MECHANISM DRAFTING AND DESIGN

A Workbook

John G. Nee
Industrial Education and Technology
Central Michigan University
Mt. Pleasant, Michigan

Prakken Publications, Inc.
416 Longshore Drive, Box 8623
Ann Arbor, Michigan 48107

Prakken Publications, Inc.
Ann Arbor, Michigan

ISBN 0-911168-45-1

Library of Congress Catalog Card No. 80-80861

Printed in the United States of America

Preface

Mechanism Drafting and Design: A Workbook combines graphical and mathematical approaches to the analysis and synthesis of both classical and modern mechanism problems. Designed for students of drafting, machine design, and mechanical technology at the postsecondary level, this workbook will be of maximum benefit when accompanied by instructor lectures and information sheets. It can also be used as a comprehensive learning module when keyed to one of the suggested reference texts noted.

An introductory "How to Use This Text" unit provides guidelines for course organization and equipment needed, as well as a list of mechanism texts and engineering reference books that offer pertinent supplementary reading. The workbook itself is divided into two major parts, followed by an appendix of detailed technical data and 87 problem sheets to be used by the student in working through each unit's drafting/calculation assignments.

Part I gives the student background information and experience in dealing with fundamental mechanism and motion concepts. Each topic provides for extensive problem solving, using basic principles of drafting, trigonometry, algebra, and technical physics. Units should be taught in the sequence presented, since each one builds on concepts developed in preceding units.

Part II presents some intriguing contemporary mechanism problems designed to stimulate interest in the application of principles learned in Part I. Unit 19, in particular, allows considerable latitude in exercising design creativity.

Specific page references to the texts cited in the "Suggested References" section accompany each unit. The student is encouraged to draw upon as many of these sources as possible in completing the problem assignments.

Varied methods of presentation have been utilized in this workbook to improve the approaches for student and instructor usage. Students can do the problems in Part I at an independent pace, although they should be completed in the sequence given in the book. Part II can be worked completely on an individualized basis. It is quite conceivable and appropriate for the instructor to have students work independently on different problems in the same class.

The text has been three-hole punched and perforated to accommodate insertion into a ring binder as well as easy removal of the problem sheets for student assignments.

The book is organized in such a way as to provide the basis for developing a reference notebook for use in an entry-level mechanical technology-related position. It is designed to have enough flexibility and instructional options for use in a one-quarter course to a two-semester sequence. The choice and number of problems selected by the instructor can be based on the background and occupational interests of the students. In a one-quarter course, the instructor may direct students to complete Part I and a limited number of problems from Part II. A one- or two-semester course would permit assignment of most if not all of Part II, supplemented with other mechanical design work of the instructor's choice.

Students using this text should have had at least two semesters of high school drafting and preferably one semester of technical school, community college, or university-level technical drafting. Also, basic courses in algebra, trigonometry, and technical physics with an emphasis on mechanics are necessary prerequisites.

Acknowledgments

Many kind and generous individuals and organizations have contributed to the completion of this book. They are listed here for instructor and student reference:

Allis-Chalmers Corporation, Milwaukee, Wisconsin

AM Bruning Division, Addressograph Multigraph Corporation, Schaumburg, Illinois

American Manufacturing Company, Inc., Tacoma, Washington
American Sprocket Chain Manufacturers Association, Park Ridge, Illinois
Amoco Oil Company, Chicago, Illinois
Beloit Corporation, Beloit, Wisconsin
Boston Gear, Quincy, Massachusetts
E. Raymond Brooke, Milton Keynes, England
Browning Manufacturing Company, Maysville, Kentucky
Burro Crane, Inc., Chicago, Illinois
Chevrolet Division, General Motors Corporation, Detroit, Michigan
Cincinnati Milacron, Cincinnati, Ohio
Commercial Cam Division, Emerson Electric Company, Chicago, Illinois
Curtis Universal Joint Company, Inc., Springfield, Massachusetts
De-Sta-Co Division, Dover Corporation, Detroit, Michigan
Niels Diffrient, Alvin Tilley, Joan Bardagjy, *Human Scale 1/2/3*, New York, New York
Kenneth Erickson, University of Wisconsin - Stout, Menomonie, Wisconsin
Fairbanks Morse Engine Division, Colt Industries, Beloit, Wisconsin
Ferguson Machine Company Division, UMC Industries, Inc., St. Louis, Missouri
Ford Motor Company, Dearborn, Michigan
General Motors Corporation, Detroit, Michigan
Geneva Mechanisms Corporation, Tampa, Florida
Henry Dreyfuss Associates, New York, New York
Invo Spline, Inc., Warren, Michigan
ISI Manufacturing, Inc., Fraser, Michigan
JLG Industries, Inc., McConnellsburg, Pennsylvania
Lapeer Manufacturing Company, Lapeer, Michigan
MIT Press, Cambridge, Massachusetts
National Broach and Machine Division, Lear Siegler, Inc., Detroit, Michigan
New Departure - Hyatt Division, General Motors Corporation, Detroit, Michigan
Parker Hannifin Corporation, Cleveland, Ohio
Penton/IPC, Inc., Cleveland, Ohio
Randolph Auston Company, Manchaca, Texas
Southworth, Inc., Portland, Maine
The Fellows Gear Shaper Company, Springfield, Vermont
Thern, Inc., Winona, Minnesota
Unimation, Inc., Danbury, Connecticut
Up-Right Scaffolds Division, Up-Right, Inc., Berkeley, California
Watson-Guptill Publications, New York, New York
Whitney Library of Design, New York, New York

Numerous excellent suggestions were made by university, technical institute, and community college faculty and students. Engineers, technicians, and company representatives were consulted throughout the book's preparation. Many educator colleagues and industrial friends gave generously of their time and resources in developing this book. A note of appreciation is due Theresa Cotter and Pam Sponseller for their untiring efforts and assistance in the preparation of the manuscript. Finally, this book is dedicated to Kathy, Chris, and Lynn Nee.

About the Author

Currently Professor in the Department of Industrial Education and Technology at Central Michigan University, John G. Nee holds a diploma from the William Hood Dunwoody Industrial Institute, bachelor's and master's degrees from the University of Wisconsin-Stout, and an Ed.D. degree from the University of Minnesota. A Certified Senior Engineering Technician and a Certified Manufacturing Engineer, Dr. Nee has work experience in machine design at the 3M Corporation, the Beloit Corporation, and various consulting engineering firms. His teaching experience includes 14 years at the community college, technical institute, and university levels. His publications include *Jig and Fixture Design and Detailing* (Prakken Publications) and numerous technical and educational articles and monographs. He has had administrative experience in directing occupational program development and institutional research, and consulting in corporate training.

Contents

How to Use This Text

THIS book is designed to present experiences which will enable you, the mechanical technology student, to mathematically and graphically analyze and design mechanical mechanisms. Problems have been selected to familiarize you with contemporary mechanisms which are being planned and constructed in industry. The design and analysis activities presented will enable you to conceptualize and communicate using engineering graphics.

A number of mathematical problems have been included in selected units to highlight technical information pertinent to the mechanism problems. A thorough understanding of the concepts presented by these problems is just as important for job growth as are good drafting skills.

You should have had at least two semesters of high school drafting and preferably one semester of technical school, community college, or university-level technical drafting in order to satisfactorily complete most of the problems found in this book. You should be relatively familiar with principles of algebra, trigonometry, and basic technical physics, with an emphasis on mechanics. Engineering graphics reference texts are listed under "Suggested References," following. Extensive use of one or more of these texts and handbooks is recommended in completing the problems and questions.

The Appendix provides various technical data and conversion tables needed for working through the mechanism design problems. You should become very familiar with the material contained in the Appendix.

Suggested References

The following mechanism references are recommended for use in completing the graphical analysis and design problems presented in this workbook. Answers to the problems can be drawn from these sources; specific page references for each source can be found immediately after the "Problem Assignment" section in each unit.

Esposito, Anthony. *Kinematics for Technology.* Columbus, Ohio: Charles E. Merrill Publishing Co., 1973.

Greenwood, Douglas C., ed. *Engineering Data for Product Design.* New York: McGraw-Hill Book Co., 1961.

———. *Mechanical Details for Product Design.* New York: McGraw-Hill Book Co., 1964.

———. *Product Engineering Design Manual.* New York: McGraw-Hill Book Co., 1959.

Hardison, Thomas B. *Introduction to Kinematics.* Reston, Va.: Reston Publishing Co., 1979.

Hinkle, Rolland T. *Kinematics of Machines.* 2nd ed. Englewood Cliffs, N.J.: Prentice-Hall, 1960.

Hirschhorn, Jeremy. *Dynamics of Machinery.* New York: Barnes & Noble, 1967.

Kepler, Harold B. *Basic Graphical Kinematics.* 2nd ed. New York: McGraw-Hill Book Co., 1973.

Lent, Deane. *Analysis and Design of Mechanisms.* 2nd ed. Englewood Cliffs, N.J.: Prentice-Hall, 1970.

Michels, Walter J., and Wilson, Charles E. *Mechanism: Design-Oriented Kinematics.* Chicago: American Technical Society, 1969.

Oberg, Erik; Jones, Franklin D.; and Horton, Holbrook L. *Machinery's Handbook.* 20th ed. New York: Industrial Press, 1976.

Patton, William J. *Kinematics.* Reston, Va.: Reston Publishing Co., 1979.

Tao, D.C. *Fundamentals of Applied Kinematics.* Reading, Mass.: Addison-Wesley Publishing Co., 1967.

The engineering graphics and technical drawing sources listed below are recommended as general references. You would be well served by having some of these available for use in completing many drafting-related mechanism problems.

Brown, Walter C. *Drafting for Industry.* South Holland, Ill.: Goodheart-Willcox Co., Inc., 1978.

Earle, James H. *Engineering Design Graphics.* 3rd ed. Reading, Mass.: Addison-Wesley Publishing Co., 1977.

French, Thomas E., and Vierck, Charles J. *Engineering Drawing and Graphic Technology.* 11th ed. New York: McGraw-Hill Book Co., 1972.

Giesecke, Frederick E., et al. *Engineering Graphics.* 2nd ed. New York: Macmillan Co., 1975.

———. *Technical Drawing.* 6th ed. New York: Macmillan Co., 1974.

Jensen, Cecil, and Helsel, Jay. *Engineering Drawing and Design.* 2nd ed. New York: McGraw-Hill Book Co., 1979.

Levens, A.S. *Graphics Analysis and Conceptual Design.* 2nd ed. New York: John Wiley & Sons, 1968.

Luzadder, Warren J. *Fundamentals of Engineering Drawing.* 7th ed. Englewood Cliffs, N.J.: Prentice-Hall, 1977.

Rising, J.S., et al. *Engineering Graphics.* 5th ed. Dubuque, Iowa: Kendall/Hunt Publishing Co., 1977.

Course Organization

Lectures, visual presentations, student design group analysis and research, guest lecturers, and field trips should be used if this book is to be of maximum benefit.

Your instructor will select suitable problems appropriate for a given individual or group ability level.

All of your drawings, completed discussion questions, references, and class notes should be maintained in a suitable location within this book. This book could serve as a valuable resource for advanced courses and as a beginning reference for your first job as a drafter or mechanical designer.

Your course grade will be based on the following requirements:

Number Required	*Course Requirements*	*Percentage Weighting*
______	Individual Projects	______ %
______	Team Design Project	______ %
______	Quizzes and Tests	______ %
______	Reports/Presentations	______ %
______	Checking Assignments	______ %
______	______	______ %
______	______	______ %

Equipment List

The following drawing instruments and supplies are typical of those used by mechanical design layout and detail drafters in developing mechanism drawings. Your instructor will specify the equipment and materials you will need in the drafting lab and for home assignments.

MATERIALS AND EQUIPMENT NEEDED

Description	*Catalog Identification*	*Source*
( ) Drafting machine	______	______
( ) Drafting set	______	______
( ) T-square	______	______
( ) Drawing board	______	______
( ) Scales	______	______
(decimal/metric/and	______	______
mechanical engineers')	______	______
( ) Adjustable triangle	______	______
( ) 45° Triangle (8″ to 10″)	______	______
( ) 30°-60° Triangle	______	______
(10″ to 12″)	______	______
( ) Eraser	______	______
( ) Tape, 1/2″ roll	______	______
( ) Erasing shield	______	______
( ) Sanding pad	______	______
( ) Lead Holder (0.5 mm,	______	______
0.7 mm, 0.9 mm,	______	______
2 mm, ______)	______	______
( ) Leads (6H, 4H, 2H,	______	______
H, F, HB, ______)	______	______
( ) Lead pointer	______	______
( ) Compass (6″ to 8″)	______	______
( ) Divider	______	______
( ) Irregular curve	______	______
( ) Drafting brush	______	______
( ) Circle template	______	______
(decimal/fractional/	______	______
metric)	______	______
( ) Protractor	______	______
( ) Vellum (sizes A, B,	______	______
C, D, ______)	______	______
( ) Needlepoint	______	______
( ) Magnifying glass	______	______

Design Teams

In recent years there has been an almost universal pattern of emphasis on creative design, with a corresponding de-emphasis on time-consuming artistic drafting skills. While drafting skills should not be disregarded because of accuracy requirements, some savings in drafting time is possible with the use of special templates and newly designed drafting instruments.

The proper focus of engineering graphics in mechanism design layout and detailing is to serve as an aid in, and as a means of, communicating ideas to others. This capacity involves more than mere technical or manipulative competence. It involves your willingness to attack a problem never before seen or studied and for which data are often incomplete.

The mechanism design layout problems presented in this book are intended to bridge the gap between the purely mental aspect of three-dimensional problem solving and its actual industrial application. By using this workbook format, you will learn the fundamentals and techniques of engineering graphics and, at the same time, gain experience in applying these principles to the solution of mechanism design layout problems.

An efficiently organized mechanical design department of any industrial concern is often as complex as the human characteristics of those individuals employed within it. In these departments, design personnel are assigned to groups; a senior designer serves as the group leader, with designers, detailers, lead drafters, checkers, and technical clerks making up the complement of each group.

To set up a facsimile of a mechanical design department found in industry, your class can be divided into a number of "design teams." Rather than attempt to duplicate all of the specific industrial positions, teams should be limited to three to five students each, with an appointed group leader to coordinate activities of the group. Problems assigned by the instructor can then be approached with a team effort, following the problem-solving steps outlined next.

DESIGN GROUP ASSIGNMENTS

GROUP A	GROUP B
Group Leader ________	Group Leader ________
____________________	____________________
____________________	____________________
____________________	____________________
____________________	____________________

GROUP C	GROUP D
Group Leader ________	Group Leader ________
____________________	____________________
____________________	____________________
____________________	____________________
____________________	____________________

Design Problem Solving

As with creativity, an emphasis on problem solving is imperative in a mechanical technology program. Various stages in the problem-solving process are identified below. Each of these steps should be covered when working through every mechanism design problem:

1. *Problem statement.* This can be general or specific depending on the desired outcome, but it should reflect a need relative to a design problem. Examples include: "Design a linkage system that will produce six-inch, straight-line travel of a point" or "design a mechanism which will guide a point in a true elliptical path."
2. *Analysis.* This raises specific questions about the mechanism problem.
3. *Data gathering.* This provides a collection of relevant information. It should include your contacts with vendors and various industries through field trips, guest speakers, and the reviewing of mechanism design references.
4. *Proposed solutions.* This step consists of experiments, design alternatives, sketches, presentations, and evaluations.
5. *Final mechanism design solution.* This consists of your finished working drawings and calculations.

After finishing the data gathering and proposed solution steps, each design team should give a presentation to the class that deals with the bits and pieces of information uncovered and the justification for a tentative mechanism design solution. After this phase, your team should propose a final solution consisting of finished drawings. The solution can then be evaluated by another design group which would, in turn, provide you with check-prints of the final design.

Introduction

MECHANISMS, which make up the basis of machine design, consist of a number of moving elements or parts that provide motions and forces necessary to a machine's function. The study of mechanisms involves the analysis and synthesis of these various, connected moving parts. In everyday life we are surrounded by linkage systems that could serve as examples for analysis: hinges, typewriters, automobile suspension systems, locking devices in kitchen appliances, common mechanics' and woodworking tools, and construction machinery, to name but a few.

When studying kinematic concepts, you are encouraged to use both graphical and analytical techniques, and to seek out examples of potential mechanism design problems from devices with which you are familiar. You should develop constant concern for precision linework through the use of the protractor, compass, and dividers to determine and transfer angular and linear displacements and acceleration and velocity vectors. The engineers' and metric scales should be used to ensure accuracy to the nearest .01 inch or 0.2 mm, respectively.

A machine designer is primarily concerned with the forces, displacements, velocities, and accelerations set up in modern industrial mechanisms, many of which are simply variations of classical mechanisms developed by mathematicians centuries ago. The following schematic drawings illustrate the analysis of a few simple classical mechanisms used in industry. Some of the linkage symbols used in developing schematic representations of mechanisms are illustrated in Fig. 3. Many engineers and technicians use templates with these or similar symbols. Some of these symbols are arranged in a schematic to represent a complete mechanism as shown in Fig. 4. A pictorial of the same mechanism is illustrated in Fig. 5 to give a better idea of the linkage relationships.

Most of the mechanisms found in machinery consist of variations of the simple four-bar linkage. Fig. 6 depicts this basic linkage in a few of its several forms. You will analyze basic four-bar linkage arrangements during this course to determine various link displacements. Fig. 6*a* shows the basic four-bar linkage, made up of two crank links (one driver and one follower), a connecting link, and a line between the fixed centers (a

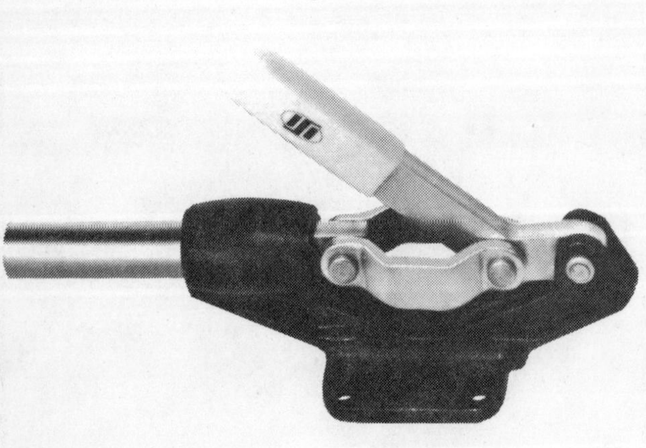

Courtesy of ISI Manufacturing, Inc.

Fig. 1. Simple toggle clamping mechanism.

Courtesy of Ford Motor Co.

Fig. 2. Complex multi-mechanism engine.

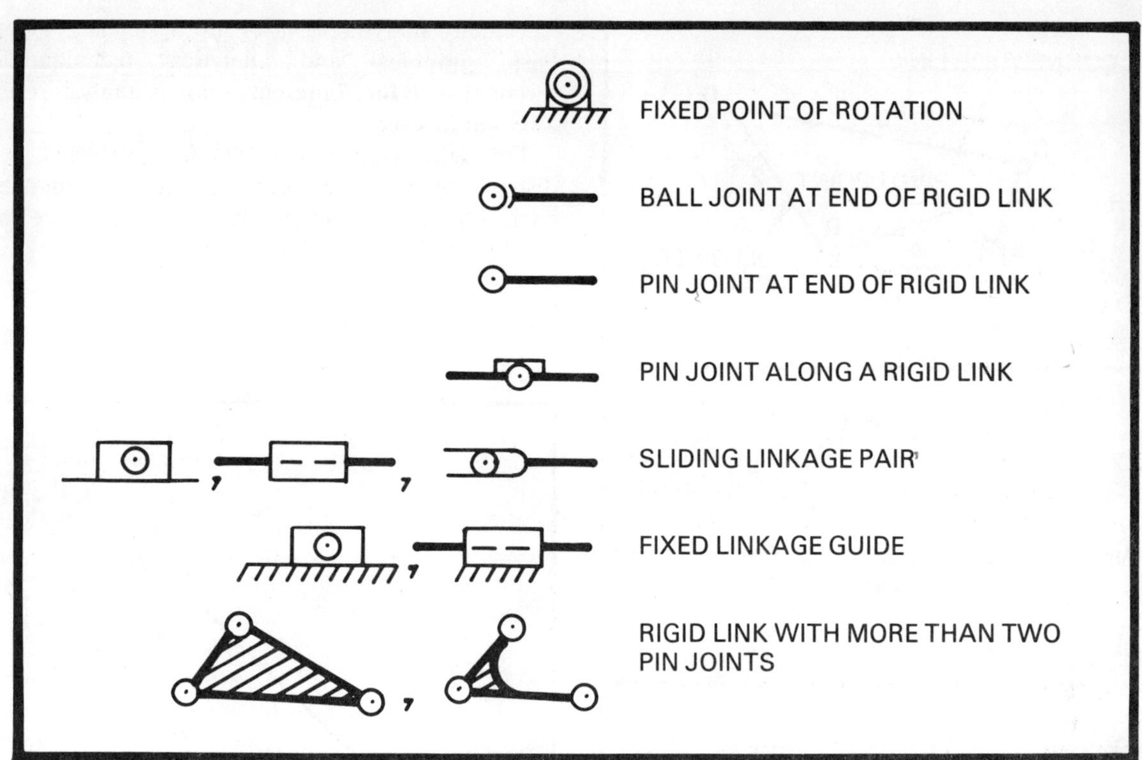

Fig. 3. Schematic mechanism symbols.

ground or fixed link). The cranks can rotate if link 1 is smaller than links 2 or 3 or 4. Link motion can easily be predicted through graphical techniques. Fig. 6*b* presents a crank and rocker variation whose operation requires the link relationships: $1 + 2 + 3 > 4$; $1 + 3 - 2 > 4$; $3 - 1 + 2 > 4$; and $1 + 4 + 2 > 3$. A four-bar linkage with a sliding member is illustrated in Fig. 6*c*. One crank has been replaced with a circular slot having an effective crank distance of link 2. The effective crank and pivot are actually replaced with the slot.

Fig. 4. Schematic of a straight-line mechanism.

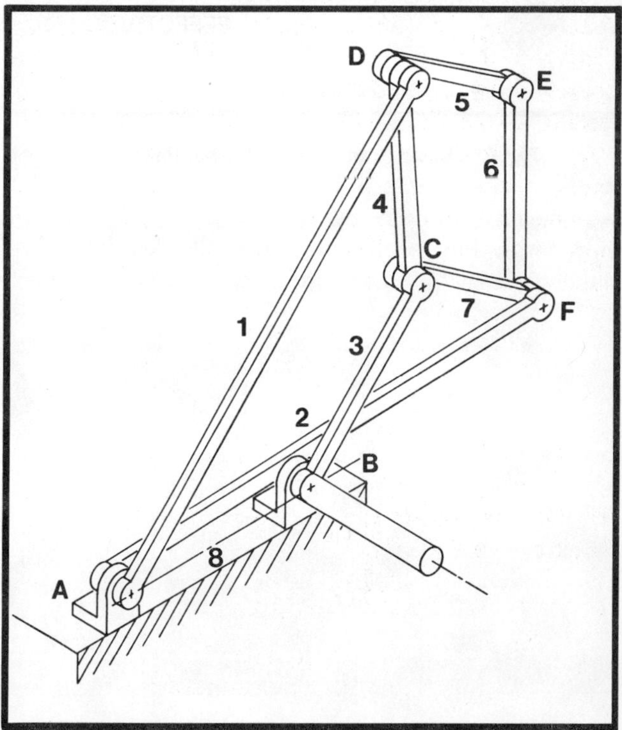

Fig. 5. Pictorial of a straight-line mechanism.

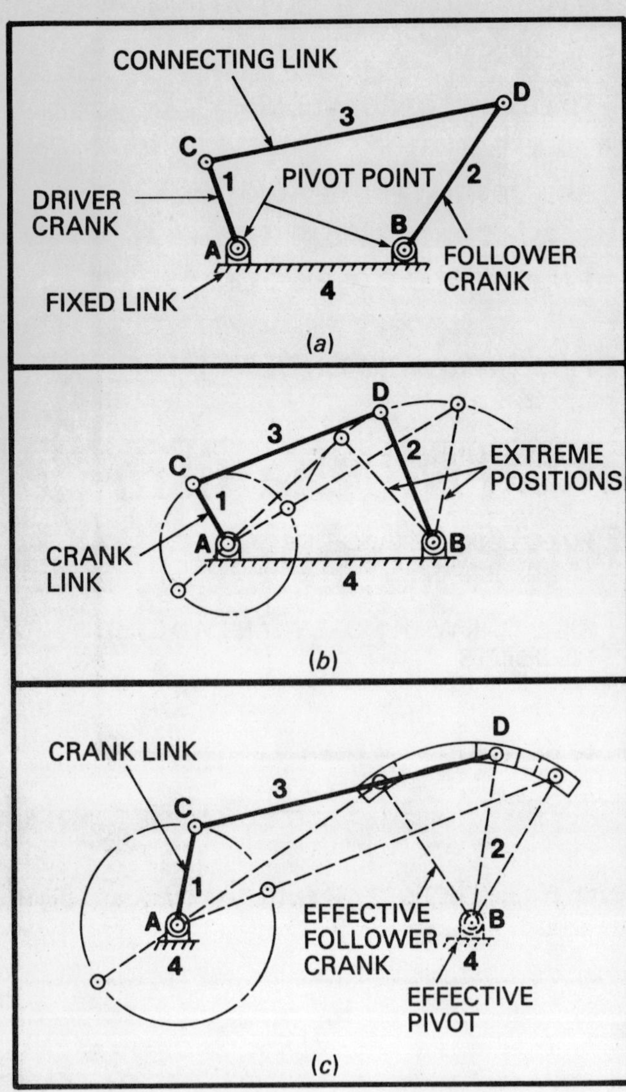

Fig. 6. Basic examples of four-bar linkages.

Velocity analysis provides another means of applying basic graphical and analytical techniques. The mechanism in Fig. 7 presents velocity analysis results for a six-bar linkage.

Through exposure to mechanism problems, you will discover the ways in which drafting, mathematics, and basic physics help to analyze and synthesize moving machine parts.

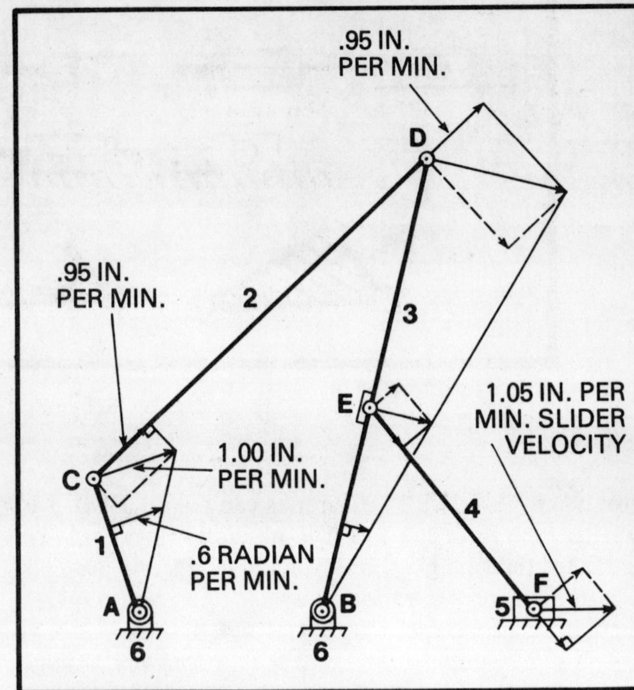

Fig. 7. Velocity analysis results for a six-bar linkage.

PART I Fundamental Mechanism and Motion Concepts

Graphical Analysis of Mechanism Forces

1

composition and resolution of forces

MECHANICAL linkage systems have to be analyzed for forces and stresses as well as for velocities and accelerations. In completing a linkage layout, the designer must determine forces of tension, compression, and shear. Much of the analysis can be completed by the use of graphical methods in conjunction with mathematics.

Graphical methods provide important advantages over a totally mathematical solution. First, vector quantities can usually be presented with greater visible clarity than other methods offer. Second, the rapid graphical solution is often more efficient because less time and fewer calculations are required. Also, calculations can be easily checked.

An understanding of vector geometry terminology is required before the techniques of graphical solution can be applied. Basic terms found in graphical vector analysis include:

Scalar Quantity. A quantity having magnitude only and whose measure is completely described by a single number. Quantities such as volume, temperature, and pressure are scalar.

Vector Quantity. A quantity having both magnitude and direction which can be represented by a directed line segment and which can be added graphically. Velocity, displacement, and force are vector quantities.

Vector. Any directed straight line segment whose length is proportional to the magnitude of the vector quantity.

Force. A vector quantity, considered as a push or pull, which changes or tends to change the state of rest or motion of a body.

Vector Chain. A series of vectors drawn head to tail in proper continuity of direction, forming a closed circuit when the vector system is in equilibrium.

Equilibrium. The resulting state of rest or of uniform motion when the sum of all vector quantities acting on a body equals zero.

Resultant. The simplest single vector quantity which will produce the same effect when substituted for the vector quantities acting on a body.

Equilibrant. The vector quantity which will produce equilibrium. The magnitude of the equilibrant is equal to that of the resultant, but acts in the opposite direction.

Composition. The process of reducing a vector system to a simpler system.

Component. Any one of two or more vector quantities into which a given vector can be resolved.

Resolution. The process of replacing a vector quantity by its components.

The following are definitions and examples of vector systems having different types of force vectors:

Coplanar. All forces lie in the same plane.

Noncoplanar. All forces do not lie in the same plane.

Concurrent. The action lines meet at a common intersection.

Nonconcurrent. The action lines do not meet at a common intersection.

Parallel. The action lines of the forces are parallel.

Nonparallel. The action lines of the forces are not parallel.

Example 1. *Noncoplanar, nonconcurrent, and nonparallel vector system*: Forces F_1, F_2, F_3, and F_4 are not on the same plane. They do not intersect at a common point and they are not parallel.

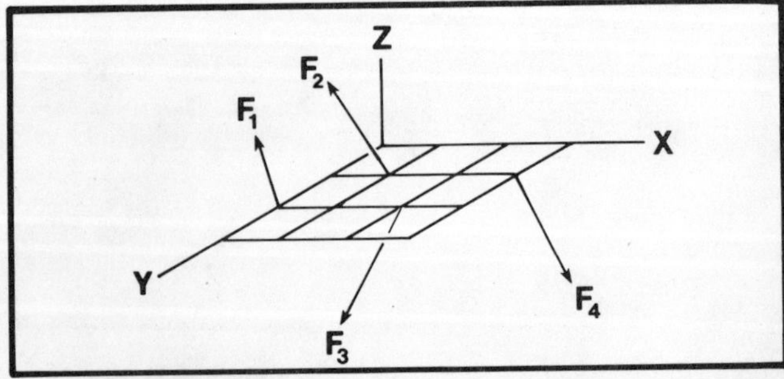

Fig. 1-1. Noncoplanar, nonconcurrent, and nonparallel vector system.

Example 2. *Noncoplanar, nonconcurrent, and parallel vector system*: Forces F_1, F_2, F_3, and F_4 are not on the same plane. They are parallel, but they do not intersect at a common point.

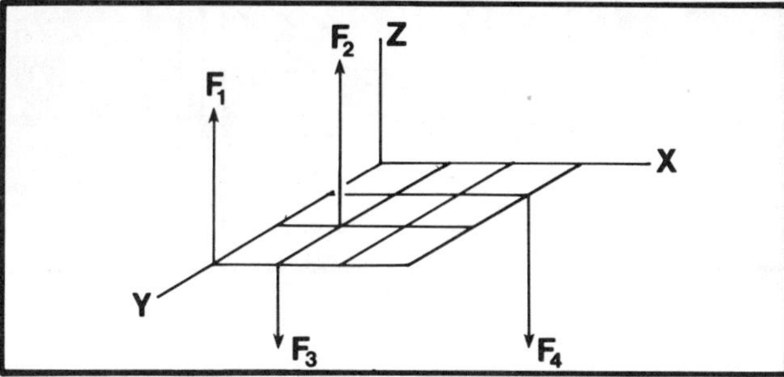

Fig. 1-2. Noncoplanar, nonconcurrent, and parallel vector system.

Example 3. *Coplanar, nonconcurrent, and parallel vector system*: Forces F_1, F_2, and F_3 act on the body, but they do not intersect at a common point. They are parallel and also in the same plane.

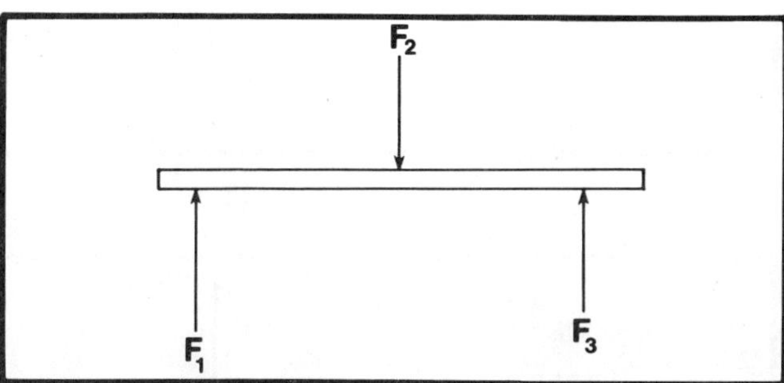

Fig. 1-3. Coplanar, nonconcurrent, and parallel vector system.

Example 4. *Coplanar, nonconcurrent, and nonparallel vector system*: Forces F_1, F_2, and F_3 act on the body but do not intersect at a common point. They are in the same plane but are not parallel.

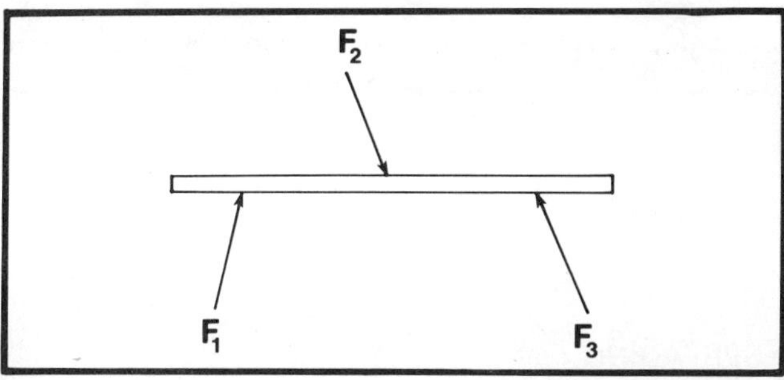

Fig. 1-4. Coplanar, nonconcurrent, and nonparallel vector system.

Example 5. *Noncoplanar, concurrent, and nonparallel vector system*: Forces F_1, F_2, F_3, F_4, and F_5 act on a common point. They are not parallel, and all of the forces do not act in a single plane. Any two vectors can be considered coplanar when isolated in pairs.

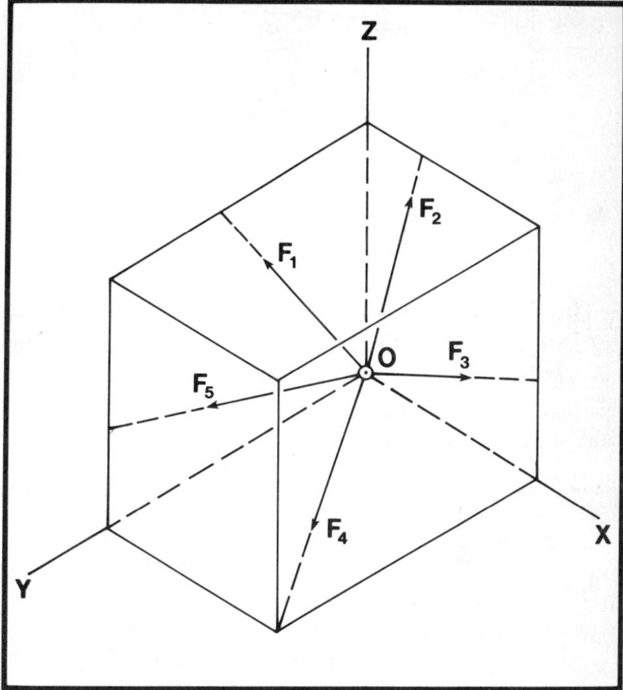

Fig. 1-5. Noncoplanar, concurrent, and nonparallel vector system.

The composition and resolution of coplanar and concurrent forces can take various forms, as depicted in Figs. 1-6 through 1-23. The parallelogram, triangle, and polygon methods are three common means of determining the resultant of force vectors. In Fig. 1-6*c*, note that force $F_?$ is reversed.

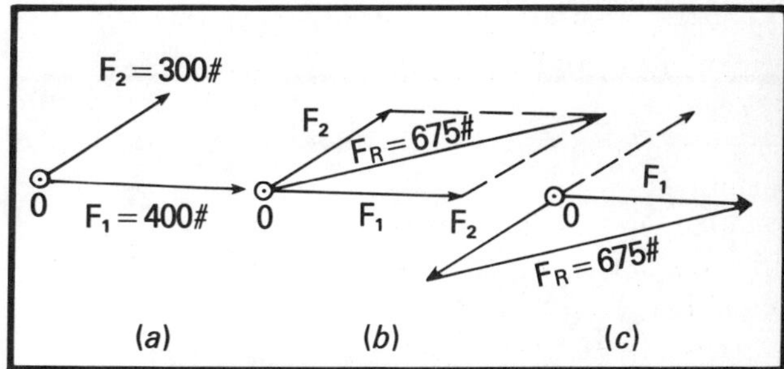

Fig. 1-6. Composition of two force vectors. (*a*) Vector components. (*b*) By parallelogram method. (*c*) By triangle method.

In Fig. 1-7, the solution is obtained by finding the resultant of F_1 and F_2 first, then determining the resultant of $F_1 + F_2$ and F_3. Forces may be combined in any order.

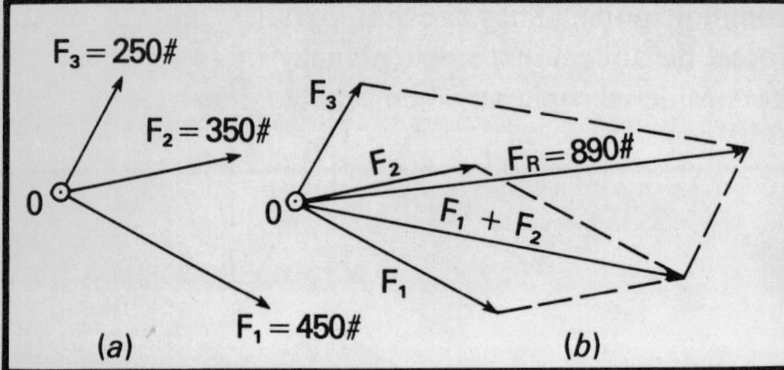

Fig. 1-7. Composition of three force vectors. (*a*) Vector components. (*b*) By parallelogram method.

The composition of multiple force vectors by the polygon method is illustrated in Fig. 1-8. After placing vectors head-to-tail, the resultant is equal to the length of the line XY required to complete the polygon. Note that vector XY has its head placed at the head of F_5.

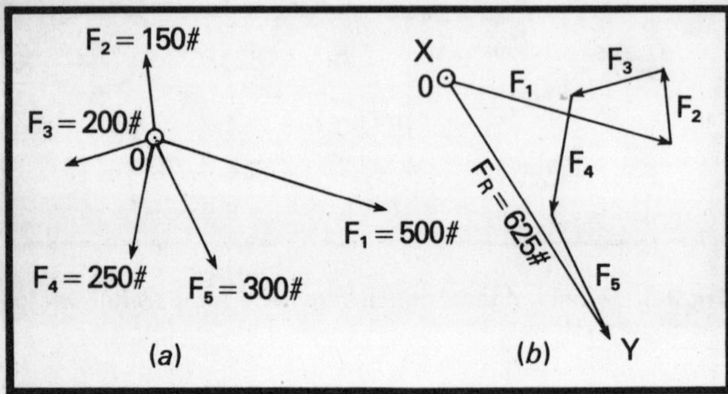

Fig. 1-8. Composition of multiple force vectors. (*a*) Vector components. (*b*) By polygon method.

Figs. 1-9 through 1-15 show the resolution of a load for several basic structures. These examples are fundamental in the analysis and design of structures and constrained linkages.

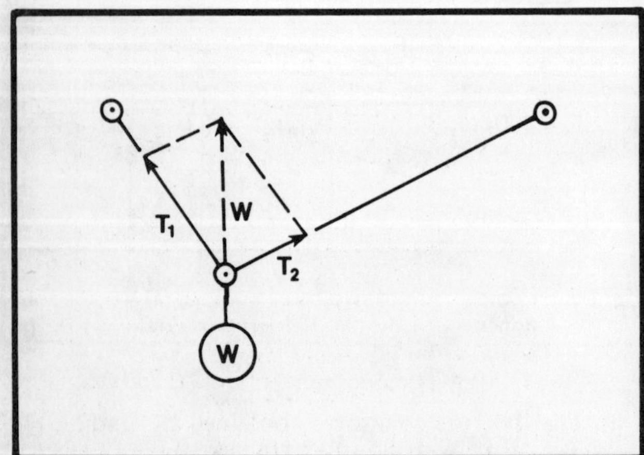

Fig. 1-9. Resolution of load *W*.

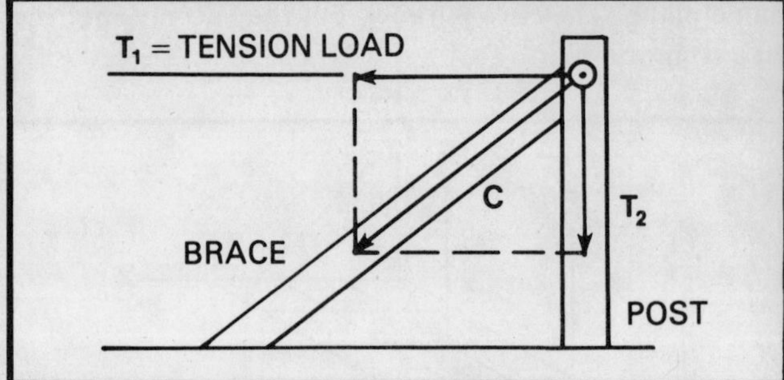

Fig. 1-10. Resolution of load *T*.

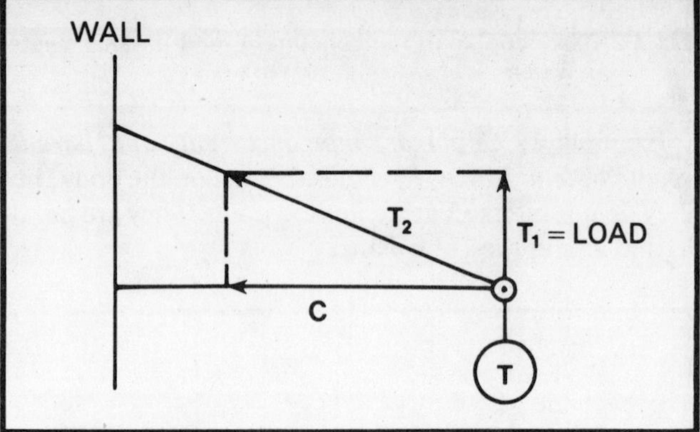

Fig. 1-11. Resolution of load *T*.

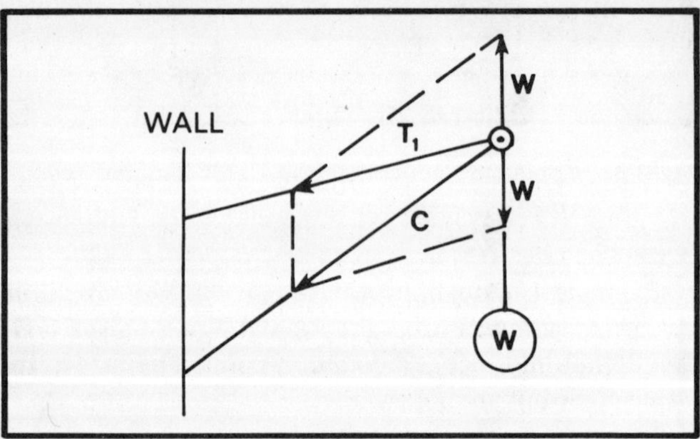

Fig. 1-12. Resolution of load *W*.

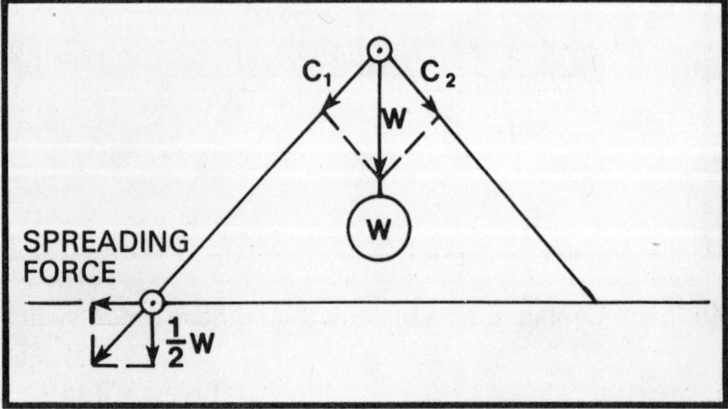

Fig. 1-13. Resolution of load *W*.

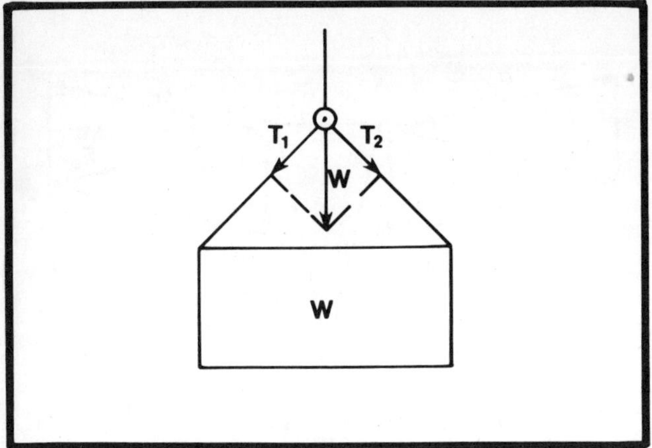

Fig. 1-14. Resolution of load W.

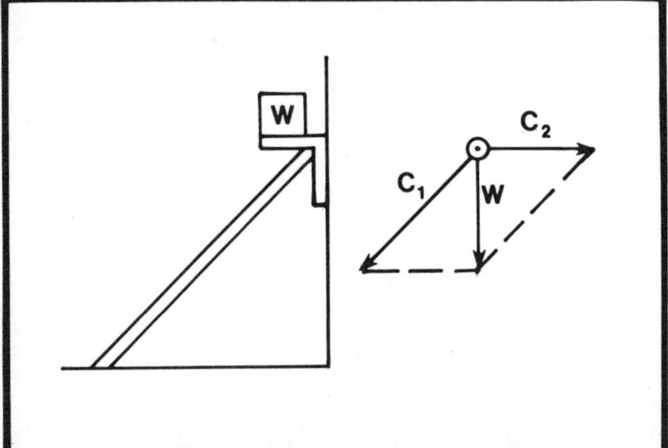

Fig. 1-15. Resolution of load W.

Fig. 1-16 illustrates the resolution of force W. E_1 and E_2 are efforts applied to the weight; T_1 and T_2 are tensions in the line.

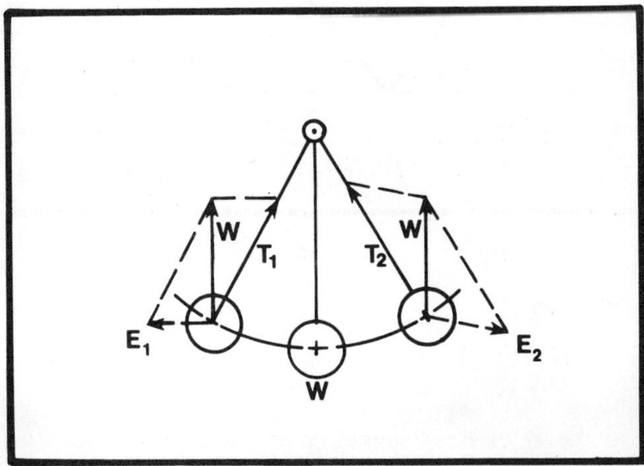

Fig. 1-16. Resolution of force W.

Fig. 1-17 shows the composition of forces W and E. E is the effort applied to weight W and N is the load normal (90°) to the incline.

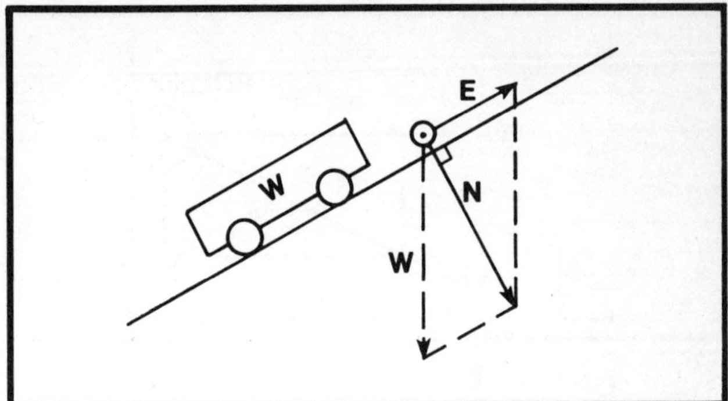

Fig. 1-17. Composition of forces W and E.

In Fig. 1-18, F_1 is the force on the piston, F_2 is the force on the connecting rod, and F_3 is the side thrust. EL is the effective length of the crank.

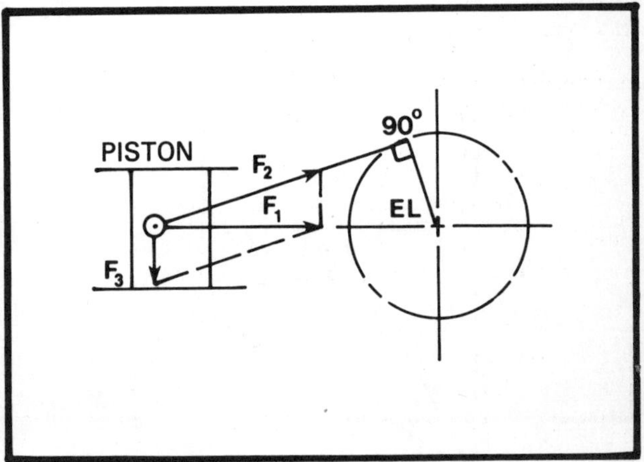

Fig. 1-18. Resolution of force F_1.

In Fig. 1-19, G is the gravity or weight, CF is the centrifugal force acting on the car and passengers, and R is the resultant force on the car and passengers.

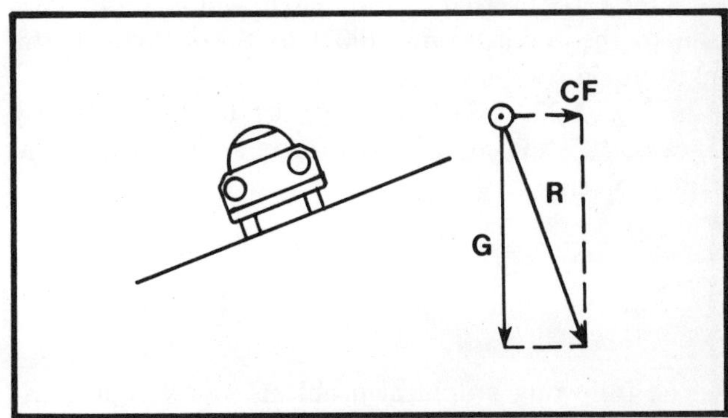

Fig. 1-19. Composition of forces G and CF.

In Fig. 1-20, E is the force applied to the roller, F is the force applied to the resistance R, and C is the compression to which the roller is subjected.

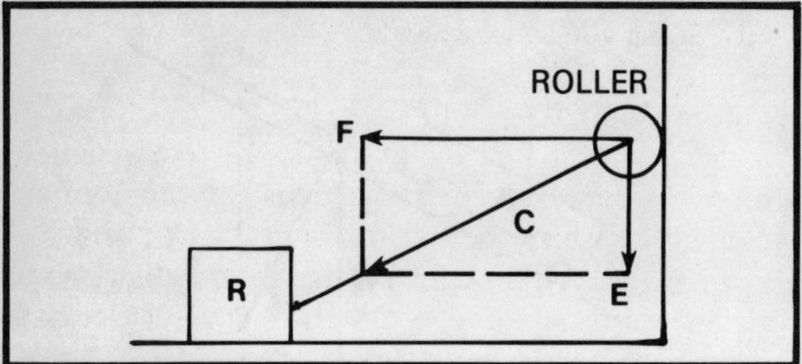

Fig. 1-20. Composition of forces F and E.

In Fig. 1-21, E is the effort applied to the weight W, tangent to the path of motion. G is the gravity or weight, and C is the compression on the arm.

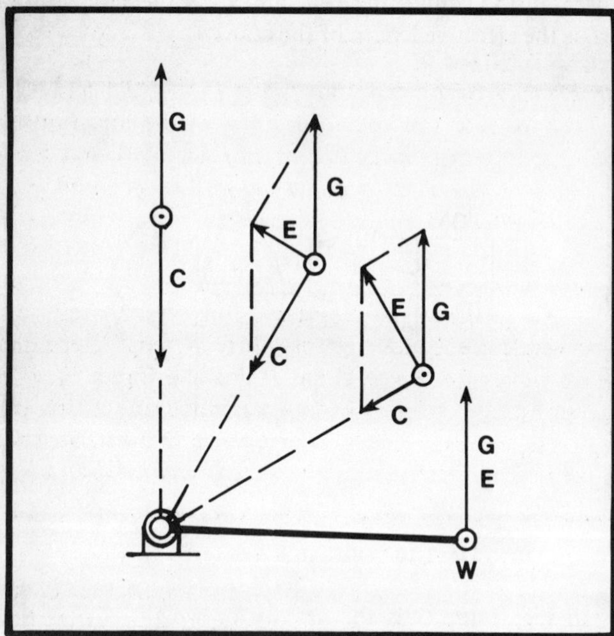

Fig. 1-21. Resolution of force G.

In Fig. 1-22, F_b is the force of the blow on the wedge, F_w is the force exerted by the wedge, and F_n is the reaction to the force normal (90°) to the incline on the opposite side of the wedge.

In Fig. 1-23, F_1 is the force set up by the screw threads, F_2 is the compression force, and F_3 is the total lifting force.

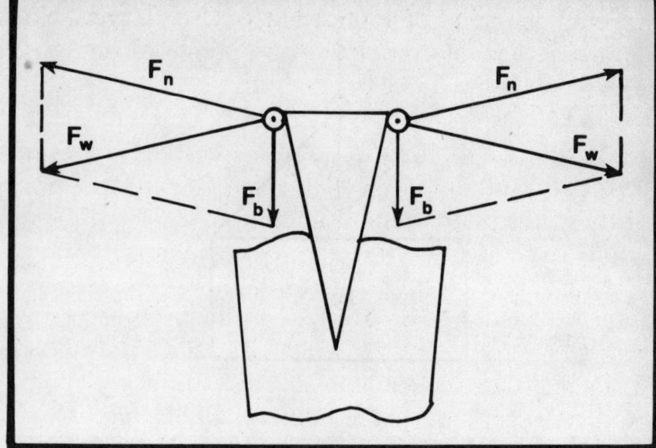

Fig. 1-22. Resolution of force F_b.

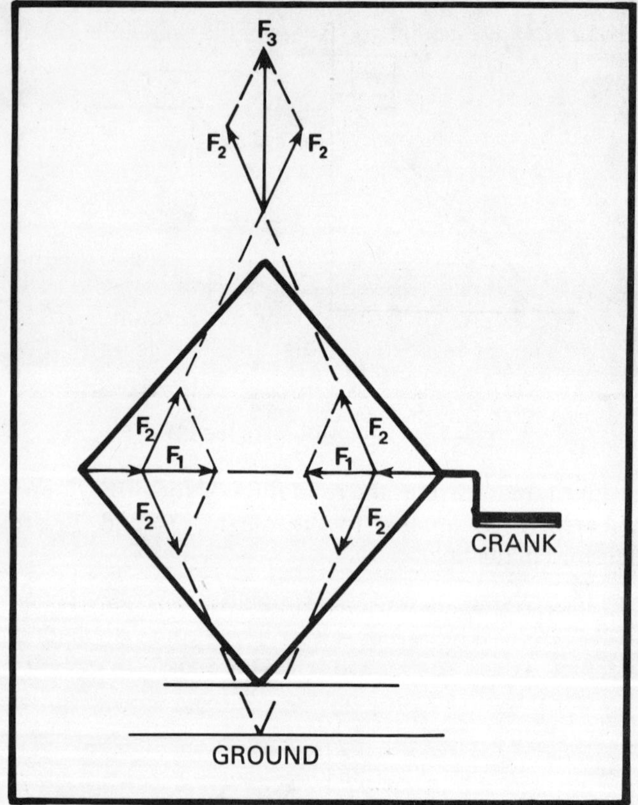

Fig. 1-23. Resolution of force F_1.

Problem Assignment

The following graphical problems and design questions are intended to provide you with experience in graphically analyzing and calculating basic mechanism constraints. Select, with instructor direction, the problem-sheet assignments suitable for you. Then complete the numbered problems in ordered sequence because each solution builds on preceding problems. Be sure to use the references suggested by your instructor (*list follows*) as sources for working through the assignment. Problem sheets are located at the back of the book.

Problem Sheet 1A

Problem 1. Find the resultant force R necessary to replace forces F_1, F_2, and F_3. Indicate the angle that the resultant makes with a horizontal datum drawn through the point of application, 0. Use the parallelogram method in completing the graphical analysis. (Scale: .25 in. = 1,000 lbs.)

Problem 2. Determine the resultant R necessary to replace forces F_1, F_2, F_3, F_4, and F_5. Indicate the angle that the resultant makes with a vertical datum drawn through the point of application, 0. Use the parallelogram method in completing the graphical analysis. (Scale: 1.00 in. = 10 lbs.)

Problem 3. Graphically find the resultant R necessary to replace forces F_1, F_2, F_3, F_4, F_5, and F_6. Indicate the angle that the resultant makes with a horizontal datum drawn through the point of application, 0. Use both the parallelogram and polygon methods in completing the analysis. (Scale: 1.0 mm = 20.0 kg.)

Problem Sheet 1B

Problem 1. Determine the forces in the rigid links 2 and 3. Indicate compression (+) and tension (−). F_1 shows direction only and is not drawn to scale. F_1 equals 4,000 lbs. (4 kips). Use the triangle method as well as mathematics to determine the forces. (Scale: .50 in. = 2 kips.)

Problem 2. Graphically find the forces in links 3, 4, and 5 when the mechanism is statically balanced in the position shown. Indicate compression (+) and tension (−). The air cylinder produces a total of 2,000 lbs. compression in link 2. Use the triangle method in your analysis. If 120 psi air pressure is introduced into the piston end of the cylinder, what would its diameter need to be in order to meet the existing design criteria? Select your own scale for the analysis.

Problem Sheet 1C

Problem 1. If three nonparallel forces are in equilibrium, they must be concurrent. Force F_1 equals 500 lbs. Determine the necessary force to be applied at point A, and the reaction at pin-point B, in order to keep the linkage in equilibrium. Utilize a force diagram and the triangle method in completing the analysis. Select your own scale.

Problem Sheet 1D

Problem 1. In the knuckle press shown, determine the force F_1 necessary to keep the linkage in equilibrium. Graphically solve for the reactions R_1, R_2, and R_3, and the forces in links 3 and 4. Use the dimensions given to calculate and verify the values determined graphically. Select your own scale.

Problem 2. The linkage shown is in static balance. Graphically determine the forces F_1 and F_2 necessary to balance the 1,000-lb. load. Also, determine the reactions R_1 and R_2. Use the dimensions given to calculate and verify the values determined graphically. Select your own scale.

Problem Sheet 1E

Problem 1. Graphically determine F_1, F_2, F_3, and F_4. Find the forces in links 2, 3, and 4 of the constrained linkage system. Indicate compression (+) and tension (−). Select the appropriate analysis method and scale in completing the problem.

Problem Sheet 1F

Problem 1. Given the frontal and horizontal views of the concurrent, noncoplanar force system, determine the values for the resultant R and the forces F_1, F_2, and F_3. Use the parallelogram method and a first or primary auxiliary view to determine R. Illustrate the method for determining the values for F_1, F_2, and F_3. (Scale: 1.00 in. = 1,000 lbs.)

Problem Sheet 1G

Problem 1. Given the frontal and horizontal views of the concurrent, noncoplanar force system, determine the values for the resultant R and the forces F_1, F_2, and F_3. Use the polygon method and first or primary auxiliary views to determine the resultant and the forces. (Scale: 1.0 mm = 50.0 kg.)

References

The following source and page references can be used in developing the graphical solutions and calculations for the problem assignments.

Esposito, Anthony. *Kinematics for Technology.* Columbus, Ohio: Charles E. Merrill Publishing Co., 1973, pp. 19-28.

Hardison, Thomas B. *Introduction to Kinematics.* Reston, Va.: Reston Publishing Co., 1979, pp. 1-30.

Hinkle, Rolland T. *Kinematics of Machines.* 2nd ed. Englewood Cliffs, N.J.: Prentice-Hall, 1960, pp. 1-10.

Hirschhorn, Jeremy. *Dynamics of Machinery.* New York: Barnes & Noble, 1967, pp. 45-72.

Kepler, Harold B. *Basic Graphical Kinematics.* 2nd ed. New York: McGraw-Hill Book Co., 1973, pp. 13-27.

Lent, Deane. *Analysis and Design of Mechanisms.* 2nd ed. Englewood Cliffs, N.J.: Prentice-Hall, 1970, pp. 39-49.

Michels, Walter J., and Wilson, Charles E. *Mechanism: Design-Oriented Kinematics.* Chicago: American Technical Society, 1969, pp. 6-12.

Oberg, Erik; Jones, Franklin D.; and Horton, Holbrook L. *Machinery's Handbook,* 20th ed. New York: Industrial Press, 1976, pp. 291-311.

Patton, William J. *Kinematics.* Reston, Va.: Reston Publishing Co., 1979, pp. 31-58.

Tao, D.C. *Fundamentals of Applied Kinematics.* Reading, Mass.: Addison-Wesley Publishing Co., 1967, pp. 9-20, 334-42.

Graphical Velocity Analysis 2

linear motion velocity

FORCE vectors are confined to rigid bodies in equilibrium. The vectors are therefore part of a balanced force system. If these bodies are acted upon by unbalanced forces, motion will result. Many motion problems can be solved graphically by vector geometry. The following examples illustrate the graphical solutions of problems involving linear motion.

Fig. 2-1 illustrates velocity vector addition. V_p is the velocity of a plane, V_w is the wind velocity and direction, and V_r is the resultant velocity and direction of the plane.

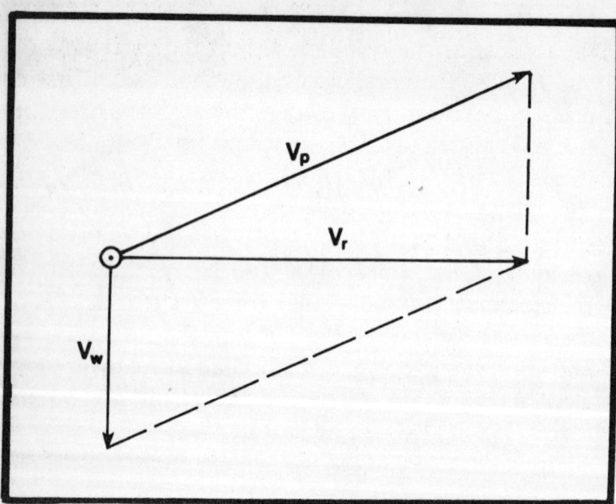

Fig. 2-1. Velocity vector addition.

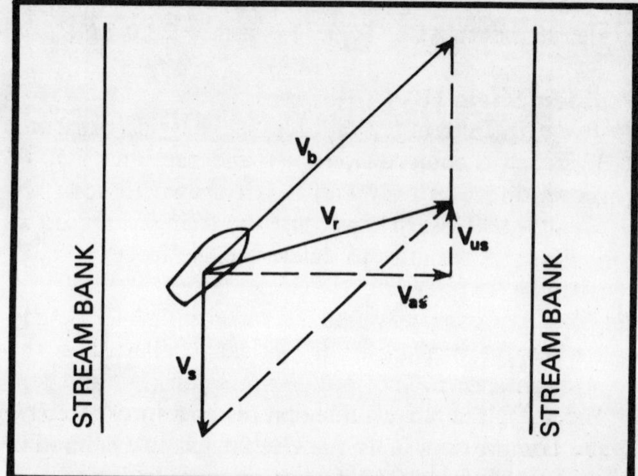

Fig. 2-2. Composition and resolution of velocity vectors.

Fig. 2-2 shows the composition and resolution of velocity vectors. V_b is the velocity of the boat, V_s is the velocity of the stream, and V_r is the resultant velocity and direction of the boat. V_{us} is the velocity upstream and V_{as} is the velocity across the stream.

In Fig. 2-3, V_w is the velocity and direction of the wind, V_{ws} is the theoretical velocity of the wind across the sail, and V_b is the velocity of the boat.

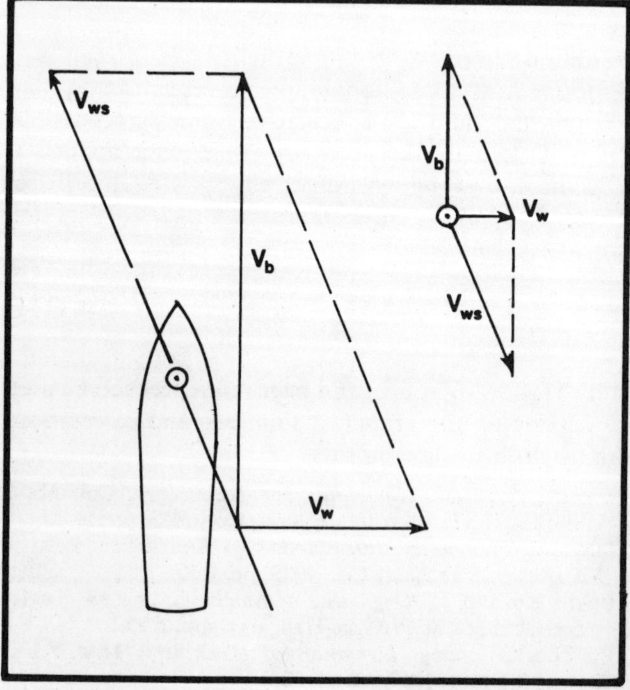

Fig. 2-3. Composition of velocity vectors V_w and V_{ws}.

A point moving along a line has *linear velocity*. When the point travels equal distances in successive equal intervals, it has *constant* or *uniform* velocity. The velocity can be calculated by determining its displacement and dividing by the given time interval:

$$V = \frac{S}{T} \qquad (2.1)$$

For example, in Fig. 2-4, if a point P travels 3 in. from X to Y in 1.5 sec., and 3 in. from Y to Z in the next 1.5 sec., its velocity is constant and equals:

$$V_p = \frac{S}{T} = \frac{3}{1.5} = 2 \text{ in./sec.}$$

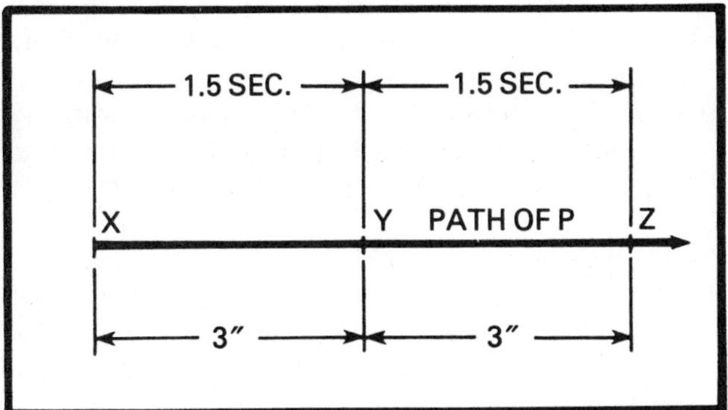

Fig. 2-4. Constant linear velocity.

When a point travels unequal distances in successive equal time intervals, its velocity is said to be *variable* and has a different value each instant. In Fig. 2-5, a point moves 6 in. from X to Y in 3 sec. and 9 in. from Y to Z in the next 3 sec. The average velocity between X and Y equals:

$$V_{av} = \frac{S}{T} = \frac{6}{3} = 2 \text{ in./sec.}$$

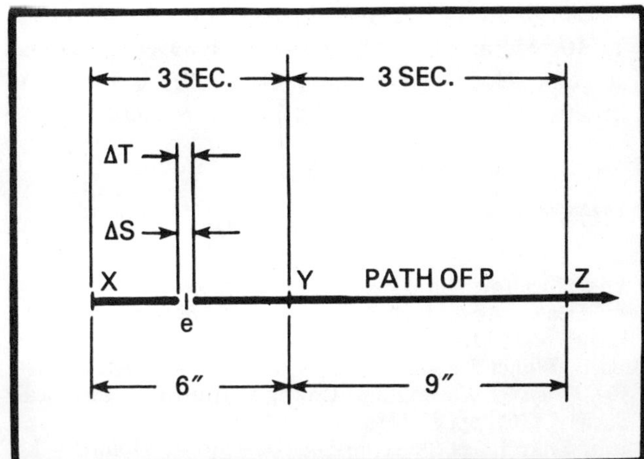

Fig. 2-5. Variable linear velocity.

This average velocity is not the velocity for all positions between X and Y, however, since the velocity is changing continuously as the point moves. It is only a theoretical number.

To determine the *instantaneous velocity* at any given point e, we consider a very small displacement ΔS and the corresponding small time interval ΔT which is required to travel this distance. As these small increments ΔS and ΔT approach zero, their ratio will equal the instantaneous value of the velocity of point e:

$$V_e = \frac{\Delta S}{\Delta T} \text{ (ΔT approaches zero).} \qquad (2.2)$$

When a point moves along a curved path, the direction of its velocity at any given position is the direction of the curved path at that point, which is best defined by the tangent drawn to the curve. In Fig. 2-6, a point traveling counterclockwise along the arc XYZ has different inclinations of velocity at each position, as shown by the tangents at X, Y, and Z. If the point travels equal lengths of arc in equal time intervals, the magnitude of the velocity at any point equals

$$V = \frac{\text{arc } XY}{T} \qquad (2.3)$$

where T equals the time required to travel from X to Y.

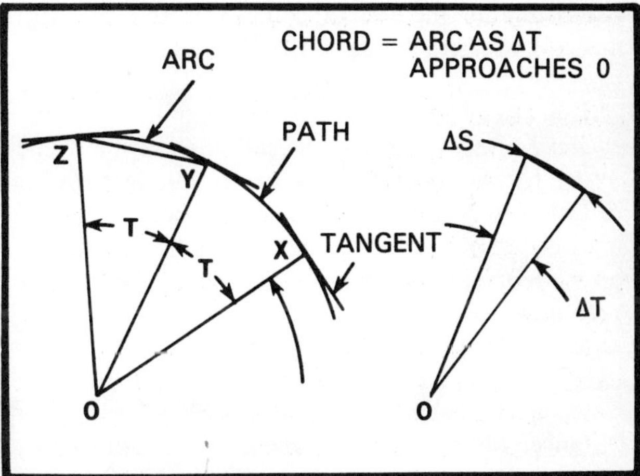

Fig. 2-6. Linear velocity on a curved path.

Since the direction of velocity is different for each location, the displacement S between X and Y (i.e., the chord XY) is not equal to the arc XY. Therefore the equation

$$V = \frac{\Delta S}{\Delta T} \qquad (2.4)$$

can only be applied when ΔS is so small as to be essentially equal to the arc subtended. When this is the case, the general equation applies:

$$V = \frac{\Delta S}{\Delta T} \text{ (ΔT approaches zero).} \qquad (2.5)$$

Problem Assignment

The following graphical problems and design questions are intended to provide you with experience in graphically analyzing and calculating basic mechanism constraints. Select, with instructor direction, the problem-sheet assignments suitable for you. Then complete the numbered problems in ordered sequence because each solution builds on preceding problems. Be sure to use the references suggested by your instructor (*list follows*) as sources for working through the assignment. Problem sheets are located at the back of the book.

Problem Sheet 2A

Problem 1. Graphically determine the resultant velocity of the airplane when there is a due-south wind velocity. Indicate the resultant's bearing. (Scale: 1.00 in. = 50 mph.)

Problem 2. Graphically determine the resultant velocity and bearing of the airplane. (Scale: 1.00 in. = 50 mph.)

Problem 3. Find the resultant velocity R necessary to replace the velocities V_1, V_2, V_3, and V_4. Indicate the angle that the resultant makes with a vertical datum drawn through the point of application, 0. Use the parallelogram method in completing the graphical analysis. (Scale: 1.0 mm = 100.0 mm/sec.)

Problem Sheet 2B

Problem 1. Determine and indicate the average velocity V_{av} for the point a as it moves to the right toward d. Calculate V_{av} for each segment as well as for the total distance. Indicate whether the point is speeding up (+) or slowing down (−). (Scale: 1.00 in. = 1.00 in.)

Problem 2. Determine and indicate the average velocity V_{av} for point a as it moves to the right toward e. Calculate V_{av} for each segment as well as for the total distance. Indicate whether the point is speeding up (+) or slowing down (−). (Scale: 1.00 in. = 2.00 in.)

Problem 3. Determine and indicate the average velocity V_{av} for the point a as it moves to the right toward e. Calculate V_{av} for each segment as well as for the total distance. Indicate whether the point is speeding up (+) or slowing down (−). (Scale: 1.00 in. = 2.00 in.)

Problem 4. Using a scale of 1.00 in. = 4.00 in., indicate the time T for distances of 6.00 in., 5.00 in., and 15.00 in. when the average velocity V_{av} is 52 in./sec. What is the total time required for movement from point a to point e?

Problem 5. Illustrate the instantaneous velocity V_e when ΔT is .0010 sec. and ΔS is 1/1,000 of the total distance point a travels. (Scale: 1.00 in. = 1.00 in.)

Problem Sheet 2C

Problem 1. Determine the linear average velocity for point a. Calculate V_{av} for each segment as well as for the total circular distance. (Scale: 1.0 mm = 20.0 mm.)

Problem 2. Determine the angular velocity (in degrees) for line a as it rotates about point 0. (Scale: 1.00 in. = 1.00 in.)

Problem 3. Determine the angular velocity (in radians) for line a as it rotates about point 0. (Scale: 1.00 in = 1.00 in.)

Problem 4. Point a has a V_{av} of 20 in./sec. Determine the value for time T in each segment and for the total circular path. (Scale: 1.00 in. = 2.00 in.)

Problem 5. Point a takes 10 sec. to travel through each segment, for a total of 30 sec. Determine the angular velocity (in degrees per second) of line a as it revolves about point 0. (Scale: 1.00 in. = 1.00 in.)

Problem 6. Point a travels through the three segments at 10, 15, and 20 sec., respectively. Determine the angular velocity (in radians per second) for line a as it rotates about point 0. (Scale: 1.00 in. = 10.00 in.)

References

The following source and page references can be used in developing the graphical solutions and calculations for the problem assignments.

Esposito, Anthony. *Kinematics for Technology.* Columbus, Ohio: Charles E. Merrill Publishing Co., 1973, pp. 57-112.

Hardison, Thomas B. *Introduction to Kinematics.* Reston, Va.: Reston Publishing Co., 1979, pp. 1-30.

Hinkle, Rolland T. *Kinematics of Machines.* 2nd ed. Englewood Cliffs, N.J.: Prentice-Hall, 1960, pp. 11-68.

Hirschhorn, Jeremy. *Dynamics of Machinery.* New York: Barnes & Noble, 1967, pp. 9-44.

Kepler, Harold B. *Basic Graphical Kinematics.* 2nd ed. New York: McGraw-Hill Book Co., 1973, pp. 99-161.

Lent, Deane. *Analysis and Design of Mechanisms.* 2nd ed. Englewood Cliffs, N.J.: Prentice-Hall, 1970, pp. 36-66.

Michels, Walter J., and Wilson, Charles E. *Mechanism: Design-Oriented Kinematics.* Chicago: American Technical Society, 1969, pp. 83-168.

Oberg, Erik; Jones, Franklin D.; and Horton, Holbrook L. *Machinery's Handbook.* 20th ed. New York: Industrial Press, 1976, pp. 325-340.

Patton, William J. *Kinematics.* Reston, Va.: Reston Publishing Co., 1979, pp. 39-42, 61-64.

Tao, D.C. *Fundamentals of Applied Kinematics.* Reading, Mass.: Addison-Wesley Publishing Co., 1967, pp. 10-46.

Basic Graphical Motion Concepts 3

position, displacement, and velocity

THE *position* of a point in a plane may be defined in terms of the length and direction of a line drawn to it from a single fixed point. The two positions B_1 and B_2 of a moving point B can be defined in terms of the lengths and directions of two lines drawn from the fixed pivot point 0. The path traced by a moving point is called a *locus.* The locus of the point may be straight or curved. The *displacement* of a moving point—its change in position—is the straight line distance between positions, usually measured in inches, feet, millimeters, or centimeters. Displacement is independent of the actual path the point takes.

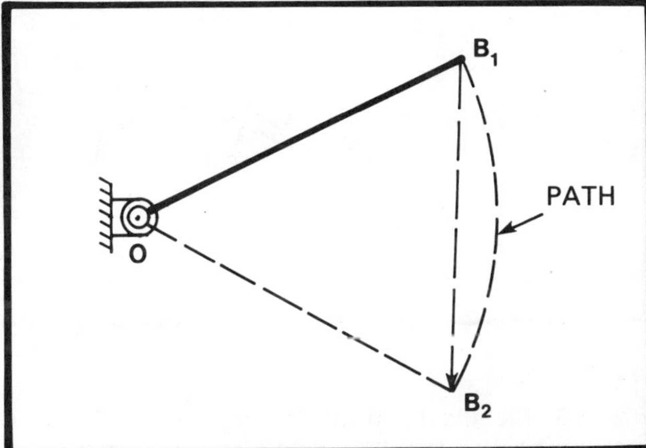

Fig. 3-1. Displacement B_1,B_2 of a point.

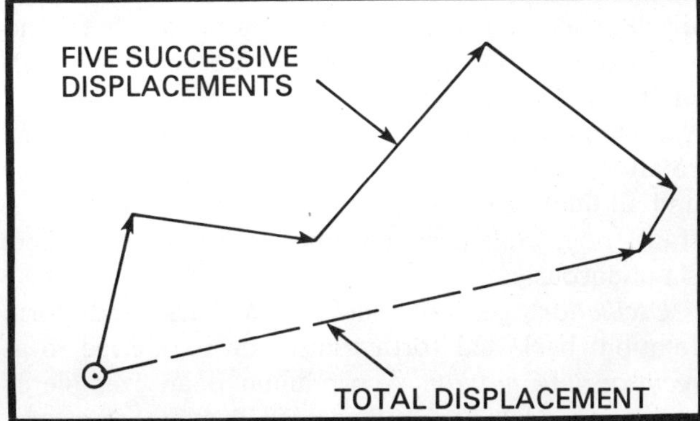

Fig. 3-2. Successive displacements.

In Fig. 3-3, A and B are two points on the moving body at the initial position. Line AB represents the initial position of the body and line A_1B_1 represents the final position.

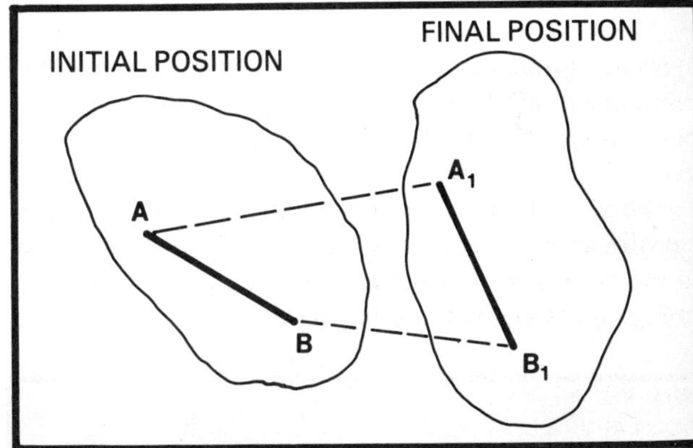

Fig. 3-3. Plane motion of a body with a line segment.

A point that is changing its position is said to be in *motion. Absolute motion* describes the movement of a point in relation to the frame. A point's motion in relation to another point or link is referred to as *relative motion.* Points moving with the same motion (parallel paths) are in *translation.* If the parallel paths are straight, the points have *rectilinear translation;*

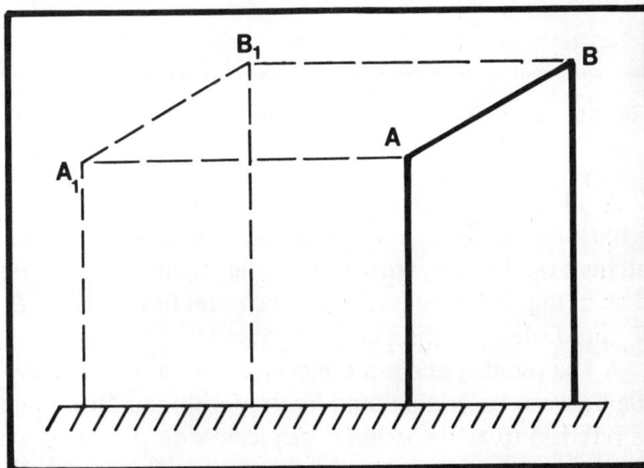

Fig. 3-4. Rectilinear translation.

curved parallel paths put the points in *curvilinear translation.*

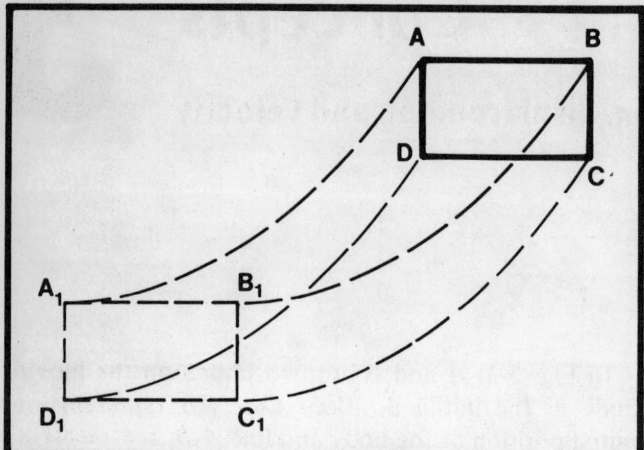

Fig. 3-5. Curvilinear translation.

Plane motion refers to motion in a single plane. *Plane mechanisms* are those which have plane motion and can be completely described in a single two-dimensional view.

The plane motion of a link can exist as rectilinear or curvilinear translation when all points on the link move in the same direction at the same speed. The link shown in Fig. 3-6 moves with both types of translation.

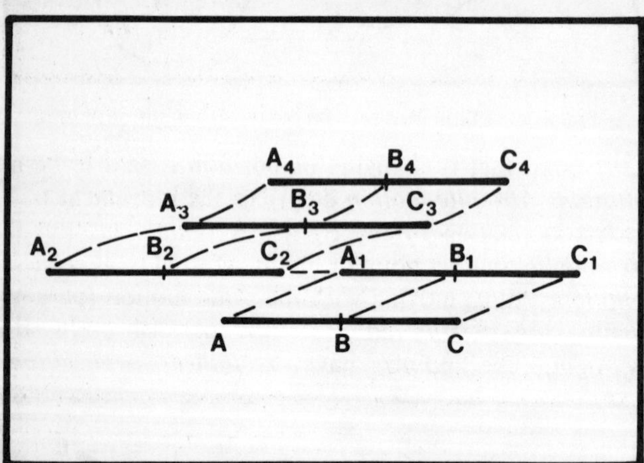

Fig. 3-6. Link translated to a number of alternative positions.

A point is considered to be rotating when it moves at a fixed radius about a *center of rotation.* In this motion, all lines on the body turn at the same angular speed. The disk in Fig. 3-7 rotates about fixed point 0. Points *A*, *B*, *C*, and *D* describe circular paths about 0.

A line rotating about a center of rotation forms an angle between its original and its new position. This angle is referred to as the *angular displacement.* The angular displacement of a rotating body has both magnitude and direction (either clockwise or counterclockwise). In

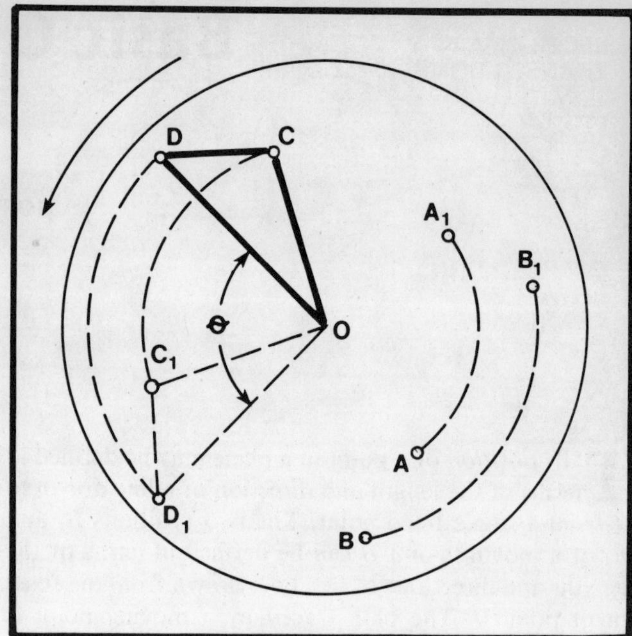

Fig. 3-7. Rotation.

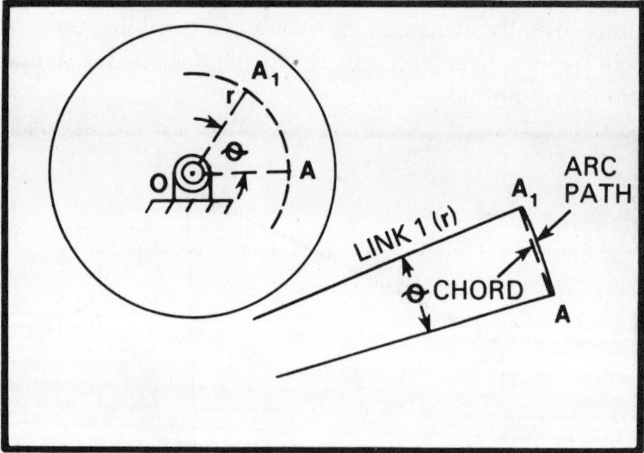

Fig. 3-8. Linear and angular displacement.

Fig. 3-8, the angular displacement of crank link 1 is measured by angle θ in a clockwise direction from its initial to its final position regardless of any possible crank oscillation between the two positions. Angular displacement is usually measured in radians but sometimes in degrees or revolutions. (One revolution is 360° or 2π rad. See Appendix C.)

Combined motion, consisting of both translation and rotation, exists when all points change their position and all lines turn. The motion of points exists only as translation, while lines can translate, rotate, or do both simultaneously.

Oscillation generally defines a back and forth rotation; back and forth translation is referred to as *reciprocating motion.* A pendulum is an example of oscillation, while the piston engine illustrates reciprocating motion.

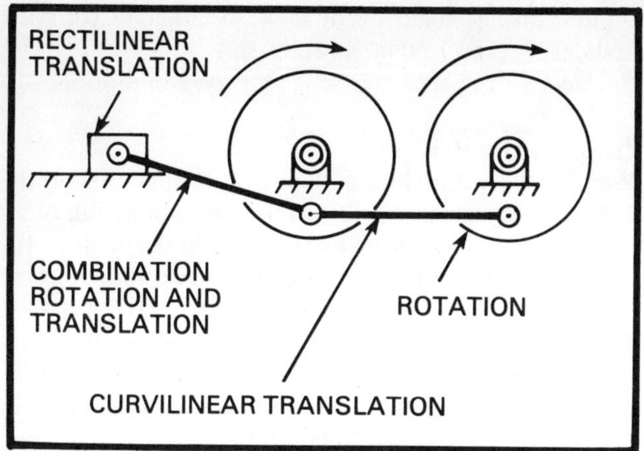

Fig. 3-9. Drive system illustrating four types of motion.

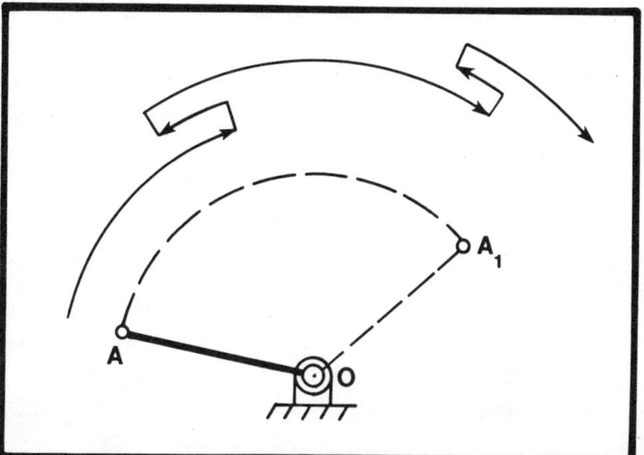

Fig. 3-10. Displacement of oscillating crank.

Spherical motion is defined as motion in space with each point on the moving member maintaining a constant distance from a common pivot point. See Fig. 3-11.

Helical motion occurs when combined motions of rotation and translation exist along the axis of rotation. A common example is the turning of a nut on a bolt.

The *velocity* of a point is the time rate of its change of position. Velocity of a point is usually given in inches per second, or in millimeters per second. Velocity of a rotating line is sometimes given in revolutions per second (or minute) or in degrees per second (or minute); however, it is generally given in radians per second (or minute).

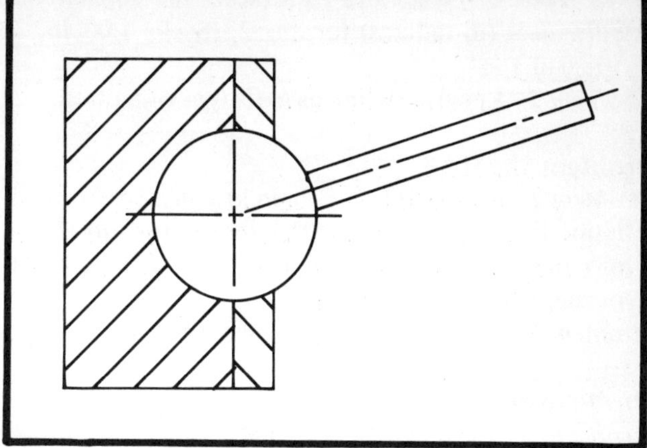

Fig. 3-11. Ball-socket joint.

The instantaneous linear velocity of a point on a rotating disk is proportional to its radius. This can be expressed as:

$$V = r\omega \qquad (3.1)$$

where V = linear velocity of point (in./sec.)
r = radius (in.)
ω = angular velocity of body (rad/sec.).

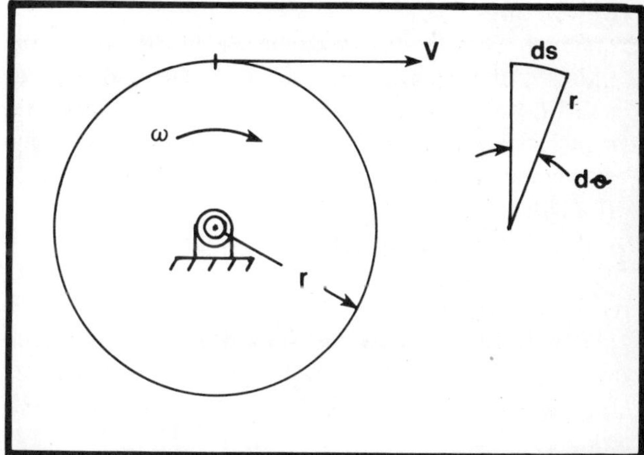

Fig. 3-12. Relationship between linear and angular velocity.

Problem Assignment

The following graphical problems and design questions are intended to provide you with experience in graphically analyzing and calculating basic mechanism constraints. Select, with instructor direction, the problem-sheet assignments suitable for you. Then complete the numbered problems in ordered sequence because each solution builds on preceding problems. Be sure to use the references suggested by your instructor (*list follows*) as sources for working through the assignment. Problem sheets are located at the back of the book.

Problem Sheet 3A

Problem 1. Rotate link 4 clockwise in 15° intervals for the illustrated portion of a transport mechanism. Plot the paths of D, E, and F. Indicate the angular displacement (in radians) for link 2. (Scale: 1.00 in. = 10.00 in.)

Problem 2. Label each link as to its type of motion.

Problem Sheet 3B

Problem 1. Rotate link 2 counterclockwise for its maximum displacement. Use 30° intervals for link 2 and plot the path for point F, indicating its displacement. (Scale: 1.00 in. = 10.00 in.)

Problem 2. Label each link as to its type of motion.

Problem Sheet 3C

Problem 1. Move point F to its highest and lowest positions by placing the piston (link 6) in successive positions; plot the path of point E. Indicate point E's displacement. Determine the ratio of the displacements for F and E. (Scale: 1.00 in. = 10.00 in.)

Problem 2. Label each link as to its type of motion.

Problem Sheet 3D

Problem 1. Move point F to trace the shape shown by dashed lines. Find the corresponding positions for points G and B. Determine the reduction ratios for points G and B to F. (Scale: 1.00 in. = 10.00 in.)

Problem 2. Label each link as to its type of motion.

Problem Sheet 3E

Problem 1. Rotate link 5 clockwise in 30° intervals. Indicate the angular displacement (in radians) of point E about point F. Show the maximum displacement of the piston (link 2). Does the maximum displacement occur at a 30° interval? (Scale: 1.00 in. = 10.00 in.)

Problem 2. Label each link as to its type of motion.

Problem Sheet 3F

Problem 1. Rotate the wheel (link 5) counterclockwise every 30° to determine the maximum displacements for the sliders (links 2, 3, 4, and 6). Does the maximum displacement occur at a 30° interval for each slider? (Scale: 1.00 in. = 10.00 in.)

Problem 2. Label each link as to its type of motion.

Problem Sheet 3G

Problem 1. Rotate link 5 clockwise so that E is on the horizontal centerline. Determine the elongation of xy for the coil tension spring (link 4) and the angle Ø. Indicate the load in the spring for the position shown. (Scales: 1.00 in. = 10.00 in.; .25 in. = 100 lbs.)

Problem 2. Rotate link 2 in either direction to determine its maximum displacement. Indicate the angular displacements of links 4 and 6. Plot the path of point E. (Scale: 1.00 in. = 10.00 in.)

Problem Sheet 3H

Problem 1. Rotate link 2 into positions X, Y, and Z. Points C and B move on vertical planes, link 4 revolves about axis D, and link 2 revolves about axis A. Illustrate the mechanism in the three views for the three new positions. Graphically determine true angles for β. (*Hint*: Use auxiliary views first to show plane $ABCD$ as an edge and then as a true-shape.)

Problem Sheet 3I

Problem 1. Input shaft A rotates at 1,750 rpm. Determine the rpm and direction of transfer disk D and the output shaft for B when D is revolved about point C to positions x, y, and z. (Scale: 1.00 in. = 5.00 in.)

Problem Sheet 3J

Problem 1. The transport mechanism shown is used for materials handling. The motion of the transport member (link 2) gives intermittent advancement of the materials (3, etc.). To move each individual article it is necessary that all points on link 2 move in similar and equal paths. Plot the path of link 2 by starting with point D. Rotate link 6 with 24 equal intervals. Make sure that the top of link 2 drops below surface 1. (Scale: 1.00 in. = 10.00 in.)

References

The following source and page references can be used in developing the graphical solutions and calculations for the problem assignments.

Esposito, Anthony. *Kinematics for Technology.* Columbus, Ohio: Charles E. Merrill Publishing Co., 1973, pp. 19-55.

Hardison, Thomas B. *Introduction to Kinematics.* Reston, Va.: Reston Publishing Co., 1979, pp. 31-55.

Hinkle, Rolland T. *Kinematics of Machines.* 2nd ed. Englewood Cliffs, N.J.: Prentice-Hall, 1960, pp. 11-33.

Hirschhorn, Jeremy. *Dynamics of Machinery.* New York: Barnes & Noble, 1967, pp. 18-44.

Kepler, Harold B. *Basic Graphical Kinematics.* 2nd ed. New York: McGraw-Hill Book Co., 1973, pp. 29-48.

Lent, Deane. *Analysis and Design of Mechanisms.* 2nd ed. Englewood Cliffs, N.J.: Prentice-Hall, 1970, pp. 1-62.

Michels, Walter J., and Wilson, Charles E. *Mechanism: Design-Oriented Kinematics.* Chicago: American Technical Society, 1969, pp. 1-82.

Oberg, Erik; Jones, Franklin D.; and Horton, Holbrook L. *Machinery's Handbook.* 20th ed. New York: Industrial Press, 1976, pp. 325-358.

Patton, William J. *Kinematics.* Reston, Va.: Reston Publishing Co., 1979, pp. 1-29.

Tao, D.C. *Fundamentals of Applied Kinematics.* Reading, Mass.: Addison-Wesley Publishing Co., 1967, pp. 7-20.

Instant Center Analysis 4

location of centros

A *machine* is a combination of resistant bodies so arranged that they transmit mechanical forces. They produce some effect or work, accompanied by certain determinate forces. A *mechanism* is a combination of rigid bodies so arranged that the motion of one compels the motion of the others.

A mechanism *link* may be defined as a rigid body which serves to transmit force from one piece to another or to cause or control motion. A mechanism *crank* link may be defined in a general way as an arm rotating or oscillating about an axis.

Four-bar mechanisms consist of a ground or fixed link, two cranks, and pin joints, connected by a floating link or coupler. A *constrained* mechanism is one with links so attached that all links move in a predictable manner, and in which all points on the various links will repeat their respective paths with each cycle of the mechanism. *Nonconstrained* mechanisms contain points on one or more links that have nonpredictable motion.

An *instant center* (*centro*) is a point about which any mechanism member rotates at a given instant of time. No matter how complex the motion of a mechanism member, it exhibits pure rotation about a unique point called a centro. The instant center of two links is the point at which these two bodies have no motion with respect to each other; the two links are revolving about a common center at that instant.

The characteristics of instantaneous centers make their use in velocity analysis very important. Their special features provide a precise and quick method for analyzing velocities in a complex mechanism. Important facts about instant centers include:

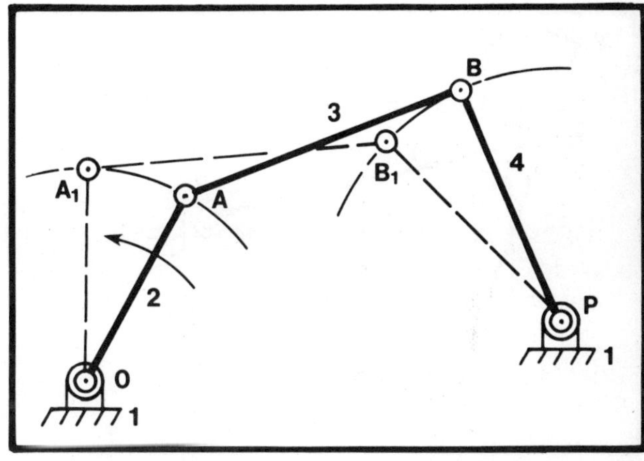

Fig. 4-2. Position analysis for constrained motion.

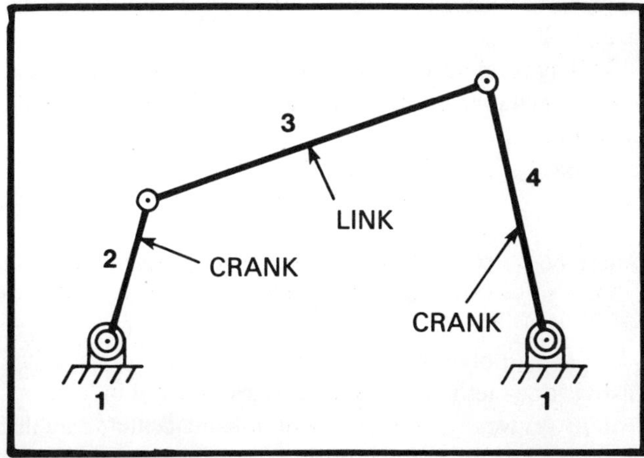

Fig. 4-1. Four-bar mechanism with constrained motion.

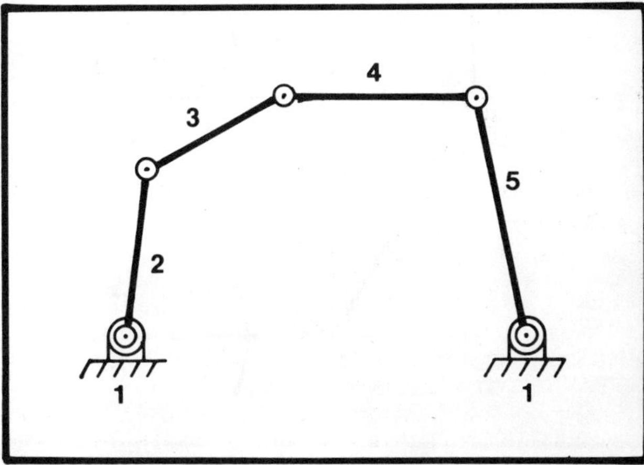

Fig. 4-3. Five-bar mechanism with nonconstrained motion.

1. There is one instant center for each floating link in a mechanism.

2. There is no one common instant center for all links in a mechanism.

3. Instant centers change as the mechanism moves.

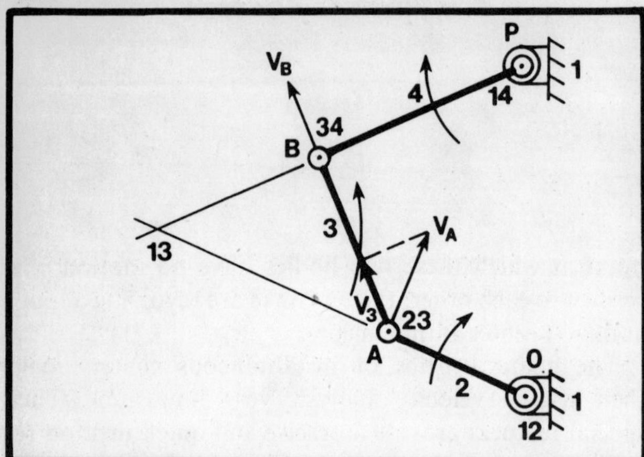

Fig. 4-4. Direction analysis and instant center location.

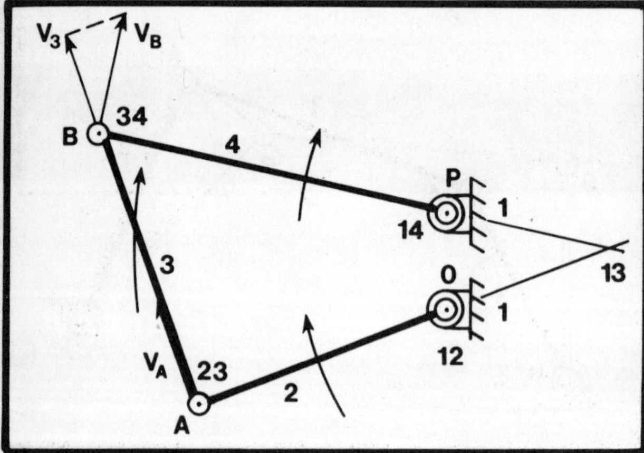

Fig. 4-5. Direction analysis and instant center location.

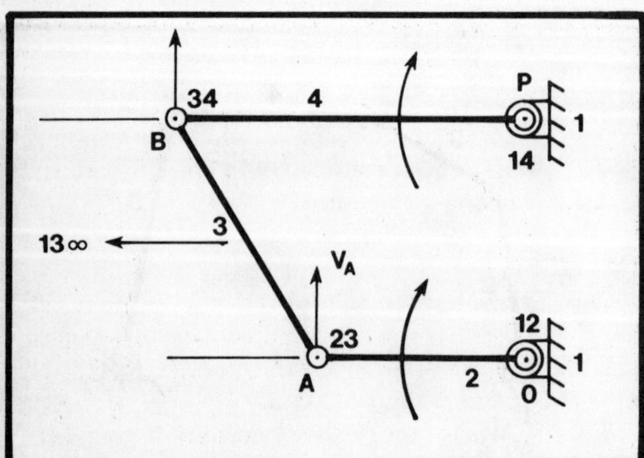

Fig. 4-6. Direction analysis and instant center location.

The linear velocity of a point on a crank link is determined by its distance from the center of rotation of the crank and the angular velocity of the crank. The velocities of two points on the same link are directly proportional to their distances from the link's center of rotation. This concept serves as the basis for the construction of a *proportional velocity triangle*, as illustrated in Fig. 4-7. The distance from the center of rotation is the only variable in the equation for linear velocity of the two points.

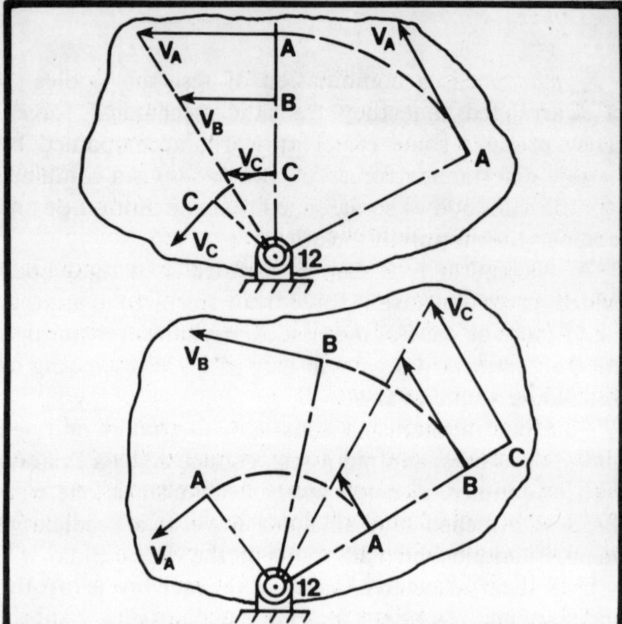

Fig. 4-7. Proportional velocity triangle for points on a rotating body.

Instant centers can be located by:

1. Observation.

2. Kennedy's Theorem, which states that any three bodies having plane motion relative to each other have only three instant centers which lie along the same straight line.

3. The total number of instant centers belonging to a given mechanism can be determined by the following equation:

$$N_C = \frac{N_L(N_L - 1)}{2} \tag{4.1}$$

where N_L = total number of fixed and moveable links
N_C = total number of instant centers (centros).

Creating polygons to represent the links of a mechanism is one method often used as a visual aid in locating instant centers. In Fig. 4-8, the instant centers for the four-bar linkage are found by first marking the easily identifiable joints on the mechanism as instant centers

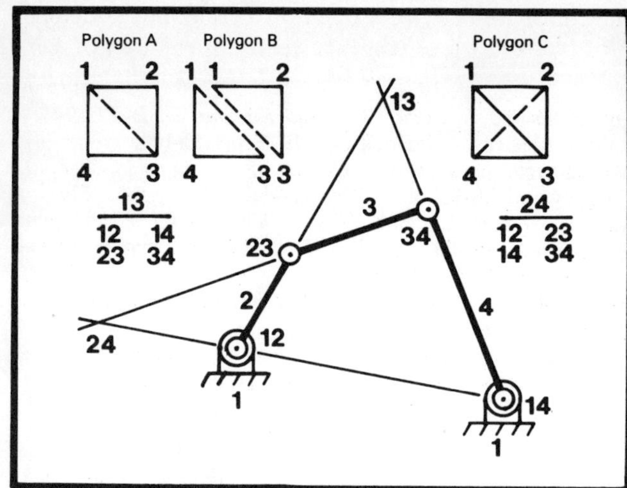

Fig. 4-8. Locating instant centers.

12, 23, 34, and 14 (numbers formed by combining the connecting link numbers). Four-cornered Polygon A is then drawn, with corners numbered in order—one for each link on the mechanism. The dashed line indicating instant center 13 (which must be located on the mechanism) divides the polygon into triangles with sides 12 and 23, and 14 and 34. These triangles are separated for clarity in Polygon B, and their sides recorded in the list below it. (There will be four pairs of numbers associated with locating any one instant center on the mechanism.) Instant center 13 on the mechanism will thus lie on 12, 23, 14, and 34. Polygon C shows the development of instant center 24; line 13 has been drawn solid after locating it on the mechanism, and the same procedure is used to form triangles with common line 24.

You can now apply the preceding approach to a six-bar mechanism (Fig. 4-9). At first glance this may appear to be a difficult problem, but the centro method is very straightforward and easy to apply.

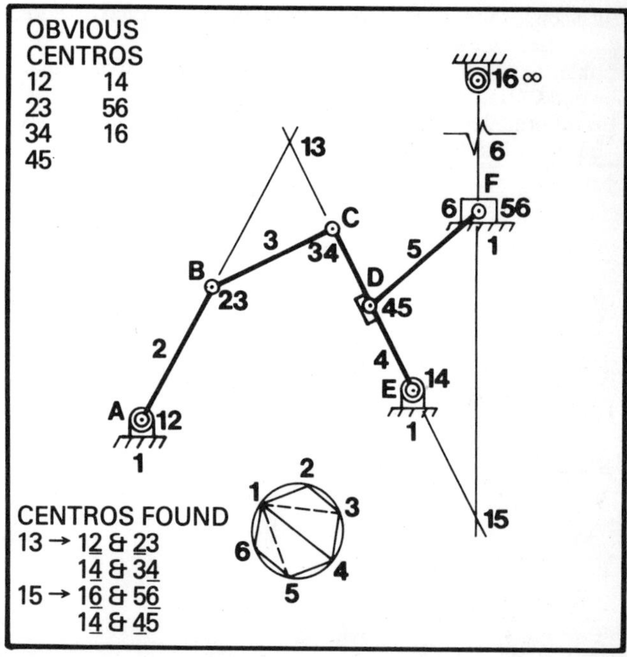

Fig. 4-9. Instant center analysis of a six-bar mechanism.

Problem Assignment

The following graphical problems and design questions are intended to provide you with experience in graphically analyzing and calculating basic mechanism constraints. Select, with instructor direction, the problem-sheet assignments suitable for you. Then complete the numbered problems in ordered sequence because each solution builds on preceding problems. Be sure to use the references suggested by your instructor (*list follows*) as sources for working through the assignment. Problem sheets are located at the back of the book.

Problem Sheet 4A

Problem 1. Locate the instant center for the slider crank mechanism.

Problem 2. Locate the instant centers for *AB* and for *BCD.*

Problem Sheet 4B

Problem 1. Locate the instant center for *CBF* and for *AB.*

Problem 2. Locate the instant center for *DC.*

Problem Sheet 4C

Problem 1. Locate the instant centers for *FCx* and for *BCD.*

Problem Sheet 4D

Problem 1. Determine the relative velocities for points *A*, *B*, *C*, and *E* when the velocity of point *D* is known. (Scale: 1.0 mm = 100.0 mm/sec.)

Problem Sheet 4E

Problem 1. Solve for the relative velocities of points *B* and *C* and prove your answer mathematically. Determine the angular velocity of the crank. (Scales: 1.00 in. = 800 fpm; 1.00 in. = 1.00 in.)

Problem 2. Solve for the relative velocities of points *B*, *C*, and *D* and prove your answer mathematically. Determine the angular velocity of the cam. (Scales: 1.00 in. = 50.00 in./sec.; 1.00 in. = 2.00 in.)

References

The following source and page references can be used in developing the graphical solutions and calculations for the problem assignments.

Esposito, Anthony. *Kinematics for Technology.* Columbus, Ohio: Charles E. Merrill Publishing Co., 1973, pp. 72-89.

Hardison, Thomas B. *Introduction to Kinematics.* Reston, Va.: Reston Publishing Co., 1979, pp. 57-84.

Hinkle, Rolland T. *Kinematics of Machines.* 2nd ed. Englewood Cliffs, N.J.: Prentice-Hall, 1960, pp. 34-43.

Hirschhorn, Jeremy. *Dynamics of Machinery.* New York: Barnes & Noble, 1967, pp. 10, 28, 29.

Kepler, Harold B. *Basic Graphical Kinematics.* 2nd ed. New York: McGraw-Hill Book Co., 1973, pp. 73-100.

Lent, Deane. *Analysis and Design of Mechanisms.* 2nd ed. Englewood Cliffs, N.J.: Prentice-Hall, 1970, pp. 55, 57.

Michels, Walter J., and Wilson, Charles E. *Mechanism: Design-Oriented Kinematics.* Chicago: American Technical Society, 1969, pp. 148-162.

Patton, William J. *Kinematics.* Reston, Va.: Reston Publishing Co., 1979, p. 71.

Tao, D.C. *Fundamentals of Applied Kinematics.* Reading, Mass.: Addison-Wesley Publishing Co., 1967, pp. 48-60.

Velocity Analysis of Linkages

5

instant centers and velocity triangles

An understanding of velocity is one of the most important concepts in the study of mechanisms. Acceleration and displacement are directly related to velocity. You should develop a strong foundation of velocity analysis in order to work the majority of mechanism problems which you will encounter.

The preceding units have presented material on motion concepts, handling of velocities, and location of instantaneous centers of rotation. You are now asked to transfer the velocity of a crank to other links and points in a mechanism. In Fig. 5-1, the velocity of point *A* is equal to 10 in./sec. in the direction shown; you are to find the direction and velocity of points *B* and *C* and the angular velocity of link 4.

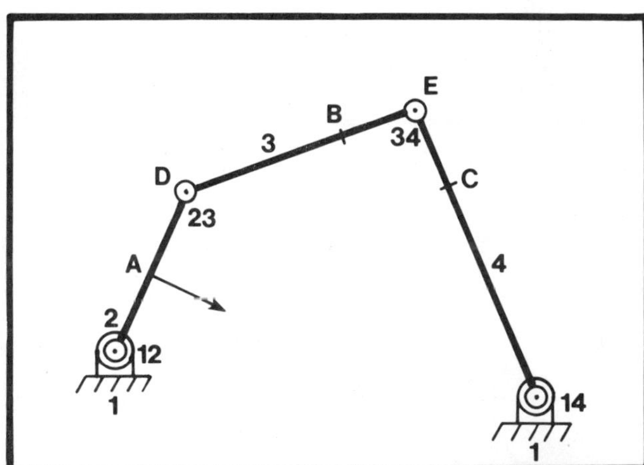

Fig. 5-1. Velocity analysis of a four-bar linkage.

Since point *A* lies on link 2 only, instant center 12 must be used to transfer this velocity to point *D* on link 2. *D* is common to link 3—the link on which point *B* lies. Therefore, the first step is to find the velocity of *D* and to locate instant center 13. (The required construction is shown in Fig. 5-2.) Once these are known, you can find the velocity of any other point on link 3 by rotating the desired point to form a new *proportional velocity triangle.* In this way, velocities for points *B* and *E* can be determined. In Fig. 5-3, *B* and *E* were rotated

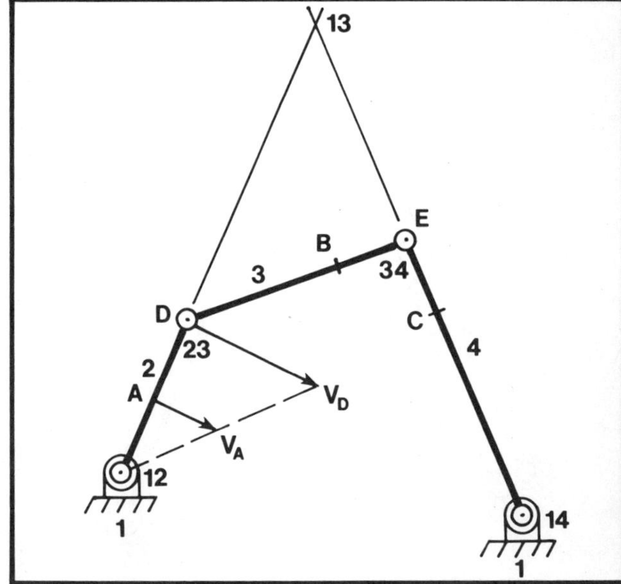

Fig. 5-2. Proportional velocity triangle and instant center construction.

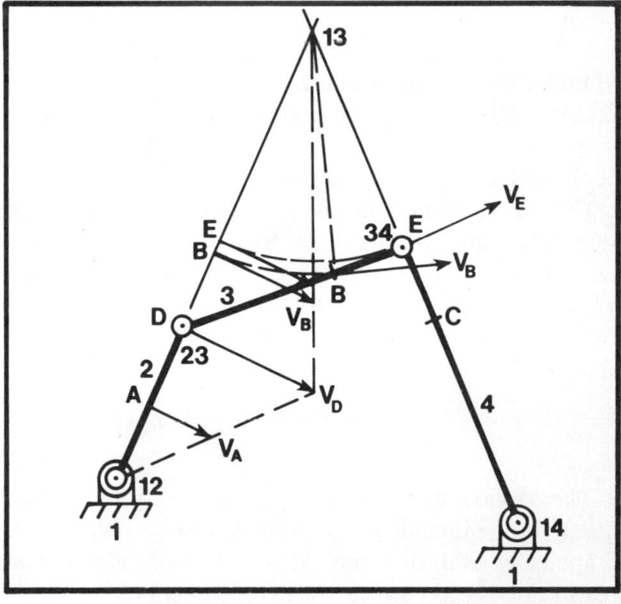

Fig. 5-3. Proportional velocity triangle and vector rotation.

to form proportional velocity triangles; their velocity magnitudes were then determined, and the points rotated back to their original positions. Velocity magnitude can be scaled from either position of the vector.

There is now a known velocity for a point on link 4—namely, *E*. By using instant center 14, you can draw a proportional velocity triangle for link 4 and find the velocity magnitude and direction of point *C*. The angular velocity for link 4 can be shown by measuring one unit out from instant center 14. Calculations will verify the graphical angular velocity solution.

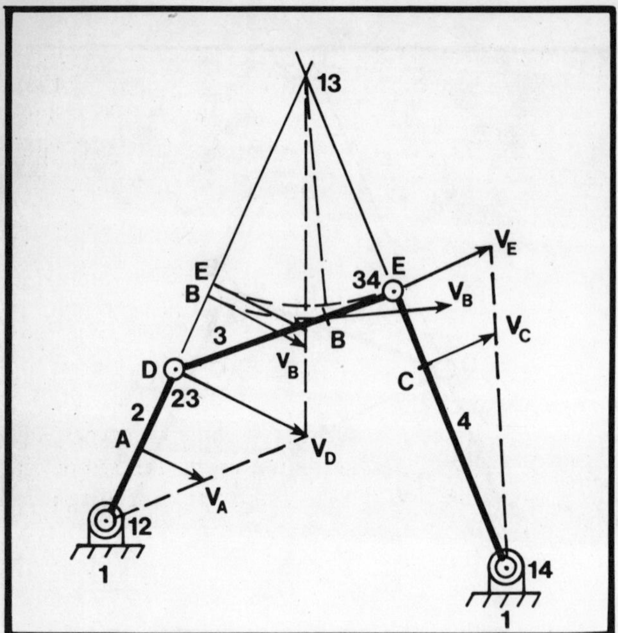

Fig. 5-4. Angular velocity determination by use of instant centers.

You can now apply the preceding approach to a more complicated six-bar linkage, as in Fig. 5-5. If the angular velocity of link 2 is known, you can construct proportional velocity triangles to determine the velocity of piston F in the position shown.

The component velocity method is one other graphical technique often used to analyze velocities in a mechanism. This technique has been applied to the four-bar linkage in Fig. 5-6 and the six-bar linkage in Fig. 5-7.

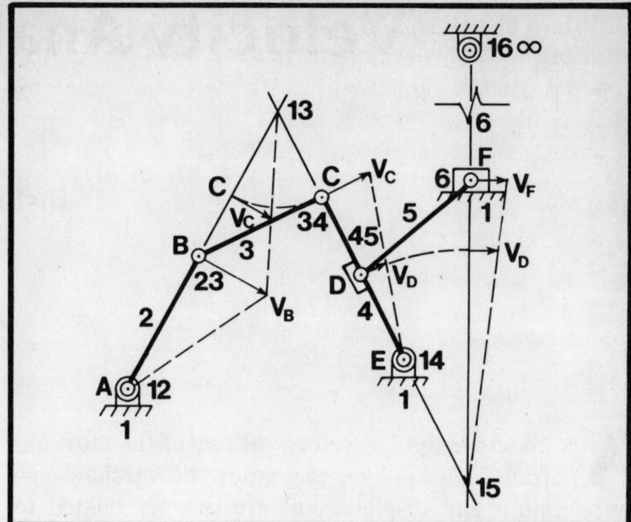

Fig. 5-5. Velocity analysis of a six-bar linkage.

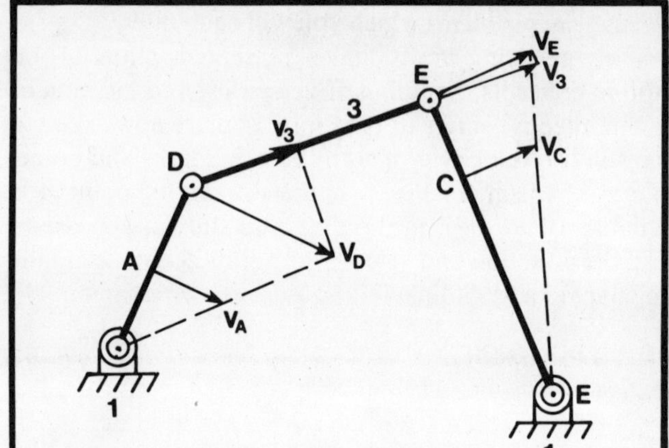

Fig. 5-6. Component velocity method (four-bar linkage).

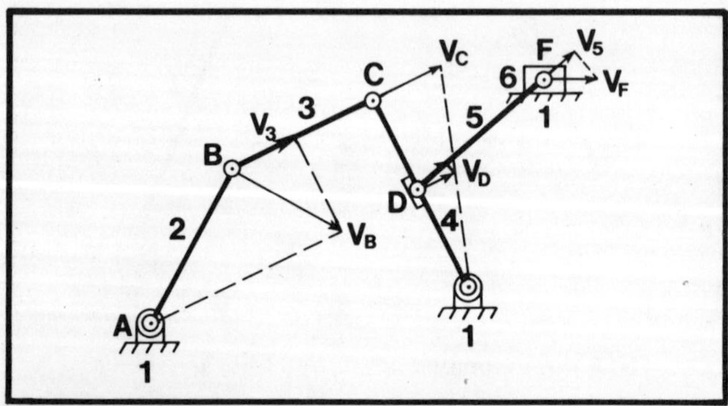

Fig. 5-7. Component velocity method (six-bar linkage).

Problem Assignment

The following graphical problems and design questions are intended to provide you with experience in graphically analyzing and calculating basic mechanism constraints. Select, with instructor direction, the problem-sheet assignments suitable for you. Then complete the numbered problems in ordered sequence because each solution builds on preceding problems. Be sure to use the references suggested by your instructor (*list follows*) as sources for working through the assignment. Problem sheets are located at the back of the book.

Problem Sheet 5A

Problem 1. The velocity of slider C is 1.75 in./sec. along its constrained path to the left. Determine the velocities of link 3 and point B. Also determine the angular velocity of link 2. Use the component velocity method in completing the solution. (Scales: 1.00 in. = 1.00 in./sec.; 1.00 in. = 2.00 in.)

Problem 2. The velocity of slider D is 1.5 in./sec. along its constrained path in a downward direction. Determine the velocities of points A, B, and C by use of instant centers. Verify your answers mathematically. (Scales: 1.00 in. = 1.00 in./sec.; 1.00 in. = 2.00 in.)

Problem Sheet 5B

Problem 1. Wheel 5 turns at 1.5 rad/sec. counterclockwise about 0. Determine the velocities of points C, D, E, and F and links 7, 8, 9, and 10. Use the component velocity method in completing the solution. (Scales: 1.0 mm = 1.0 mm/sec.; 1.0 mm = 1.0 mm.)

Problem Sheet 5C

Problem 1. Link 4 revolves with a clockwise angular velocity of 2.0 rad/sec. Determine the velocities of points C, D, E, and F and the angular velocity of link 2. Use instant centers in completing the solution; verify your answers mathematically. (Scales: 1.00 in. = 1.00 in./sec.; 1.00 in. = 2.00 in.)

Problem Sheet 5D

Problem 1. Slider 6 (point F) is moving downward at a velocity of 1.5 in./min. Determine the angular velocities of links 2 and 4 and the velocity of point E. (Scales: 1.00 in. = 1.00 in./sec.; 1.00 in. = 2.00 in.)

Problem Sheet 5E

Problem 1. Point C has a downward vertical velocity of 100.00 in./sec. Determine the velocities of points B and D. Also determine the angular velocity of link 5. (Scales: 1.00 in. = 100.00 in./sec.; 1.00 in. = 2.00 in.)

Problem 2. Determine the velocities of points x, B, and C when link 4 rotates counterclockwise with an angular velocity of 10.0 rad/sec. Select the appropriate velocity scale. (Scale: 1.00 in. = 1.00 in.)

Problem Sheet 5F

Problem 1. Crank link 2 turns clockwise about A at 1.25 rad/sec. Determine the velocity of point F. (Scales: 1.00 in. = 1.00 in./sec.; 1.00 in. = 1.00 in.)

Problem Sheet 5G

Problem 1. Crank link 2 turns clockwise about A at 1.12 rad/sec. Determine the velocity of point x. (Scales: 1.00 in. = 1.00 in./sec.; 1.00 in. = 1.00 in.)

Problem Sheet 5H

Problem 1. Link 2 rotates about point A at 1.00 rad/sec. Determine the velocity of point C when point B is in the B, X, Y, and Z positions. (Scales: 1.00 in. = 1.00 in./sec.; 1.00 in. = 1.00 in.)

Problem Sheet 5I

Problem 1. Link 5 rotates counterclockwise about A at 1.00 rad/sec. Determine the velocities of D and E. (Scales: 1.00 in. = 1.00 in./sec.; 1.00 in. = 1.00 in.)

Problem Sheet 5J

Problem 1. Point F travels at a constant velocity of 1.00 in./sec. along the path shown. Determine the angular velocities for links 2 and 5 when F is at locations 1, 4, and 9. (Scales: 1.00 in. = 1.00 in./sec.; 1.00 in. = 1.00 in.)

Problem Sheet 5K

Problem 1. Link 5 rotates about point A with an angular velocity of 10.00 rad/sec. Determine the velocities of points C and E along centerline 4 (spring centerline). Select an appropriate velocity scale. (Scale: 1.00 in. = 2.00 in.)

Problem 2. Link 2 rotates clockwise about point A with an angular velocity of 20.00 rad/sec. Determine the velocity of point F. Select an appropriate velocity scale. (Scale: 1.00 in. = 2.00 in.)

References

The following source and page references can be used in developing the graphical solutions and calculations for the problem assignments.

Esposito, Anthony. *Kinematics for Technology.* Columbus, Ohio: Charles E. Merrill Publishing Co., 1973, pp. 57-112.

Hardison, Thomas B. *Introduction to Kinematics.* Reston, Va.: Reston Publishing Co., 1979, pp. 57-84.

Hinkle, Rolland T. *Kinematics of Machines.* 2nd ed. Englewood Cliffs, N.J.: Prentice-Hall, 1960, pp. 34-68.

Hirschhorn, Jeremy. *Dynamics of Machinery.* New York: Barnes & Noble, 1967, pp. 9-44.

Kepler, Harold B. *Basic Graphical Kinematics.* 2nd ed. New York: McGraw-Hill Book Co., 1973, pp. 99-161.

Lent, Deane. *Analysis and Design of Mechanisms.* 2nd ed. Englewood Cliffs, N.J.: Prentice-Hall, 1970, pp. 39-93.

Michels, Walter J., and Wilson, Charles E. *Mechanism: Design-Oriented Kinematics.* Chicago: American Technical Society, 1969, pp. 83-168.

Oberg, Erik; Jones, Franklin D.; and Horton, Holbrook L. *Machinery's Handbook.* 20th ed. New York: Industrial Press, 1976, pp. 325-330.

Patton, William J. *Kinematics.* Reston, Va.: Reston Publishing Co., 1979, pp. 31-76.

Tao, D.C. *Fundamentals of Applied Kinematics.* Reading, Mass.: Addison-Wesley Publishing Co., 1967, pp. 62-108.

Acceleration Analysis 6

rate of change of mechanism velocities

MECHANICAL designers must be concerned with acceleration of mechanisms. Forces acting within and as a result of their acceleration often must be determined. The basic equation you will work with is

$$F = ma \qquad (6.1)$$

where F = force
m = mass
a = acceleration

Acceleration is the rate of change of velocity with respect to time. An automobile moving away from a stop sign has an initial or original velocity of zero. Within a short distance it has obtained a speed of 60 mph. If the driver were to accelerate at a constant rate and to move the automobile that distance in a time period of 6 sec., then the car would have a linear acceleration rate of 14.65 ft./sec².

The change of linear velocity per unit of time is called *linear acceleration.* A body in translation has points that move with linear acceleration. It is a vector quantity defined as

$$a = \frac{\Delta V}{\Delta T} = \frac{V_f - V_i}{T_f - T_i} \qquad (6.2)$$

where a = linear acceleration (in./sec², ft./min²)
V_f = final velocity
V_i = initial velocity
T_f = final time
T_i = initial time

Equation (6.2) applies when the acceleration is constant or when ΔT is very small. If the rate of acceleration is changing, then the acceleration calculated is the average acceleration. The average acceleration for a given time interval n is (see Fig. 6-1):

$$a_n^{av} = \frac{\Delta V}{\Delta T} = \frac{V_f - V_i}{\Delta T} \qquad (6.3)$$

When the velocity does not equal zero at $T = 0$, the following equation is used (see Fig. 6-2):

$$V_f = a(T_f - T_i) + V_i \qquad (6.4)$$

There is a specific parabolic relationship between displacement and time. Mathematically this is shown as

$$S = \left(\frac{V_i + V_f}{2}\right) \Delta T \qquad (6.5)$$

where S = displacement.
Rearranging these terms gives us

$$V_f = \frac{2S}{\Delta T} - V_i \qquad (6.6)$$

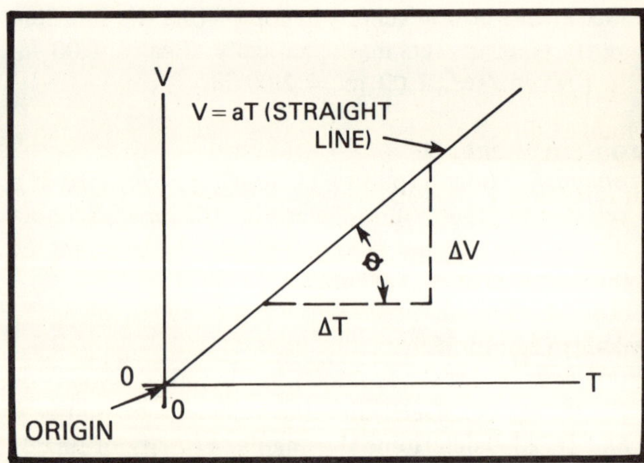

Fig. 6-1. Constant acceleration motion ($V = 0$ when $T = 0$).

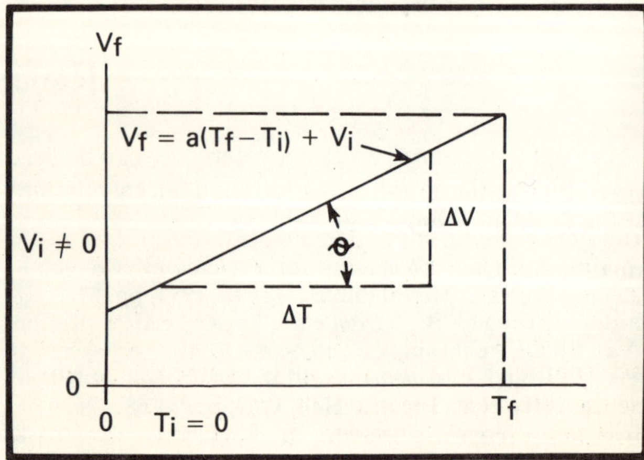

Fig. 6-2. Constant acceleration motion ($V \neq 0$ when $T = 0$).

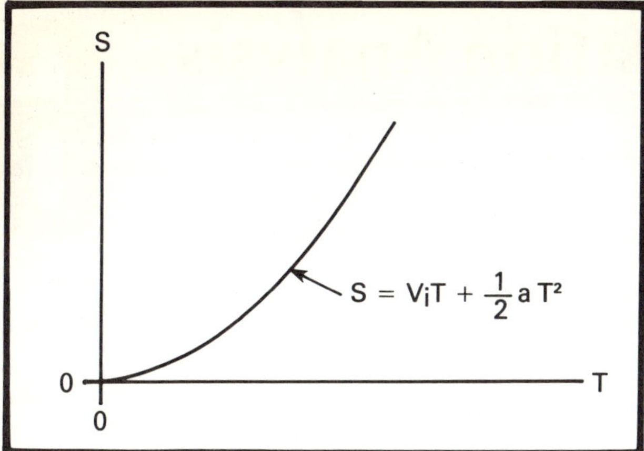

Fig. 6-3. Parabolic motion.

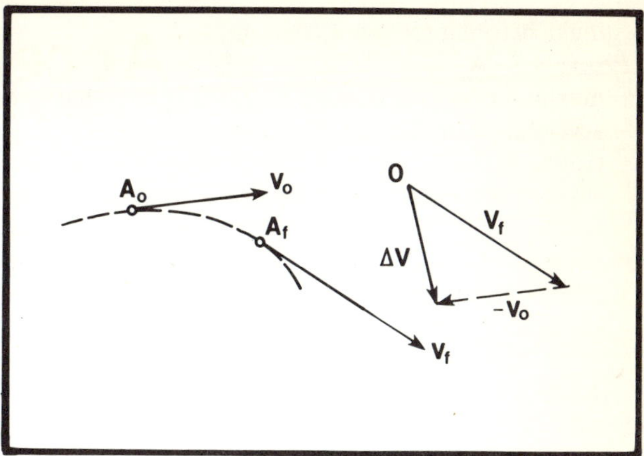

Fig. 6-4. Accelerating point on a curved path.

But since

$$V_f = V_i + a\Delta T \qquad (6.7)$$

then equating the two V_f terms results in

$$\frac{2S}{\Delta T} - V_i = V_i + a\Delta T \qquad (6.8)$$

or, upon rearrangement,

$$S = V_i\Delta T + \frac{a(\Delta T)^2}{2} \qquad (6.9)$$

When a point travels a curved path, its velocity changes direction and magnitude. The velocity change ΔV becomes a vector difference between V_0 (the velocity at zero) and V_f. Acceleration is then expressed as

$$a = \frac{V_f - \rightarrow V_0}{\Delta T} \qquad (6.10)$$

Uniform angular acceleration is the change in angular velocity ($\Delta\omega$) for a given time interval, and is written as

$$\alpha = \frac{\Delta\omega}{\Delta T} = \frac{\omega_f - \omega_i}{\Delta T} \qquad (6.11)$$

where α = angular acceleration
ω = angular velocity

Therefore

$$\omega_f = \omega_i + \alpha\Delta T \qquad (6.12)$$

Angular displacement of a line during a given time interval equals the average angular velocity during that interval multiplied by the time:

$$\theta = \omega_n^{av} \times \Delta T \qquad (6.13)$$

where θ = angular displacement.

Note that the following equations as well as those listed above are similar to those of uniform and variable linear acceleration.

$$\omega_n^{av} = \frac{\omega_i + \omega_f}{2} \qquad (6.14)$$

$$\omega_n^{av} = \omega_i + \frac{\alpha\Delta T}{2} \qquad (6.15)$$

$$\theta = \omega_i\Delta T + \frac{\alpha(\Delta T)^2}{2} \qquad (6.16)$$

$$(\omega_f)^2 = (\omega_i)^2 + 2\alpha\theta \qquad (6.17)$$

Problem Assignment

The following graphical problems and design questions are intended to provide you with experience in graphically analyzing and calculating basic mechanism constraints. Select, with instructor direction, the problem-sheet assignments suitable for you. Then complete the numbered problems in ordered sequence because each solution builds on preceding problems. Be sure to use the references suggested by your instructor (*list follows*) as sources for working through the assignment. Problem sheets are located at the back of the book.

Problem Sheet 6A

Problem 1. An automobile starting from a stop sign reaches a speed of 55 mph in 20 sec. If the acceleration of the automobile is uniform, how far does it travel before reaching 55 mph? What is the average acceleration during this time?

Problem 2. A six-cylinder engine accelerates uniformly from 0 rpm to a speed of 4,200 rpm. It requires 0.4 sec. to accelerate from 1,000 rpm to 2,000 rpm. What is its acceleration? How many revolutions must it make before it reaches 4,200 rpm?

Problem 3. A 12-in.-diameter grinding wheel has a maximum surface velocity of 2,000 fpm. What is its acceleration at that speed? If the grinding wheel is operating at 1,500 rpm and decelerating at 16 rad/sec², what is the total acceleration of a point on its surface?

Problem 4. A flywheel has a clockwise angular velocity of 5 rad/sec. and a counterclockwise angular acceleration of 20 rad/sec². The radius of the flywheel is 0.5 m. Determine the total acceleration of a point on the circumference of the flywheel.

Problem 5. A machine part drops from rest at a constant acceleration of 32 ft./sec². Determine the distance it falls during each successive period of 1 sec., for a total of 3 sec. What is the ratio of the distance covered in the first period to that covered in the second period, and the ratio of the distance in the first to that in the third?

Problem Sheet 6B

Problem 1. Wheel 2 turns counterclockwise about 0 with an angular velocity of 2.5 rad/sec. It has an angular acceleration of 10 rad/sec.² in a clockwise direction. Determine the linear accelerations of points *A, B, C, D,* and *E* on the wheel. Select an appropriate acceleration scale. (Scale: 1.00 in. = 1.00 in.)

Problem Sheet 6C

Problem 1. Crank link 2 turns about point *A* with constant angular velocity of 75 rad/sec. Determine the accelerations of points *x* and *y* in the position shown. Select an appropriate acceleration scale. (Scale: 1.00 in. = 1.00 in.)

Problem 2. The absolute acceleration of pin *A* on link 3 is 2.00 in./sec.² toward the left. The angular acceleration of link 3 is 2.5 rad/sec.² in a clockwise direction. Find the absolute acceleration of *B*. Select an appropriate acceleration scale. (Scale: 1.00 in. = 1.00 in.)

References

The following source and page references can be used in developing the graphical solutions and calculations for the problem assignments.

Esposito, Anthony. *Kinematics for Technology.* Columbus, Ohio: Charles E. Merrill Publishing Co., 1973, pp. 115-144.

Hardison, Thomas B. *Introduction to Kinematics.* Reston, Va.: Reston Publishing Co., 1979, pp. 85-104.

Hinkle, Rolland T. *Kinematics of Machines.* 2nd ed. Englewood Cliffs, N.J.: Prentice-Hall, 1960, pp. 69-94.

Hirschhorn, Jeremy. *Dynamics of Machinery.* New York: Barnes & Noble, 1967, pp. 19-29, 78, 356.

Kepler, Harold B. *Basic Graphical Kinematics.* 2nd ed. New York: McGraw-Hill Book Co., 1973, pp. 163-215.

Lent, Deane. *Analysis and Design of Mechanisms.* 2nd ed. Englewood Cliffs, N.J.: Prentice-Hall, 1970, pp. 94-149.

Michels, Walter J., and Wilson, Charles E. *Mechanism: Design-Oriented Kinematics.* Chicago: American Technical Society, 1969, pp. 169-215.

Oberg, Erik; Jones, Franklin D.; and Horton, Holbrook L. *Machinery's Handbook.* 20th ed. New York: Industrial Press, 1976, pp. 325-328.

Patton, William J. *Kinematics.* Reston, Va.: Reston Publishing Co., 1979, pp. 43-46, 65-71, 94-99.

Tao, D.C. *Fundamentals of Applied Kinematics.* Reading, Mass.: Addison-Wesley Publishing Co., 1967, pp. 110-150.

Cam Analysis 7

cam profiles, nomenclature, and displacement diagrams

CAMS are very versatile mechanisms. Through cam design you can take simple input motion and develop a large number of different and complex output motions. A few applications of cams are sewing machines, bottling and canning machines, weaving machines, machine tools, office machines, automobile engines, printing machines, and many household appliances.

The function of a cam is to impart motion to a follower by direct contact. The follower will then actuate some other link in a larger mechanism. Important factors in the motion of a follower include:

1. type of cam,
2. type and position of the follower,
3. the cam profile.

Cam profiles are developed around a *base circle.* Given an angular rotation of the cam, the follower will be moved a required distance along its axis or about a pivot point with a desired type of displacement. Cam design requires that the follower displacement be as gradual as possible in order to prevent large acceleration forces in the mechanism. Traditional motions consisting of

1. uniform velocity,
2. simple harmonic,
3. uniform acceleration (parabolic),
4. modified uniform velocity, and
5. cycloidal

are generally used but special motion may be designed into the cam.

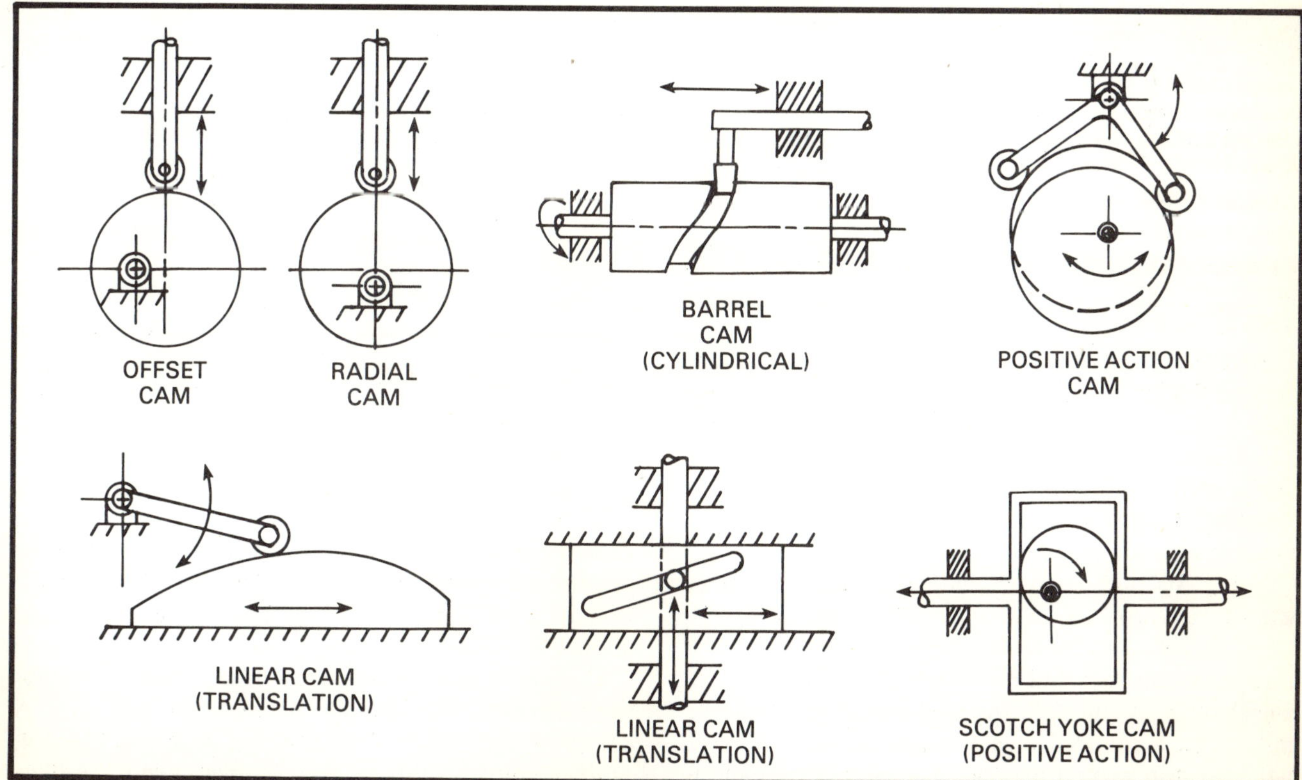

Fig. 7-1. Examples of cams.

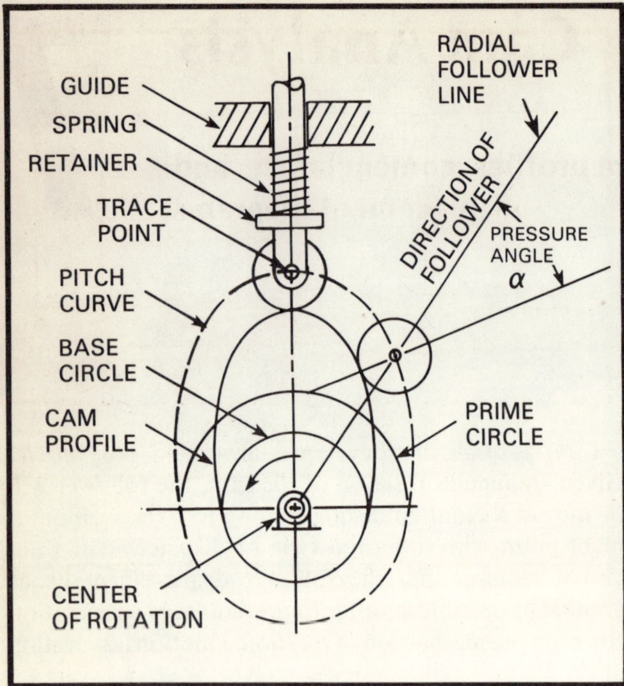

Fig. 7-2. Cam terminology.

Before you start to design cams, you should have a clear understanding of basic terminology. Fig. 7-2 and the list following indicate the main terms:

Trace Point. The end point of a knife-edge follower, or the center of the roller for a roller-type follower.

Cam Profile. Actual shape of the cam.

Pitch Curve. The path of the trace point.

Base Circle. The smallest circle drawn tangent to the cam profile with its center on the axis of the camshaft. The size of the base circle determines the size of the cam.

Prime Circle. The smallest circle drawn tangent to the pitch curve with its center on the axis of the camshaft.

Pressure Angle. The angle between the normal to the pitch curve and the direction of motion of the follower at the point of contact.

Direction of Motion. Positive or negative displacement of the follower away from the base circle or base line.

Follower. The mechanism link that is activated or displaced by the force developed by the cam profile.

Guide. A bearing or limiting surface designed to maintain a constant axis of follower motion.

Spring. Its purpose is to maintain follower contact with the cam profile.

The configuration and position of cam followers vary considerably. A cam follower may consist of a point or knife edge, a rounded or a flat surface. The follower path may also vary; it may be straight or curved and it may pass through the cam center or be offset.

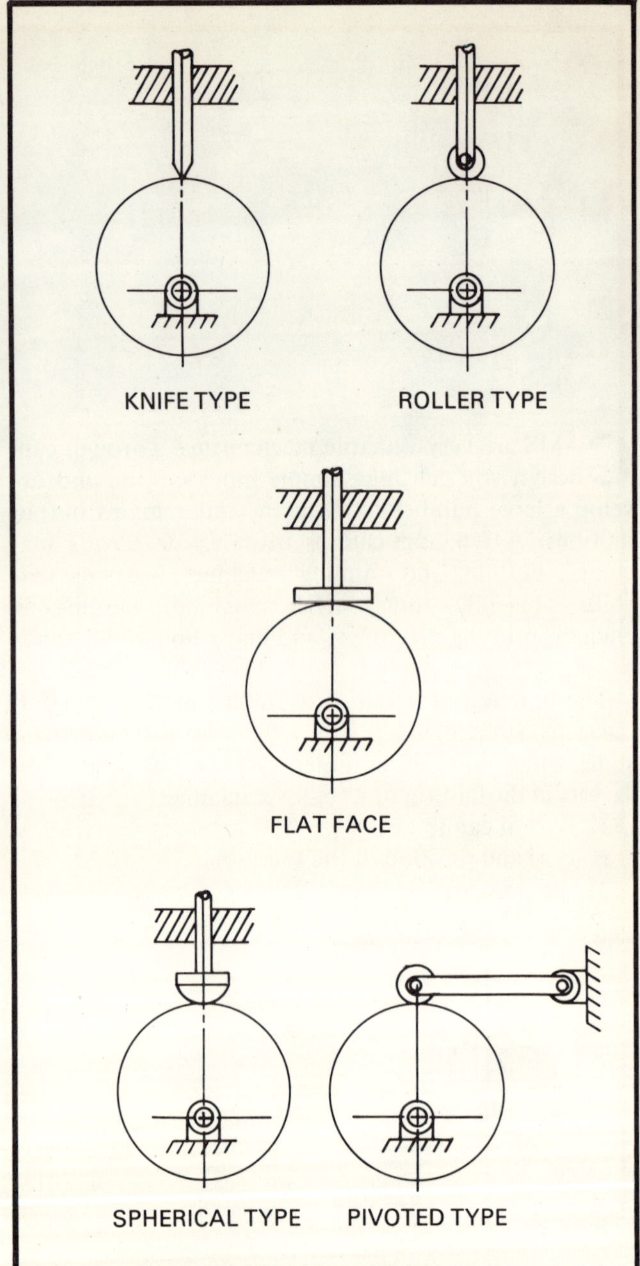

Fig. 7-3. Forms and paths of followers.

The profile of a high-speed cam (greater than 1,000 rpm) should be determined mathematically with the aid of a computer. The greatest number of cams operate at relatively low rpm, however, and their profiles can be determined by making graphical layouts. The construction of most cam profiles basically follows rules of applied geometry.

Motions of cam followers can be expressed by various displacement, velocity, and acceleration diagrams. A comparison of follower motions can be seen in Fig. 7-4; graphic constructions of their displacements are illustrated in Figs. 7-5 through 7-9. (Also see Appendix E.)

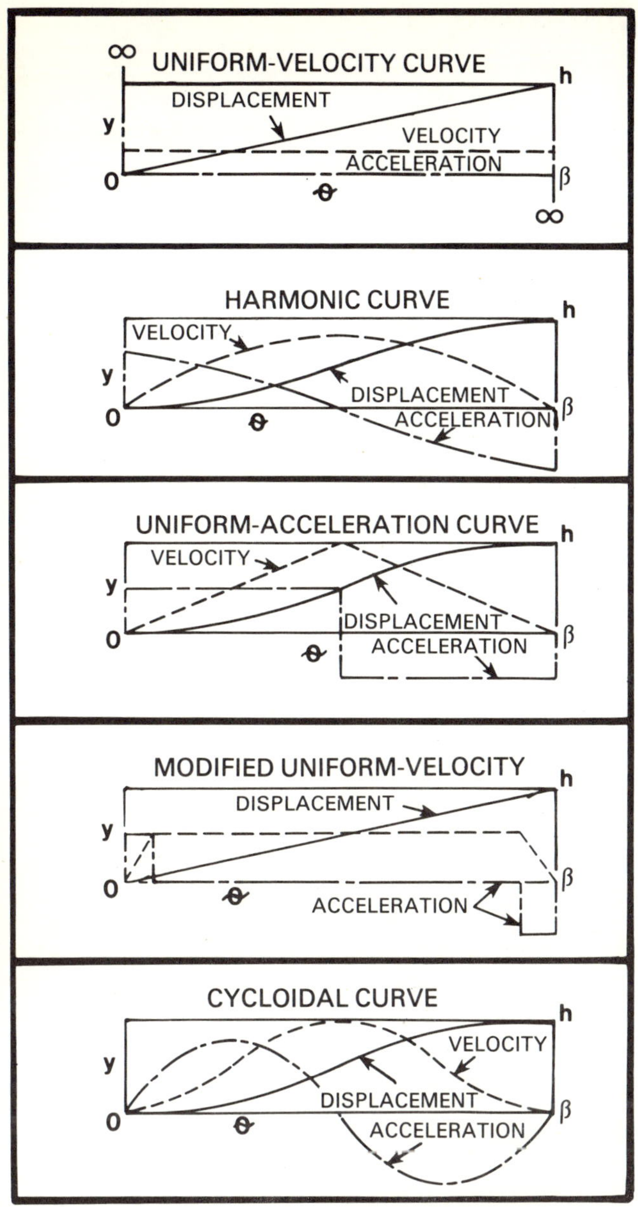

Fig. 7-4. Comparison of motion diagrams.

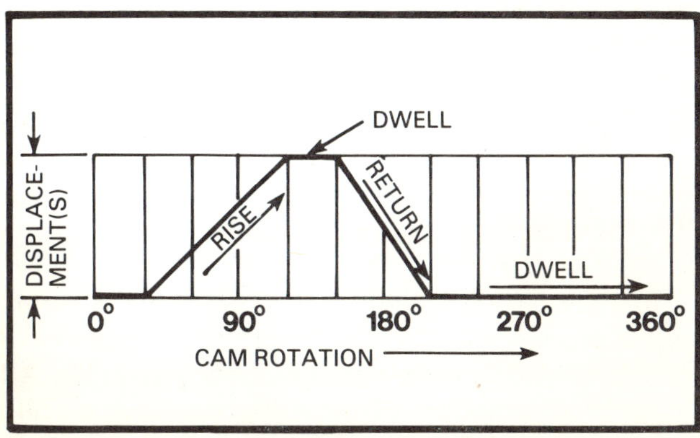

Fig. 7-5. Uniform velocity.

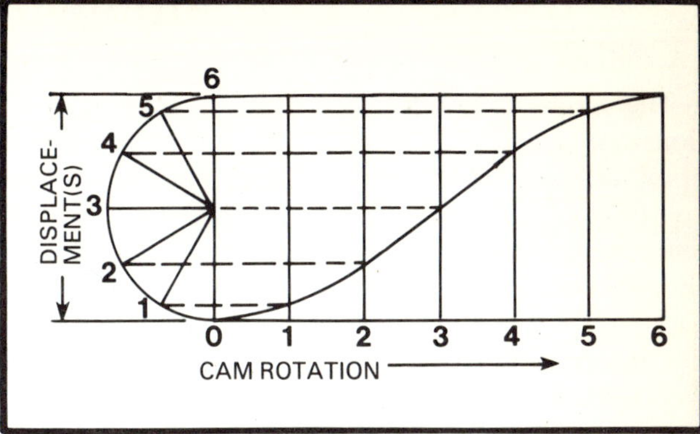

Fig. 7-6. Simple harmonic motion.

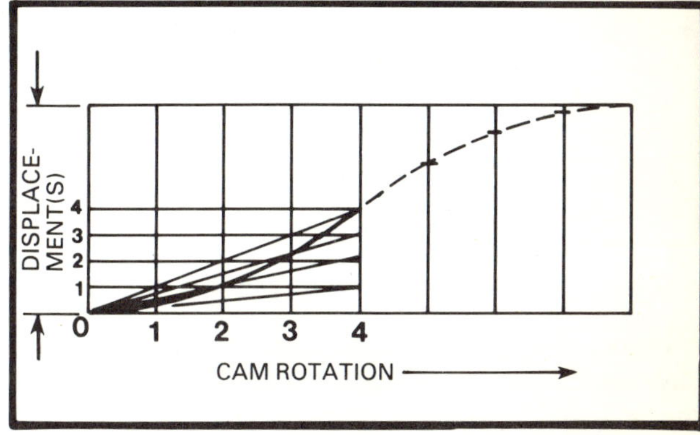

Fig. 7-7. Uniform acceleration (parabolic).

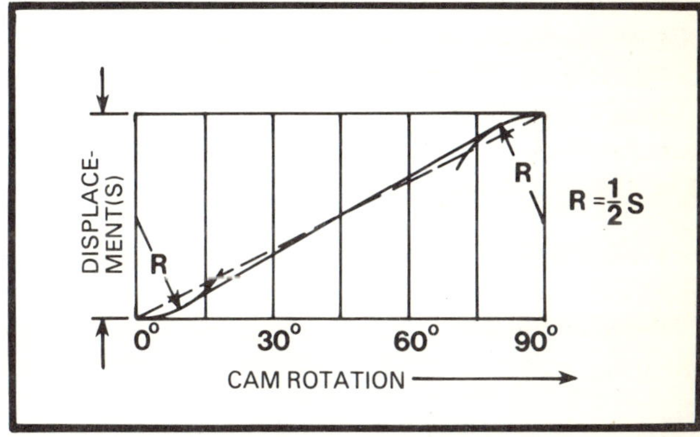

Fig. 7-8. Modified uniform velocity.

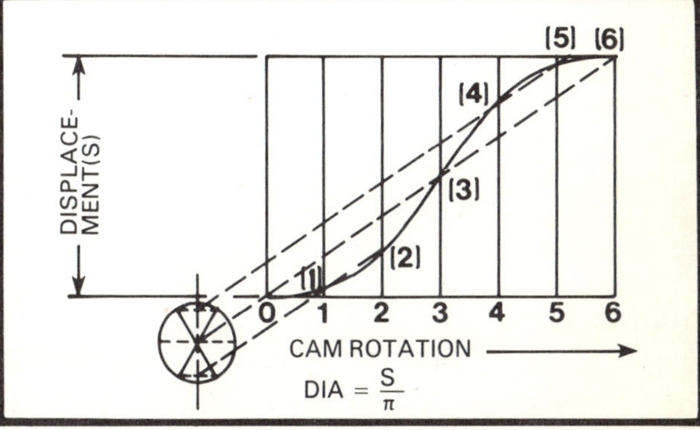

Fig. 7-9. Cycloidal.

Problem Assignment

The following graphical problems and design questions are intended to provide you with experience in graphically analyzing and calculating basic mechanism constraints. Select, with instructor direction, the problem-sheet assignments suitable for you. Then complete the numbered problems in ordered sequence because each solution builds on preceding problems. Be sure to use the references suggested by your instructor (*list follows*) as sources for working through the assignment. Problem sheets are located at the back of the book.

Problem Sheet 7A

Problem 1. In the space provided, construct a displacement diagram for the following motions:
- A. Uniform velocity throughout.
- B. Rise of 2.00 in. in 120°.
- C. Dwell for 60°.
- D. Fall for 90°.
- E. Dwell for 90°.

Problem 2. In the space provided, construct a displacement diagram for the following motions:
- A. Modified uniform velocity throughout.
- B. Rise of 1.88 in. in 100°.
- C. Dwell for 100°.
- D. Fall for 60°.
- E. Dwell for 100°.

Problem 3. In the space provided, construct a displacement diagram for the following motions:
- A. Simple harmonic motion throughout.
- B. Rise of 1.75 in. in 90°.
- C. Dwell for 45°.
- D. Fall for 90°.
- E. Dwell for 135°.

Problem Sheet 7B

Problem 1. In the space provided, construct a displacement diagram for the following motions:
- A. Uniform acceleration and deceleration throughout.
- B. Rise of 3.50 in 150°.
- C. Dwell for 50°.
- D. Fall for 75°.
- E. Dwell for 85°.

Problem 2. In the space provided, construct a displacement diagram for the following motions:
- A. Cycloidal motion throughout.
- B. Rise of 3.75 in. in 180°.
- C. Fall for 180°.

Problem Sheet 7C

Problem 1. In the space provided, construct a displacement diagram for the following motions:
- A. Rise of 3.00 in. in 90° with simple harmonic motion.
- B. Dwell for 15°.
- C. Rise of 3.00 in. in 90° with uniform accelerated and decelerated motion.
- D. Dwell for 30°.
- E. Fall of 6.00 in. in 120° with cycloidal motion.
- F. Dwell.

Problem Sheet 7D

Problem 1. Lay out a cam that will move the knife-edge follower 2 with the motion developed in Problem Sheet 7A, problem 1.

Problem Sheet 7E

Problem 1. Lay out a cam that will move the radial-roller follower 2 with the motion developed in Problem Sheet 7A, problem 2.

Problem Sheet 7F

Problem 1. Lay out a cam that will move the flat-face follower 2 with the motion developed in Problem Sheet 7A, problem 3. Determine the face length *yz* when .25 in. is allowed from the extreme tangents.

Problem Sheet 7G

Problem 1. Lay out a cam that will move the offset-roller follower 2 with the motion developed in Problem Sheet 7A, problem 1.

Problem Sheet 7H

Problem 1. Lay out a cam that will move the offset-roller follower 2 with the motion developed in Problem Sheet 7A, problem 2.

Problem Sheet 7I

Problem 1. Lay out a cam that will move the flat-face follower 2 with the motion developed in Problem Sheet 7A, problem 3. Determine the face length *yz* when .38 in. is allowed from the extreme tangents.

Problem Sheet 7J

Problem 1. Lay out the cam that will move the slider 3 with the motion developed in Problem Sheet 7A, problem 3. Determine the face length *yz* when .25 in. is allowed from the extreme tangents.

Problem Sheet 7K

Problem 1. Construct the outline of a cam that will transmit harmonic motion to a pivoted, flat-faced follower 3 in the following manner:
- A. Point *B* moves to *B'* while the cam turns through 120°.
- B. Dwell while the cam turns 90°.
- C. Point *B'* moves back to *B* while the cam turns

through 90°.

D. Dwell.

Problem 2. Show the limits of contact along the follower face. Label the limits *y* and *z*.

Problem Sheet 7L

Problem 1. Construct the outline of a cam that will move slider 6 from point *E* to point *F* in the following manner:

A. Rise with uniform accelerated and decelerated motion through 180°.

B. Fall with simple harmonic motion through 180°.

Problem Sheet 7M

Problem 1. Draw the cylindrical cam 3 which will impart the following motions to the follower 1:

A. Move point *B* 149.0 mm to the right with simple harmonic motion during 150° of cam rotation.

B. Dwell during 90° of cam rotation.

C. Move point *B* 149.0 mm to the left with equal, uniformly accelerated and decelerated motion during 120° of cam rotation.

References

The following source and page references can be used in developing the graphical solutions and calculations for the problem assignments.

Esposito, Anthony. *Kinematics for Technology.* Columbus, Ohio: Charles E. Merrill Publishing Co., 1973, pp. 181-226.

Greenwood, Douglas C., ed. *Engineering Data for Product Design.* New York: McGraw-Hill Book Co., 1961, p. 114.

———. *Mechanical Details for Product Design.* New York: McGraw-Hill Book Co., 1964, p. 252.

Hardison, Thomas B. *Introduction to Kinematics.* Reston, Va.: Reston Publishing Co., 1979, pp. 135-178.

Hinkle, Rolland T. *Kinematics of Machines.* 2nd ed. Englewood Cliffs, N.J.: Prentice-Hall, 1960, pp. 128-164.

Hirschhorn, Jeremy. *Dynamics of Machinery.* New York: Barnes & Noble, 1967, pp. 99-115.

Kepler, Harold B. *Basic Graphical Kinematics.* 2nd ed. New York: McGraw-Hill Book Co., 1973, pp. 217-275.

Lent, Deane. *Analysis and Design of Mechanisms.* 2nd ed. Englewood Cliffs, N.J.: Prentice-Hall, 1970, pp. 346-376.

Michels, Walter J., and Wilson, Charles E. *Mechanism: Design-Oriented Kinematics.* Chicago: American Technical Society, 1969, pp. 216-280.

Oberg, Erik; Jones, Franklin D.; and Horton, Holbrook L. *Machinery's Handbook.* 20th ed. New York: Industrial Press, 1976, pp. 717-734, 1,845-1,851.

Patton, William J. *Kinematics.* Reston, Va.: Reston Publishing Co., 1979, pp. 117-148.

Tao, D.C. *Fundamentals of Applied Kinematics.* Reading, Mass.: Addison-Wesley Publishing Co., 1967, pp. 185-223.

Gear Analysis and Design 8

principles of gears

A large number of power transmission devices exist for the purpose of transmitting rotary motion from one shaft to another. Gears, today, are widely used in a variety of different machine applications. Gears are used to transmit power, to change speeds, to change direction of rotation, to change shaft direction, and to provide a positive, nonslip drive. In many instances, gears are designed and built into an integral unit or gear drive for reducing or increasing speeds.

While there are many specialized gears peculiarly adapted to certain unusual applications, only the more common types will be discussed in this unit.

Spur Gears

Spur gears have teeth cut straight and parallel to the axis of shaft rotation. As a pair of teeth pass through engagement, contact by a combined sliding and rolling motion is made, with pure rolling contact taking place only at the pitch cylinder. Maximum sliding motion occurs at the tip and root of the teeth; these areas can develop excessive wear if proper lubrication is not provided.

Tooth engagement occurs instantaneously across the entire width of the tooth face and progresses from tip to root on one tooth and from root to tip on the mating tooth. Not more than two teeth are meshed at any one time; hence the load is transferred rapidly from one tooth to another. For this reason the use of spur gears is usually limited to low speeds; otherwise, excess noise and wear may occur. Specially designed spur gears may, however, be occasionally found in high speed applications. Spur gears are usually found where no high impact loads or severe thrust conditions are present (where quietness of operation is not a factor).

Spur gears are used in gear drives of many types, in speed reducers, speed increasers, planetary gears, and various types of gear trains.

Helical Gears

On helical gears, the teeth are disposed obliquely across the face at an angle instead of straight across, as with spur gears. Helical teeth make contact along a line across the profile from root to tip as they pass through engagement. Several teeth are in mesh at the same time, which produces uniform transfer of contact from tooth to tooth. Helical gears operate more quietly than spur gears; they also have greater strength and durability capacity due to the increased length of contact lines in the zone of action.

Single Helical Gears

A single helical gear has helical teeth which are cut with only one hand of helix on a gear. The helix angle may be varied to suit application requirements. Single helical gears develop moderate thrust loads which usually are accommodated by the bearings used for supporting the shafts.

Single helical gears are used for applications where both pinion and gear shafts are held in fixed axial positions and where driving or driven equipment may subject the gearing to thrust or axial forces. They can also be used where shaft axes are crossed or nonparallel. Under these conditions mating teeth make point contact, which limits their use to applications having low power requirements.

Herringbone Gears

Herringbone or double helical gears have teeth with right-hand and left-hand helices cut on each element. The thrust reaction from one helix balances that of the other. Herringbone gears should be mounted with all but one element of a gear train free to float axially to assure apex alignment and load distribution between helices for all gears in mesh. On many herringbone gear sets, the two helices are separated by a narrow gap at the center to provide greater accuracy without sacrifice of strength. This also eliminates entrapment of lubricant at the apex as the teeth pass through the mesh.

Bevel Gears

Bevel gears are used where rotary power is to be transmitted from one shaft to another shaft at an angle to the first. Bevel gears with straight teeth are used for right-angle drives. They are often called miter gears when the

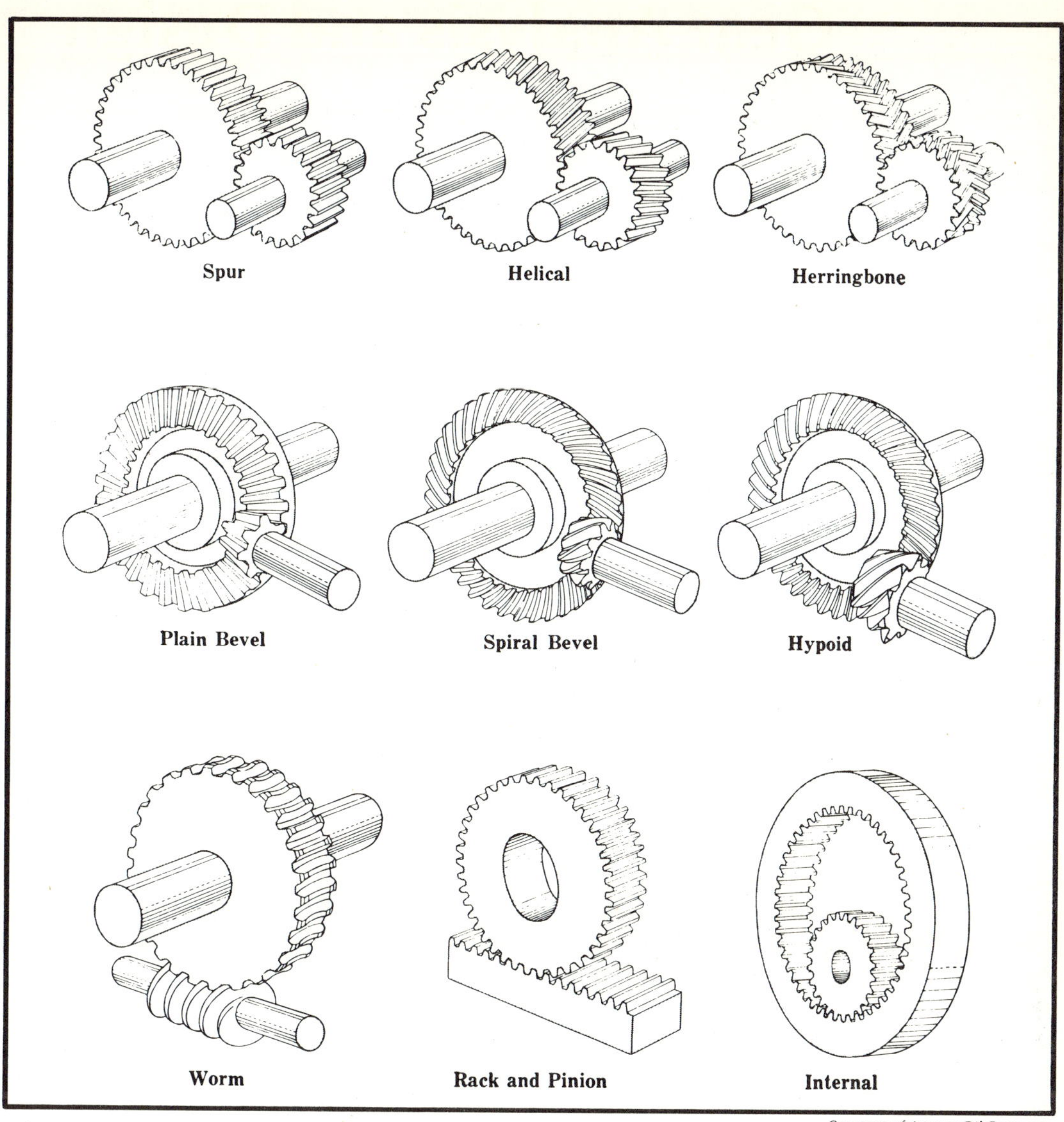

Courtesy of Amoco Oil Company.

Fig. 8-1. Principal types of gears.

gear ratio is one-to-one. To obtain the advantages of helical gear operation, the spiral bevel gear was developed. It may also be used where the shafts to be connected are not at right angles to one another.

Hypoid Gears

Hypoid gears resemble spiral bevel gears except that the axis of the pinion gear is above or below the axis of the driven gear. Because of this, both ends of the driving and driven shafts may be supported in bearings for applications requiring more rigid support, or when a continuous pinion shaft drives more than one gear. The tooth action is between that of the spiral bevel gear and the worm gear, combining the rolling action of the first and the sliding action of the latter.

Worm Gears

With worm gear drives it is possible to obtain large speed reductions in a right-angle drive. The action of worm gears is unlike that of other gear types. The worm is actually a screw which drives the worm wheel in a sliding fashion by the inclined plane principle. The worm is always connected to the power source and the wheel is always connected to the load. Worm gears can be cut single-threaded so that the wheel moves one tooth for each revolution of the worm, or they can be multiple-threaded for lower speed reduction.

The crossed axis helical gear is really a special worm gear where the tooth ratio is one-to-one and where the shafts cross at an angle. The gears are limited to light load applications. Their lubrication requirements are similar to the worm gear.

Action between the continuous worm thread and the teeth of the driven worm wheel is predominantly of a sliding rather than rolling contact. For this reason, worm gear lubrication must receive special consideration. To further reduce excessive friction from rubbing due to the sliding motion, the worm wheel is often made of bronze or cast iron, while the worm is of hardened steel. Worm gear drives for large-load or high-speed applications are often provided with air-cooling or water-cooling to keep lubricants from overheating.

Internal Gears

Internal gears have teeth cut on the inside of the rim. The internal gear drive is more compact, it runs more smoothly, and the teeth are stronger than with external spur gear drives.

With internal gears, the planetary gear arrangement is possible. When the small center pinion gear (or sun gear) rotates in a clockwise direction, the planet gears (or idler gears) rotate counterclockwise; the fixed ring gear (internal gear) causes the planet gears to rotate as a group (idler spider) in the same direction as the sun gear but at a lower speed. Input and output shafts turn in the same direction and have their axes in the same straight line. In another type of planetary drive, the planet gear group is held fixed and the ring gear, connected to the output shaft, is free to move. Planetary gear systems may make use of spur, helical, or sometimes herringbone gears.

Gears, as devices for the transmission of motion or power, are such common occurrences in mechanics and industry that a thorough knowledge of their nomenclature, calculation procedures, and design principles is essential for the engineer or engineering draftsman. This should include the ability to develop proper gear proportions through calculations and then to make a usable graphic representation or working drawing of the design.

Developing the knowledge and specific pattern of skills necessary to design gears involves a particular procedure: first, learning the nomenclature peculiar to gears, both design and application; second, gaining a working knowledge of various mathematical formulas;

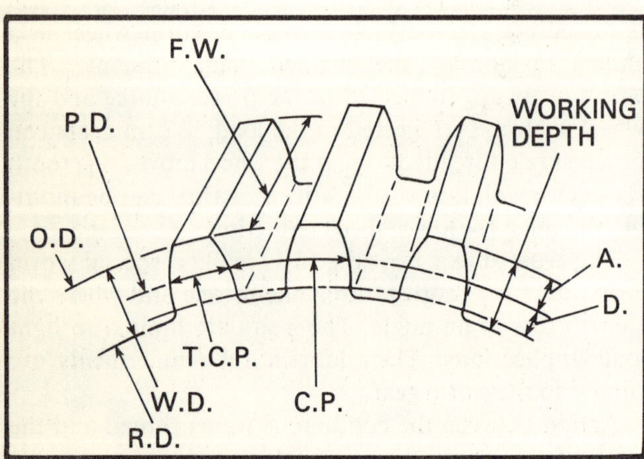

Fig. 8-2. Basic spur gear nomenclature.

Table 1
Gear Nomenclature and Abbreviations

Abbreviation	*Term*
N.	Number of teeth.
D.P.	Diametral Pitch; number of teeth for each inch of the pitch diameter.
P.D.	Pitch Diameter; diameter of the pitch circle.
C.P.	Circular Pitch; distance from a point on one tooth to the same point on the next tooth on the pitch diameter.
A.	Addendum; radial distance from the pitch circle to the outside diameter.
D.	Dedendum; radial distance between the pitch circle and the bottom of the tooth or the root diameter.
W.D.	Whole Depth; radial distance between the outside circle and the root circle.
O.D.	Outside Diameter; maximum diameter of the gear.
R.D.	Root Diameter; diameter of the root circle.
F.W.	Face Width; thickness of the gear blank or the width of the pitch surface.
P.A.	Pressure Angle.

and finally, capably using these formulas, together with accepted techniques of drafting, to graphically illustrate the gear design. (See Fig. 8-2, Table 1, and Appendix F.)

The actual working drawing of the gear consists of a front or plan view and a side view. The front view shows the gear teeth, hub, spokes, etc. In this view, three or four teeth are usually drawn by an approximate method. For the remainder of the gear, the outside diameter is made by a solid line, the pitch diameter with a dash-dot line, and the root diameter with a dashed line.

The side view is made in section, with the gear teeth not sectioned. Pitch diameter is shown with a dash-dot line, the root diameter with a solid line. See Fig. 8-3.

Dimensions that are placed on the drawing should pertain only to the gear blank, such as the shaft opening diameter and hub diameter. All calculations and dimensions pertaining to the actual gear teeth are scheduled in a separate table on industrial drawings.

The involute system of gearing is developed from a basic rack having straight-sided (flat) teeth inclined at a standard pressure angle (usually 14-1/2° or 20°); see Figs. 8-4 and 8-5. This is an interchangeable system since all gears made to operate correctly with this basic rack will operate with each other. An important advantage of involute gears is that small variations of center distance will not affect their ability to transmit uniform angular motion or velocity.

The 14-1/2° P.A. has been used for many years and

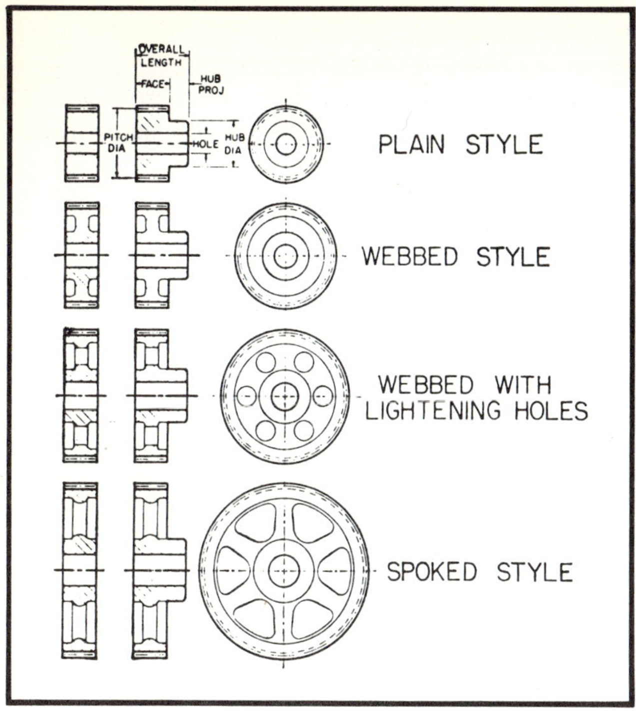

Courtesy of Boston Gear.

Fig. 8-3. Typical spur gear types.

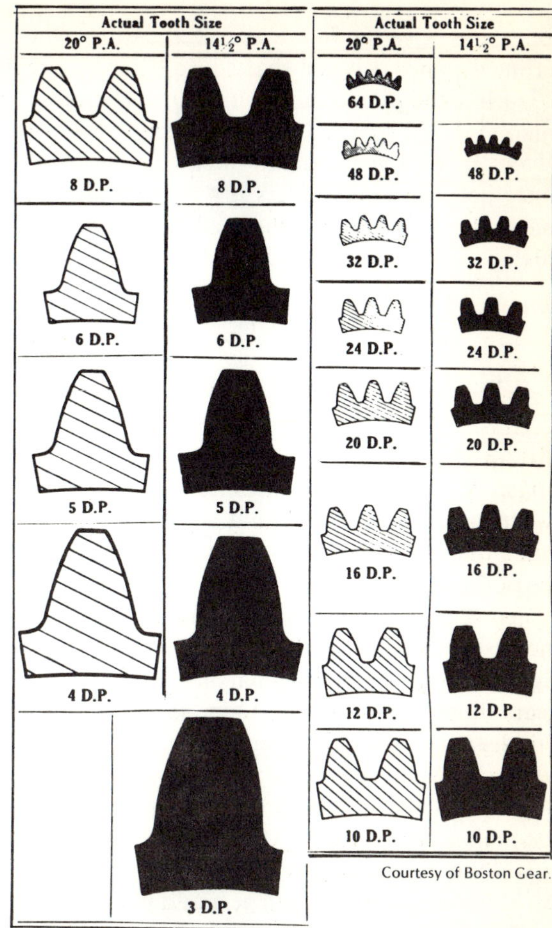

Courtesy of Boston Gear.

Fig. 8-4. Comparison of different diametral pitches.

remains useful for duplicate or replacement gearing or in situations where the control of backlash is of primary importance. The 20° P.A. has become standard for new gearing because of its smoother and quieter running characteristics, greater load-carrying ability, and fewer number of teeth affected by undercutting. Teeth having a 20° P.A. are wider at the base, and undercutting is confined to fewer than 18 teeth (compared to fewer than 32 teeth for 14-1/2° P.A.).

For best results, standard spur gears of 14-1/2° P.A. should have a minimum of 16 teeth with at least 40 teeth in a mating pair. Gears with a 20° P.A. should have a minimum of 13 teeth (26 or more in a pair).

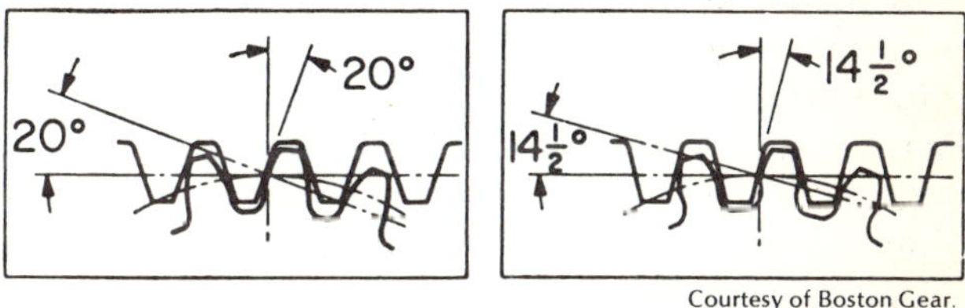

Courtesy of Boston Gear.

Fig. 8-5. Pressure angles.

Problem Assignment

The following graphical problems and design questions are intended to provide you with experience in graphically analyzing and calculating basic mechanism constraints. Select, with instructor direction, the problem-sheet assignments suitable for you. Then complete the numbered problems in ordered sequence because each solution builds on preceding problems. Be sure to use the references suggested by your instructor (*list follows*) as sources for working through the assignment. Problem sheets are located at the back of the book.

Problem Sheet 8A

Problem 1. Calculate the outside diameter, root diameter, number of teeth, and circular pitch for a gear with a pitch diameter of 22 in. and a diametral pitch of 5.5.

Problem 2. A gear with an outside diameter of 6-5/8 in. (as closely as can be measured) has 91 teeth. Determine its probable diametral pitch.

Problem 3. A gear with a diametral pitch of 6 meshes with a pinion. Distance between gear centers is 9 in.; speed ratio is 3.5:1. Calculate the following:

A. Pitch diameter of the pinion.

B. Number of teeth in the pinion and gear.

Problem 4. Calculate the outside diameter, pitch diameter, and number of teeth on each of two mating spur gears. Use the following specifications: the centerline distance between shafts is to be approximately 6-3/4 in., the diametral pitch is 10, the speed of the pinion is 1,750 rpm and that of the gear is 500 rpm.

Problem 5. Two gears are to mesh together. The distance between their centers is 4 in. One gear has to make 11 rev. while the other makes 5. The diametral pitch is 10. Calculate the following:

A. Pitch diameter of each gear.
B. Outside diameter of each gear.
C. Height of tooth.
D. Circular pitch.

Problem Sheet 8B

Problem 1. Calculate the distance between centers of two mating spur gears with a velocity ratio of 1:2.3 and a diametral pitch of 3.5. The number of teeth in the pinion is 30.

Problem 2. A gear and pinion are designed for an exact increase in velocity in the ratio of 1:4. The tooth design calls for a diametral pitch of 4. The centers of the gears should be as close to 13 in. as possible. What is this distance? How many teeth will the gear have?

Problem 3. A pair of helical gears have a normal pitch of 12 and a helix angle of 18°. The gear has 32 teeth, the pinion has 20 teeth. Approximate center distance is 2-1/4 in. Shafts are parallel. Find the following:

A. Pitch diameters of both gear and pinion.
B. Outside diameter of both gear and pinion.
C. Lead of helix.
D. Whole depth of tooth.
E. Exact center distance.
F. Normal circular pitch.

Problem 4. A bevel gear having 60 teeth and a pitch diameter of 12 in. is driven by a bevel pinion having a pitch diameter of 4 in. The shafts are at right angles. Determine the following:

A. Angular-velocity ratio of driver to driven.
B. Number of teeth on the pinion.
C. Back-cone distance for the gear.
D. Formative number of teeth for the gear.

Problem 5. In a certain worm-gear drive the worm wheel has 90 teeth, the velocity ratio is 30:1, and the circular pitch is 1.5 in. The pitch diameter of the worm is 6 in. Determine the following:

A. Center-to-center distance.
B. Lead of the worm.
C. Lead angle of the worm.

Problem Sheet 8C

Problem 1. Indicate the center-to-center distances for spur gears *A*, *B*, and *C*. Label the addendums, dedendums, and pitch diameters. Indicate the number of teeth on each gear if the diametral pitch is 10, and show the direction of rotation.

Problem Sheet 8D

Problem 1. Determine the velocity ratio of gear *A* to gear *E*; indicate the direction of rotation of gear *E*. Complete the table and show all calculations in the space provided.

Problem Sheet 8E

Problem 1. Determine the speed and direction of rotation of gear *G*. Complete as much of the table as possible with the data given and show all calculations in the space provided.

Problem Sheet 8F

Problem 1. Determine the velocity ratio of gear *A* to gear *L* and indicate the direction of rotation of *L*. Complete as much of the table as possible with the data given and show all calculations on another sheet.

Problem Sheet 8G

Problem 1. A pair of spur gears have 16 and 22 teeth, respectively, and a diametral pitch of 2; the tooth is a 20° full depth. The pinion is below the gear and drives in a clockwise direction. Point of contact is at *P*. Construct tooth profiles in contact. Determine the following:

A. Pitch circle radii.
B. Base circle radii.
C. Circular pitch.

Problem Sheet 8H

Problem 1. Graphically determine the pitch radius of the smallest pinion with a diametral pitch of 2 that will mesh with a rack without interference for:

A. 14-1/2° full-depth tooth.
B. 20° full-depth tooth.

Calculate the pitch radius, the corresponding number of teeth, and the minimum number of whole teeth for A and B.

Problem Sheet 8I

Problem 1. A set of gears have 14 and 16 teeth, respectively, a diametral pitch of 2, and a 14-1/2° pressure angle; they are full depth. The pinion is above the gear. Graphically determine the amounts that have to be removed from the addendums in order to eliminate the interference.

Problem Sheet 8J

Problem 1. Two gears have 24 and 36 teeth, respectively, a diametral pitch of 6, and a 20° pressure angle. The gear is below the pinion. Determine the center distance of the axes. Also, graphically determine the pressure angle that results when the center distance is increased .25 in.

Problem Sheet 8K

Problem 1. Two bevel gears have 18 and 45 teeth, respectively, and a diametral pitch of 6. The addendum and dedendum at the large end of the teeth are the same as for a 20° full-depth involute tooth system. The pinion and gear are to be mounted on 1 and 1-1/2 in. diameter shafts, respectively. The hub diameter is to be 1-3/4d + 1/4 and the lengths to be 1-1/2d, where d is the respective shaft diameter. The face width is to be one third of the cone distance. Draw the gears in section after calculating the following for the gear and pinion:

A. Addendum.
B. Dedendum.
C. Pitch diameter.
D. Pitch angle.
E. Cone distance.
F. Back-cone radius.
G. Addendum angle.
H. Dedendum angle.
I. Face angle.
J. Root angle.
K. Outside diameter.
L. Face width.

References

The following source and page references can be used in developing the graphical solutions and calculations for the problem assignments.

Esposito, Anthony. *Kinematics for Technology.* Columbus, Ohio: Charles E. Merrill Publishing Co., 1973, pp. 229-297.

Greenwood, Douglas C., ed. *Engineering Data for Product Design.* New York: McGraw-Hill Book Co., 1961, pp. 202-204, 228-258.

———. *Mechanical Details for Product Design.* New York: McGraw-Hill Book Co., 1964, pp. 27, 96-99.

———. *Product Engineering Design Manual.* New York: McGraw-Hill Book Co., 1959, pp. 201-203, 275.

Hardison, Thomas B. *Introduction to Kinematics.* Reston, Va.: Reston Publishing Co., 1979, pp. 179-252.

Hinkle, Rolland T. *Kinematics of Machines.* 2nd ed. Englewood Cliffs, N.J.: Prentice-Hall, 1960, pp. 176-221.

Hirschhorn, Jeremy. *Dynamics of Machinery.* New York: Barnes & Noble, 1967, pp. 131-134, 190-197, 232, 329.

Kepler, Harold B. *Basic Graphical Kinematics.* 2nd ed. New York: McGraw-Hill Book Co., 1973, pp. 277-371.

Lent, Deane. *Analysis and Design of Mechanisms.* 2nd ed. Englewood Cliffs, N.J.: Prentice-Hall, 1970, pp. 173-303.

Michels, Walter J., and Wilson, Charles E. *Mechanism: Design-Oriented Kinematics.* Chicago: American Technical Society, 1969, pp. 281-451.

Oberg, Erik; Jones, Franklin D.; and Horton, Holbrook L. *Machinery's Handbook.* 20th ed. New York: Industrial Press, 1976, pp. 735-990.

Patton, William J. *Kinematics.* Reston, Va.: Reston Publishing Co., 1979, pp. 161-228.

Tao, D.C. *Fundamentals of Applied Kinematics.* Reading, Mass.: Addison-Wesley Publishing Co., 1967, pp. 255-319.

PART II Mechanisms for Specific Applications

Industrial Robots

9

programmable automation mechanisms

INDUSTRIAL robots can work 24 hours a day, seven days a week, doing repetitious tasks in areas subject to extreme temperatures, smells, and hazardous gases. Industrial robots of the types being manufactured by Unimation, Inc., and by Cincinnati Milacron are representative of the programmable, self-contained, movable machines available to industry. These machines can be programmed to automatically perform various jobs by the means of mechanical arms and hands.

Industrial robots are capable of providing years of round-the-clock productive service. The robots can track moving lines and, while tracking, put welds in precisely the right spots; they can pluck an assembly out of a moving welding jig and transport it high overhead onto a conveyer, tracking the two continuously moving lines independently of one another.

The industrial robot is taught new jobs by leading it through a work program. Work steps are recorded in its memory by using a button programming device; the robot will repeat these programmed instructions indefinitely. It can also monitor and, if necessary, correct its own motion. See Figs. 9-1 and 9-2.

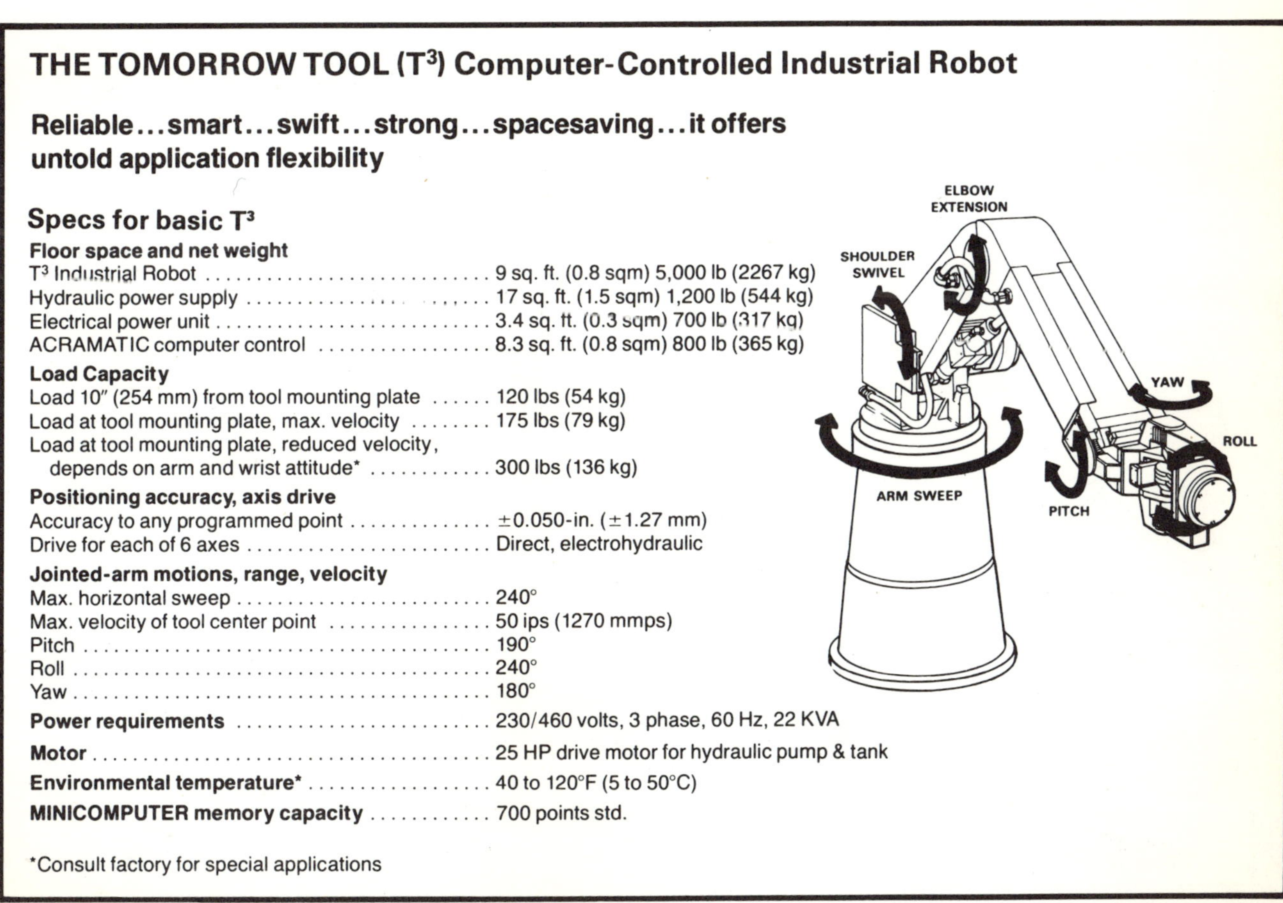

THE TOMORROW TOOL (T³) Computer-Controlled Industrial Robot

Reliable...smart...swift...strong...spacesaving...it offers untold application flexibility

Specs for basic T³

Floor space and net weight	
T³ Industrial Robot	9 sq. ft. (0.8 sqm) 5,000 lb (2267 kg)
Hydraulic power supply	17 sq. ft. (1.5 sqm) 1,200 lb (544 kg)
Electrical power unit	3.4 sq. ft. (0.3 sqm) 700 lb (317 kg)
ACRAMATIC computer control	8.3 sq. ft. (0.8 sqm) 800 lb (365 kg)
Load Capacity	
Load 10″ (254 mm) from tool mounting plate	120 lbs (54 kg)
Load at tool mounting plate, max. velocity	175 lbs (79 kg)
Load at tool mounting plate, reduced velocity, depends on arm and wrist attitude*	300 lbs (136 kg)
Positioning accuracy, axis drive	
Accuracy to any programmed point	±0.050-in. (±1.27 mm)
Drive for each of 6 axes	Direct, electrohydraulic
Jointed-arm motions, range, velocity	
Max. horizontal sweep	240°
Max. velocity of tool center point	50 ips (1270 mmps)
Pitch	190°
Roll	240°
Yaw	180°
Power requirements	230/460 volts, 3 phase, 60 Hz, 22 KVA
Motor	25 HP drive motor for hydraulic pump & tank
Environmental temperature*	40 to 120°F (5 to 50°C)
MINICOMPUTER memory capacity	700 points std.

*Consult factory for special applications

Courtesy of Cincinnati Milacron.

Fig. 9-1. Cincinnati Milacron T³ computer-controlled industrial robot specifications.

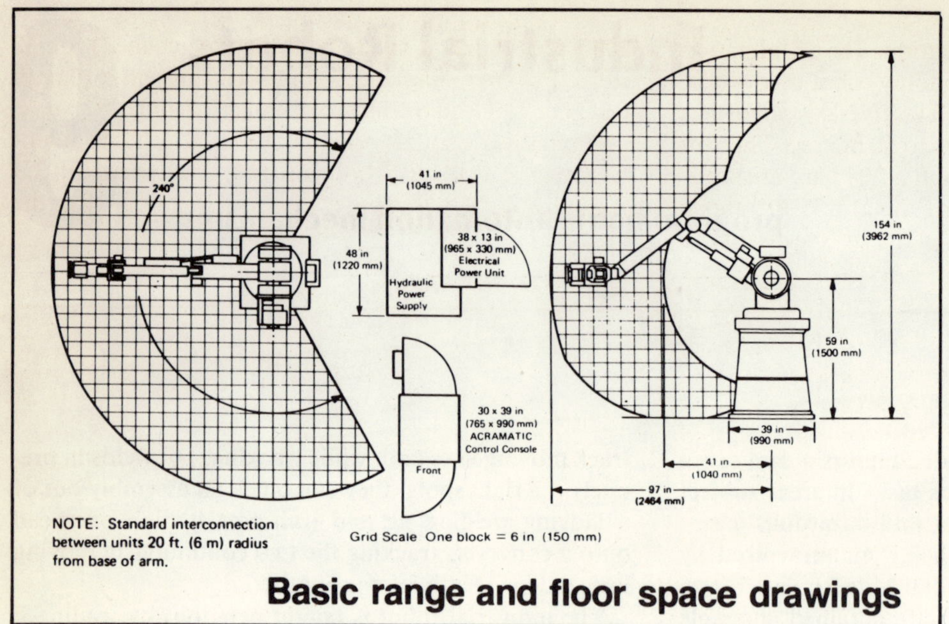

Courtesy of Cincinnati Milacron.

Fig. 9-2. Cincinnati Milacron T^3 basic range and floor space drawings. (*top*)

Fig. 9-3. Unimate Series 1000 specifications and floor space drawings. (*bottom*)

Courtesy of Unimation, Inc.

Specifications

General

Configuration	5 axes
Drive	3 arm motions, hydraulic actuation 2 wrist motions pneumatic actuation
Controller	Solid state including memory
Teaching	Lead-by-the-hand with teach pendant, point-to-point
Program capacity	32 memory steps, expandable
Controller input/output	12 signal lines (6 in, 6 out)
Power requirement	230/460V, 3Ø, 60Hz, 9KVA
Positioning repeatability	0.05 in.
Air requirement	75 cfm @ 75 psi

Load capacity	Wrist bend	500 in. — lbs.
	Wrist yaw	150 in. — lbs.
	Max. load	50 lbs.
Travel/Velocity at maximum load	Wrist bend	90°, 30°/sec
	Wrist yaw	90°, 90°/sec.
	Arm Radial	41 in., 50 in./sec.
	Arm Rotary	208°, 110°/sec.
	Arm Vertical	57°, 110°/sec.
Environment	40°F to 120°F	Humidity 0-90%

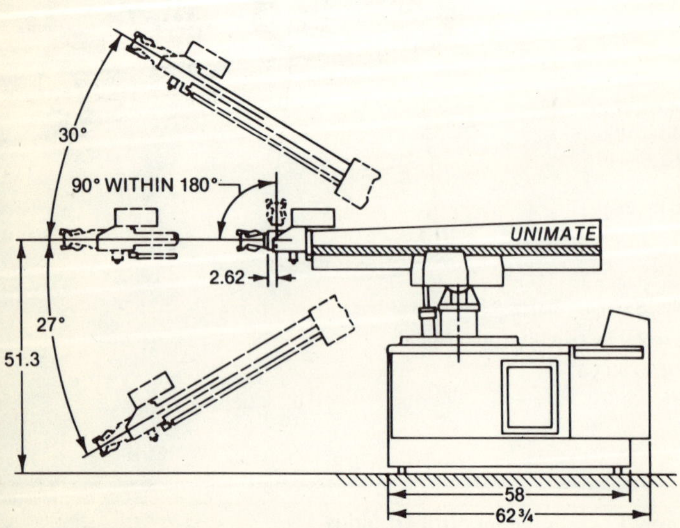

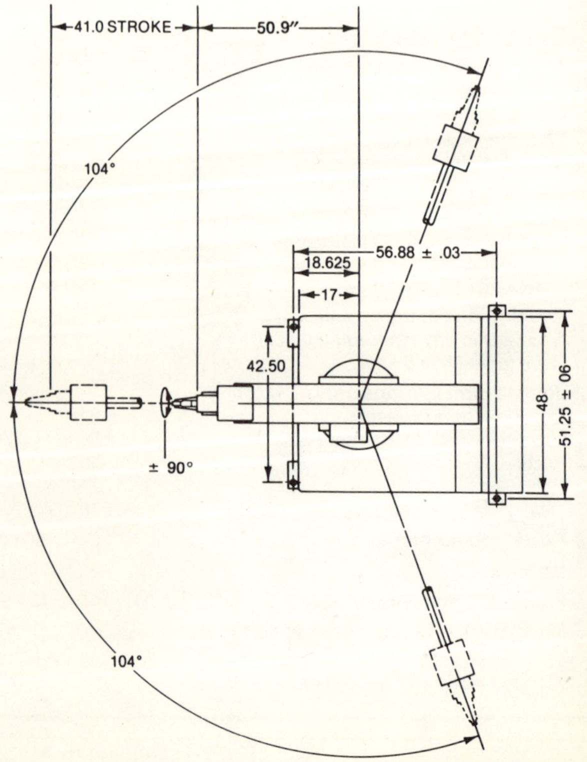

41.0 STROKE
50.9″
104°
18.625
56.88 ± .03
17
42.50
48
51.25 ± 06
± 90°
104°
30°
90° WITHIN 180°
UNIMATE
2.62
27°
51.3
58
62¾

Fig. 9-3 illustrates the specifications for a Unimation Series 1000 robot. This is an economy robot used where lifting capacity is under 50 lbs. The five-axis robot has an outreach of 41 in. at speeds up to 50 in./sec. The arm will swing in rotary travel through a 208° arc at speeds up to 110°/sec. Vertical travel of the arm is 30° above and 27° below the horizontal plane at speeds up to 100°/sec. It has a solid-state memory capacity of 32 steps. Programming is accomplished through a plug-in teaching unit lead by the programmer's hand.

Unimation also has a lighter robot which can maintain a total load of 5 lbs. and has a tip speed of 3.3 ft./sec. Fig. 9-4 illustrates the specifications for the lighter PUMA robot.

General	
Configuration	5 revolute axes
Drive	Electric servos
Controller	System computer
Teaching	By teach pendant and/or computer terminal
Program Language	VAL™
Program capacity	As required by application
External program storage	Floppy-disk (optional)
Controller input/output	16 signal lines (8 in, 8 out) TTL compatible standard
End effector control	Computer-controlled, pneumatic, 0.5 cfm @ 100 psi.
Optional accessories	CRT or TTY, floppy-disk, end effectors, alternate I/O module

Power requirement	95-130 Vac, 50-60 Hz, 750 VA max.

Performance	
Repeatability	±0.004 in.
Load capacity	5.0 lbs. (including end effector)
Tip velocity	3.3 ft./sec. max. with max. load
Static force at tip	13.2 lbs. max.

Physical characteristics	
Arm weight	120 lbs.
Controller size	12.5 in. x 19.0 in. x 20.0 in. (H x W x D)
Controller weight	80 lbs.
Controller cable length	8 ft. max.

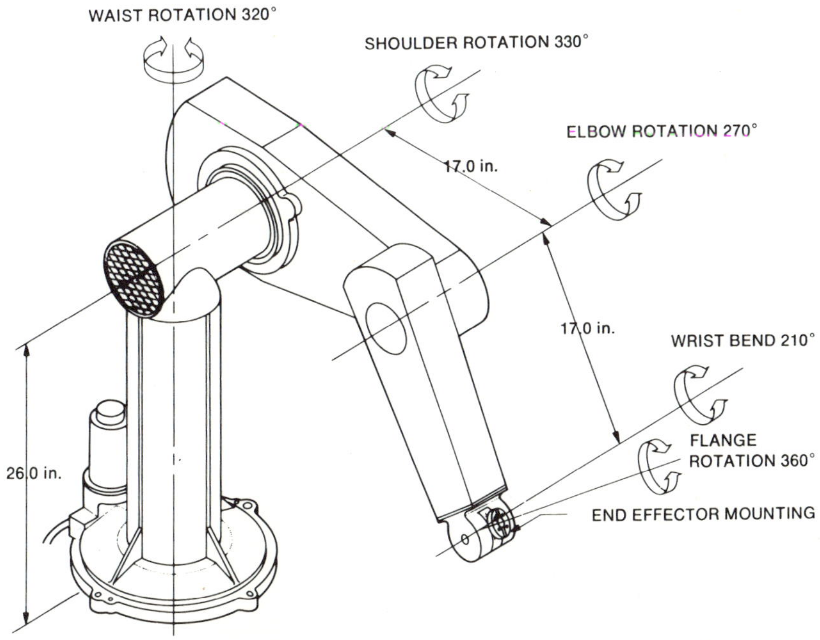

Courtesy of Unimation, Inc.

Fig. 9-4. Unimation PUMA robot specifications.

Problem Assignment

The following graphical problems and design questions are intended to provide you with experience in graphically analyzing and calculating basic mechanism constraints. Select, with instructor direction, the problem-sheet assignments suitable for you. Then complete the numbered problems in ordered sequence because each solution builds on preceding problems. Be sure to use the references suggested by your instructor (*list follows*) as sources for working through the assignment. Problem sheets ar located at the back of the book.

Problem Sheet 9A

Problem 1. The Unimate industrial robot illustrated operates with the following limitations:

A. Point C can extend 40.0 in. from its present position along the horizontal centerline shown.

B. Horizontal rotational traverse of the arm is 110° about point B in directions w from the horizontal centerline as seen in the top view.

C. Vertical traverse of the arm is limited to maximums of 30° up (y) and 27° down (z) about point A as seen in the front view.

D. The hand pivots an angular distance up 110° and down 100° (x) about C as seen in the front view.

Illustrate the total volume that point C can travel with the limitations listed above. Show the volume by cross-hatching the top, front, and right profile views. Show with a dashed line the limitations of the hand as it pivots about C. (Scale: 1:00 in. = 40.00 in.)

Problem Sheet 9B

Problem 1. The robot's fingers 5 and 6 pivot about points D and C, respectively. Movement of the hydraulic piston rod end 4 causes fingers 5 and 6 to move closer to or further away from the hand centerline. Surface E represents a fixed machine surface next to which the robot's hand must work while completing its tasks. Shapes F and G represent the outlines of two parts that the robot must pick up from the same position. The cylindrical shape F has two rectangular features, while G is a simple rectangular solid; both shapes are 20 in. high. Lay out and design the gripping surfaces that have to be mounted to the fingers in order to lift and move the two shapes. It may be necessary to design two sets of gripping surfaces. (Scale: 1.00 in. = 2.00 in.)

References

The following source and page references can be used in developing the graphical solutions and calculations for the problem assignments.

Esposito, Anthony. *Kinematics for Technology.* Columbus, Ohio: Charles E. Merrill Publishing Co., 1973, pp. 324, 325.

Patton, William J. *Kinematics.* Reston, Va.: Reston Publishing Co., 1979, pp. 231-256.

Reciprocating Piston Engine 10

slider-crank, four-bar mechanism

THE reciprocating piston engine, which is basically a *slider-crank* kinematic device, is an example of a *four-bar* mechanism, the most fundamental of mechanisms. Other examples of four-bar mechanisms include the bellcrank, rocker arm, and simple toggle linkages.

Basic kinematic terms relating to the four-bar mechanism are:

Link. A rigid body joined to other bodies in order to transmit motion or force.

Crank. A link which can make a complete revolution about a fixed pivot.

Lever. A link which oscillates but cannot make a complete revolution.

Rocker Arm. A lever pivoted near the center.

Bellcrank. A rocker arm bent at the fixed pivot.

Toggle. A slider-crank which has a tremendous mechanical advantage. A force is applied to the movable end of the crank and a clamping action is exerted by the slider. (See Unit 15.)

Let us consider a large diesel piston engine in a power generating station (Fig. 10-1) and the gas piston engine in an automobile (Figs. 10-2 and 10-3).

The cutaway drawing on the right illustrates the unique opposed piston diesel engine design. Steps in its operation include:

1. Air under blower pressure charges the cylinder via intake ports encircling top of cylinder liner.
2. Converging pistons close the exhaust and intake ports, entrapping clean air.
3. Pistons converge, compressing air trapped in the cylinder.
4. Fuel injects into the combustion chamber between the pistons.
5. Heat of compression ignites the air-fuel mixture.
6. Pressures resulting from combustion force the pistons apart, delivering power to both crankshafts, interconnected by a vertical drive.
7. Exhaust ports encircling the bottom of the cylinder liner open. Unimpeded by valves, burned gases escape. (No power is lost in operating push rods or rocker assemblies.)
8. Intake ports open, admitting clean air under blow-

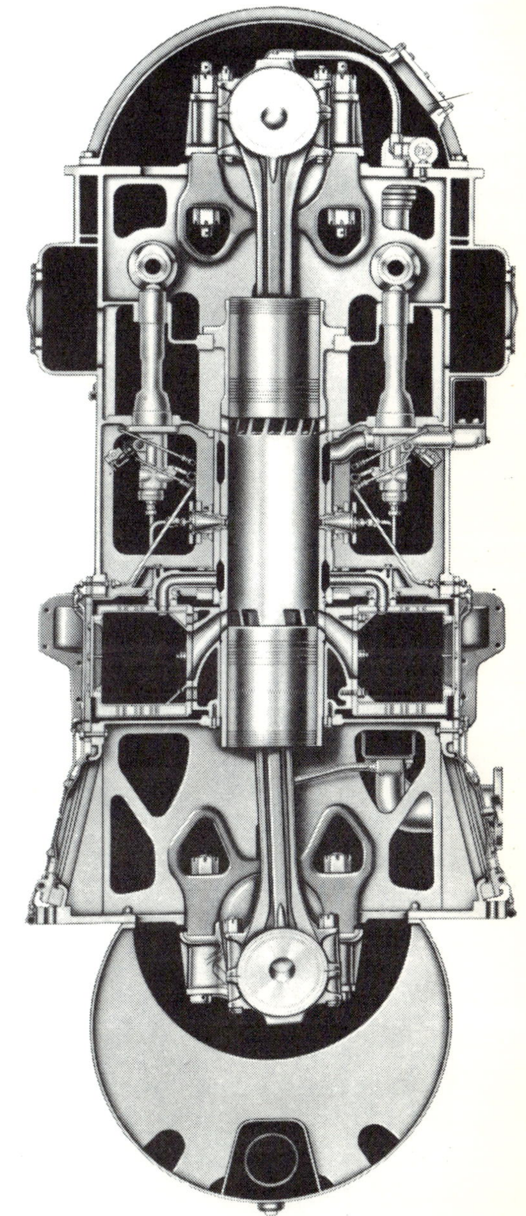

Courtesy of Fairbanks Morse Engine Division, Colt Industries.

Fig. 10-1. The opposed piston diesel engine design.

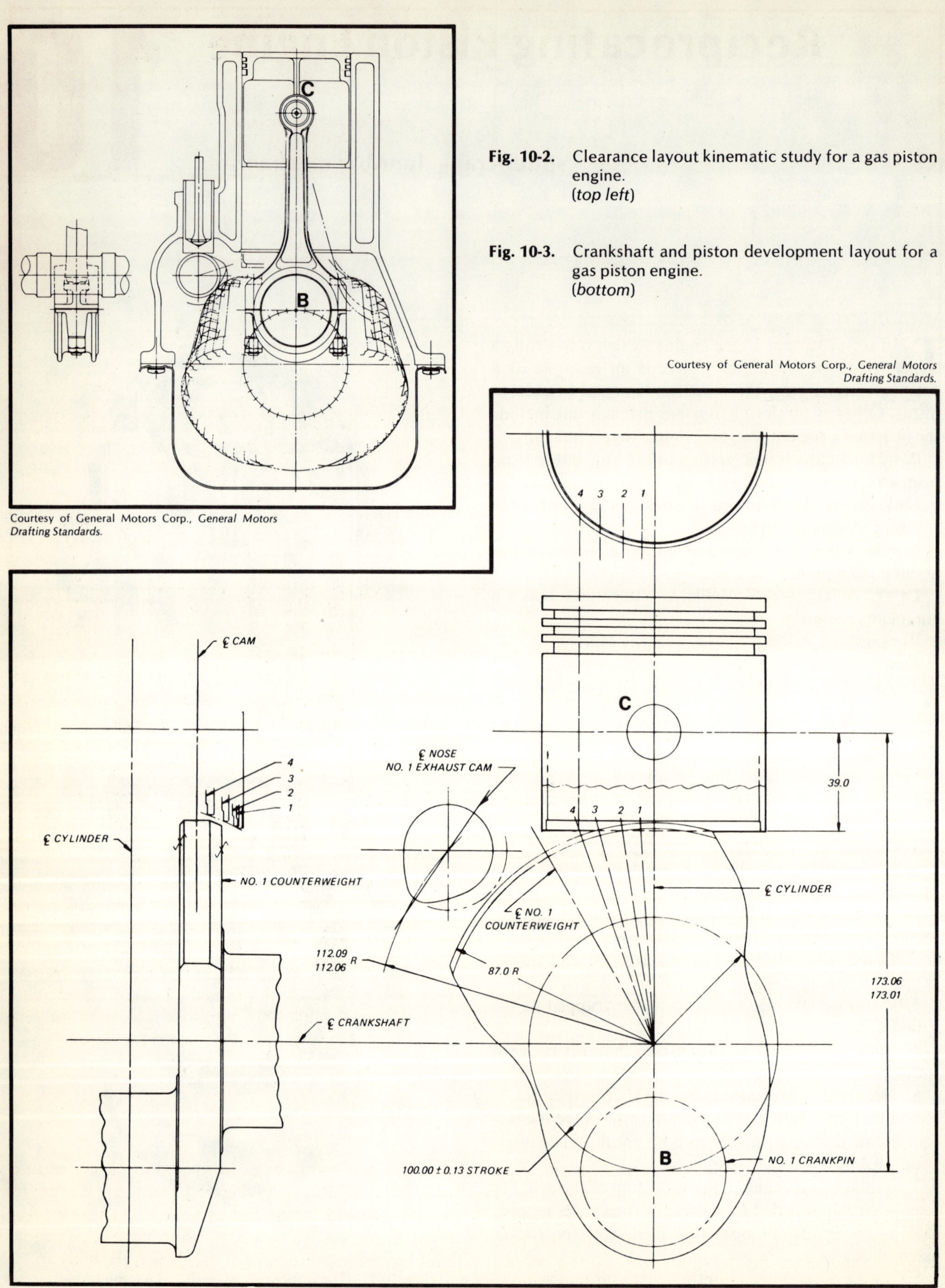

Fig. 10-2. Clearance layout kinematic study for a gas piston engine. (*top left*)

Fig. 10-3. Crankshaft and piston development layout for a gas piston engine. (*bottom*)

Courtesy of General Motors Corp., *General Motors Drafting Standards.*

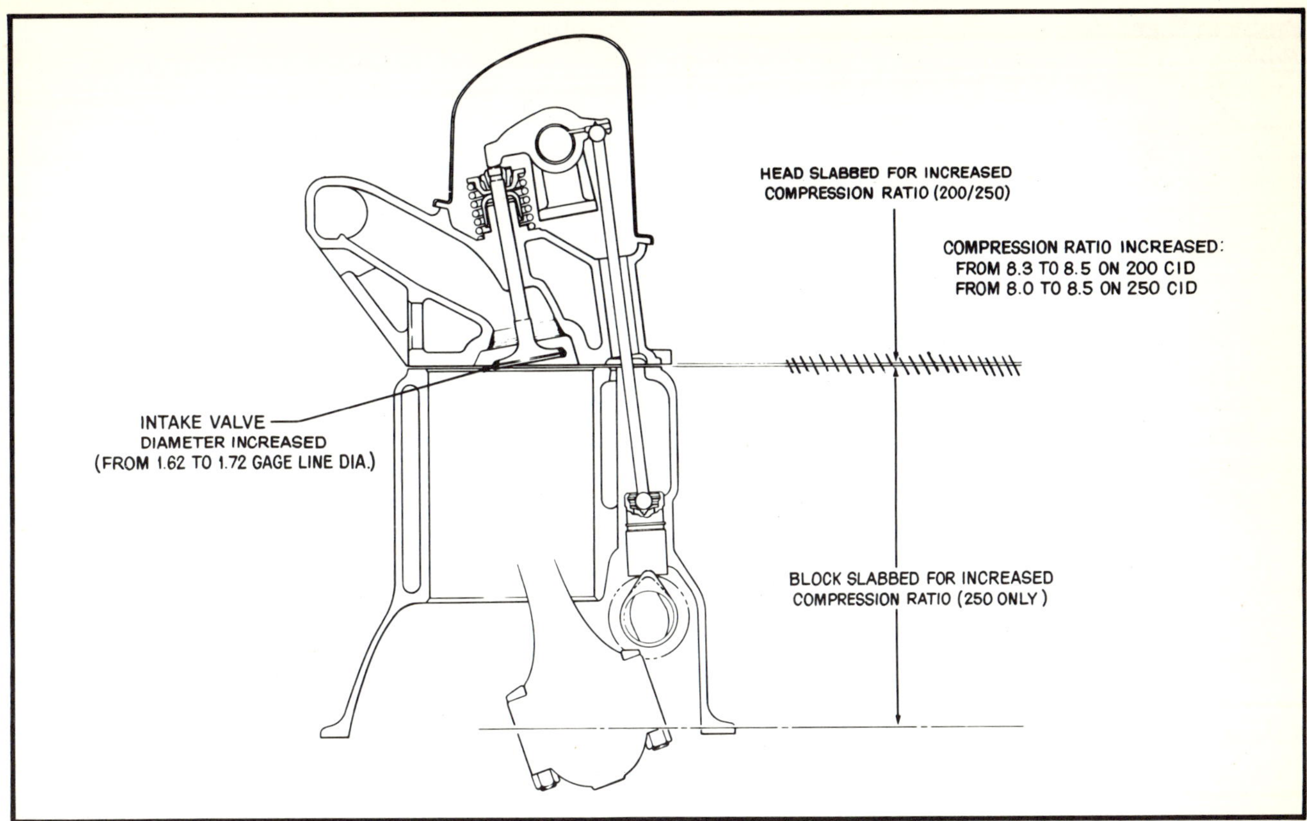

Courtesy of Ford Motor Company.

Fig. 10-4. Compression ratio variations for a gas engine.

er pressure and sweeping, in one direction, all burned gases from the compression chamber as the cycle repeats.

In the gas piston engine, the piston (slider), connecting rod (link), and crank of each cylinder have plane motion. The motion of the connecting rod is represented by the motion of line segment *BC* (Figs. 10-2 and 10-3), where points *C* and *B* are the centers of the wristpin and crankpin, respectively. The motion of any point on the connecting rod can be derived from the motion of this link.

Both the gas piston and diesel piston engines just described are examples of plane motion, since the motion of most engine members falls on a plane or can be resolved into two or more sets of plane motion.

There are three types of plane motion: *rotation*, *translation,* and *combined rotation and translation.* These types are illustrated by the running gear of the cylinders in each engine:

1. *Rotation* occurs when the crank turns about the main bearing. During the motion, any point on the crank travels in a circle and keeps its fixed radius from the axis of rotation (the centerline of the crankshaft between the main bearings).

2. The pistons shown in the illustrations travel in a straight line; thus all the points on the pistons move in straight parallel paths. The pistons are in rectilinear *translation.*

3. The wristpin of the pistons have translation, and the crankpin has rotation. Therefore, the connecting rod moves in a *combined* motion of *translation and rotation.*

Fig. 10-4 illustrates the relationships between various parts of a gas engine and the clearance concerns. Compression ratios (C.R.) are varied by changing the distances between the head and the top of the piston.

Problem Assignment

The following graphical problems and design questions are intended to provide you with experience in graphically analyzing and calculating basic mechanism constraints. Select, with instructor direction, the problem-sheet assignments suitable for you. Then complete the numbered problems in ordered sequence because each solution builds on preceding problems. Be sure to use the references suggested by your instructor (*list follows*) as sources for working through the assignment. Problem sheets are located at the back of the book.

Problem Sheet 10A

Problem 1. Plot the path that point *D* traces. Indicate the maximum velocity of *D* and mark the location(s) of point *B* where *D*'s velocity is the greatest. Does it occur when *AB* is at one of the 30° intervals? Indicate the velocity of *D* (in millimeters per second). (Scale: 1.0 mm = 2.0 mm.)

Problem 2. Calculate the total volume displaced by the piston's stroke. The piston diameter is 85.0 mm. Remember that the volume of a cylinder equals its circular area multiplied by its height ($V_{cyl} = .7854\ D^2h$). Give your answers in cubic millimeters, cubic centimeters, and cubic meters.

Problem 3. Calculate the compression ratio (C.R.) when the clearance between the bottom of the head and the top of the piston in its "up" position equals 10 mm (actual): C.R. = 1 + (stroke/clearance). State how you could increase and decrease the C.R.

Problem 4. Lay out the oil pan bottom and sides when 20.0 mm clearance is maintained from the extreme positions of the connecting rod. (Scale: 1.0 mm = 2.0 mm.)

Problem Sheet 10B

Problem 1. Plot the displacement graph for the piston wristpin, point *C*. The mechanism is drawn to a metric scale of 1.0 mm = 2.0 mm, but the graph should be drawn to a scale of 1.0 mm = 1.0 mm. Crankpin center *B* revolves in a counterclockwise direction around crankshaft center *A* at 80 rps. You are to mark 12 angular increments of 30° each, starting at *AB* about center *A*. Indicate the total stroke length (in millimeters).

Problem 2. Plot the velocity graph for the piston wristpin when point *B* revolves counterclockwise about point *A* at 80 rps. Note that as the piston moves up, the velocity is considered positive; as it moves down, the velocity is considered negative. Use a metric scale to indicate velocity (1.0 mm = 1000.0 mm/sec.). Indicate the locations where point *C* has zero velocity. What is the maximum velocity of *C*? Mark which location(s) *B* has when point *C* is at its maximum velocity.

References

The following source and page references can be used in developing the graphical solutions and calculations for the problem assignments.

Esposito, Anthony. *Kinematics for Technology.* Columbus, Ohio: Charles E. Merrill Publishing Co., 1973, pp. 82-84, 89, 90, 147-157.

Hinkle, Rolland T. *Kinematics of Machines.* 2nd ed. Englewood Cliffs, N.J.: Prentice-Hall, 1960, pp. 70, 98-110, 126.

Hirschhorn, Jeremy. *Dynamics of Machinery.* New York: Barnes & Noble, 1967, pp. 389-406.

Kepler, Harold B. *Basic Graphical Kinematics.* 2nd ed. New York: McGraw-Hill Book Co., 1973, pp. 6, 60, 61, 222-228.

Lent, Deane. *Analysis and Design of Mechanisms.* 2nd ed. Englewood Cliffs, N.J.: Prentice-Hall, 1970, p. 282.

Michels, Walter J., and Wilson, Charles E. *Mechanism: Design-Oriented Kinematics.* Chicago: American Technical Society, 1969, pp. 13, 14, 41, 51-54, 125-131.

Oberg, Erik; Jones, Franklin D.; and Horton, Holbrook L. *Machinery's Handbook.* 20th ed. New York: Industrial Press, 1976, pp. 295-305, 326-356.

Patton, William J. *Kinematics.* Reston, Va.: Reston Publishing Co., 1979, pp. 1-29, 61-114.

Tao, D.C. *Fundamentals of Applied Kinematics.* Reading, Mass.: Addison-Wesley Publishing Co., 1967, pp. 4, 7, 129, 227, 241, 242.

Steering System 11

instantaneous center of turning

THE front wheels of an automobile must be positioned to have a certain geometrical relationship to each other and to the axis of the rear wheels, so that turning while driving at low speeds results in minimal tire wear. Referring to Fig. 11-1, suppose the rear wheel axis *PQ* is extended to the left, as shown by the centerline. Turning the automobile requires that the wheels rotate about one center of rotation (instantaneous center). This center of rotation is not fixed, but changes as the front wheels are turned.

Since the rear wheel axis is fixed, the center of rotation must be somewhere on the extension of the axis

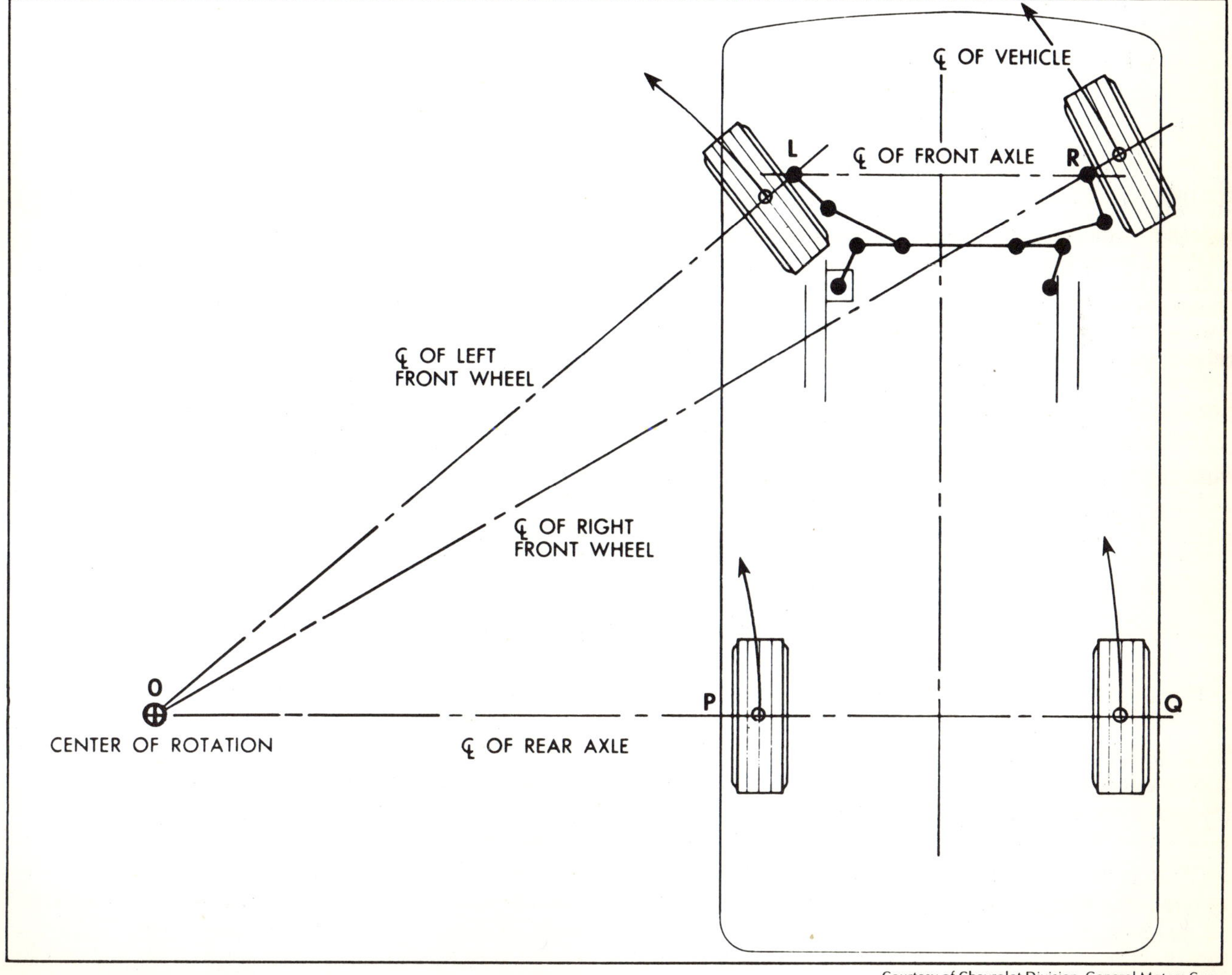

Courtesy of Chevrolet Division, General Motors Corp.

Fig. 11-1. Geometric center of rotation.

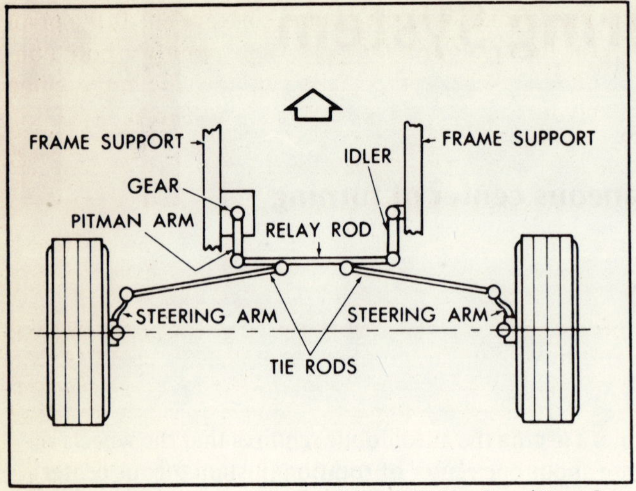

Fig. 11-2. Parallelogram linkage—front.

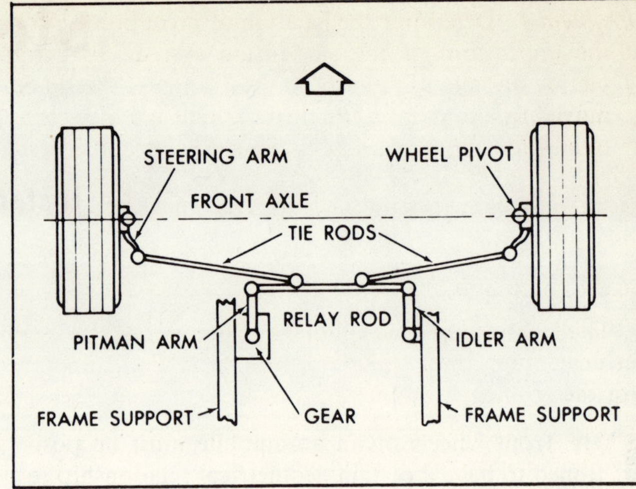

Fig. 11-3. Parallelogram linkage—rear.

PQ. If you draw the centerlines 0*L* and 0*R* at right angles to the turned positions of the front wheels, these lines, for true steering, must meet the axis *PQ* at point 0. The resultant path of the wheels is shown by the arrows.

The Ackerman steering linkage system, shown in Fig. 11-2, produces very nearly this method of front wheel steering. Problems encountered with the Ackerman system are maintaining the geometrical relationship as well as moving with the front suspension. Two basic steering linkage designs are typically used; both are the parallelogram type. One is designed in front of the axle (Fig. 11-2), the other behind the axle (Fig. 11-3).

The basic parallelogram steering linkage is made with five major components: two adjustable tie rods, one relay rod, one pitman arm, and one idler arm. The pitman and idler arms are connected to the gear and frame, respectively, and both are connected to the center link or relay rod. As the steering wheel is turned, the gear rotates the pitman arm (Fig. 11-4) which forces the relay rod to one side. The relay rod (connected to the tie rods) moves the wheels, which are pivoted by rotating the steering arms. The steering arms, when extended, will usually intersect on the rear axle.

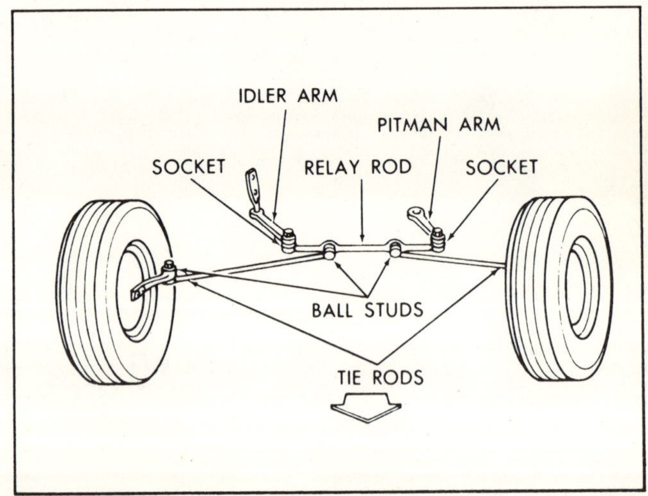

Fig. 11-4. Linkage during turn.

Problem Assignment

The following graphical problems and design questions are intended to provide you with experience in graphically analyzing and calculating basic mechanism constraints. Select, with instructor direction, the problem-sheet assignments suitable for you. Be sure to use the references suggested by your instructor (*list follows*) as sources for working through the assignment. Problem sheets are located at the back of the book.

Problem Sheet 11A

Problem 1. Complete the layout for the steering linkage illustrated when point *A* becomes the nearest instantaneous center of rotation. Locate the steering rod ends along lines 3 and 4. Select the appropriate tie rod end positions on link 5.

Problem Sheet 11B

Problem 1. Graphically determine the positions of the links in the suspension system as the vehicle hits the rut and the curb. Indicate the angles made by links 2 and 4 when the rut and curb are hit. What problem occurs in terms of the amount of side slip the tires make? (Scale: .25 in. = 1.00 in.)

Problem Sheet 11C

Problem 1. Determine the location of pivot point *E* for the upper arm of the suspension system. In completing the layout, point *D* on the tire center should be moved first to point *x* and then to point *y*. Point *B* rotates about point *A*, and triangle *BFC* is considered to be rigid in shape. Indicate the angles that links 2 and 4 make when the rut and curb are hit. Is the pivot for the upper arm the exact same point for both conditions? What compromise can be made in designing pivot bracket 1? (Scale: .25 in. = 1.00 in.)

References

The following source and page references can be used in developing the graphical solutions and calculations for the problem assignments.

Esposito, Anthony. *Kinematics for Technology.* Columbus, Ohio: Charles E. Merrill Publishing Co., 1973, pp. 72-89.

Hinkle, Rolland T. *Kinematics of Machines.* 2nd ed. Englewood Cliffs, N.J.: Prentice-Hall, 1960, pp. 34-44.

Hirschhorn, Jeremy. *Dynamics of Machinery.* New York: Barnes & Noble, 1967, p. 10.

Kepler, Harold B. *Basic Graphical Kinematics.* 2nd ed. New York: McGraw-Hill Book Co., 1973, pp. 1-12, 30-48, 73-98.

Lent, Deane. *Analysis and Design of Mechanisms.* 2nd ed. Englewood Cliffs, N.J.: Prentice-Hall, 1970, pp. 55-59.

Michels, Walter J., and Wilson, Charles E. *Mechanism: Design-Oriented Kinematics.* Chicago: American Technical Society, 1969, pp. 149-162.

Oberg, Erik; Jones, Franklin D.; and Horton, Holbrook L. *Machinery's Handbook.* 20th ed. New York: Industrial Press, 1976, p. 327.

Patton, William J. *Kinematics.* Reston, Va.: Reston Publishing Co., 1979, pp. 71-76.

Tao, D.C. *Fundamentals of Applied Kinematics.* Reading, Mass.: Addison-Wesley Publishing Co., 1967, pp. 48-77.

Packaging the Human Operator 12

two-dimensional manikin layout

HUMAN engineering and human factors are becoming more and more a part of the designer's vocabulary. The use of human factors, through graphical and other analytical techniques, should be standard in the engineering office, as one means of improving the fit between man and machine.

Human factors researchers have adopted a system of dividing the population into groups for study. They have broken down the adult population into 100 percentage groups (percentiles). Percentile 1 is the smallest possible physique, percentile 100 the largest. The top and bottom few percentiles represent people who are extremely rare. Most data charts cover percentiles 2.5 through 97.5, or 95 percent of the adult population.

Most published charts present three human figures. They represent the mean (percentile 50), plus the two extremes, percentiles 2.5 and 97.5. If a design situation fits the three theoretical physiques within reason, it should be adequate for the other percentiles.

Human factors tradeoffs in engineering are critical. If any part of the machine or human envelope is unsuitable, the overall design can strain human capabilities and impair machine performance. Human factors experts cite numerous cases where man-machine relationships have been compromised in favor of the machine, with dire consequences for the operator.

Although the human mechanism is vastly more complex and important than any man-made machinery, the basic dimensional problem of installing an operator in the crane cab, on a tractor, or in an automobile often has been given less attention than that of installing a simple fuel tank or motor. The operator, like the mechanical or electrical component, must have adequate volume, clearance, supports, and the proper surrounding environment.

Graphical and mathematical models are used to set criteria and verify specifications for operator-machine relationships. A semi-rigid model for military pilots helps to evaluate cockpit geometry; an enfleshed model helps to validate arm and leg reaches of auto drivers.

Courtesy of General Motors Corp.

Fig. 12-1. Designers working on a human factors problem.

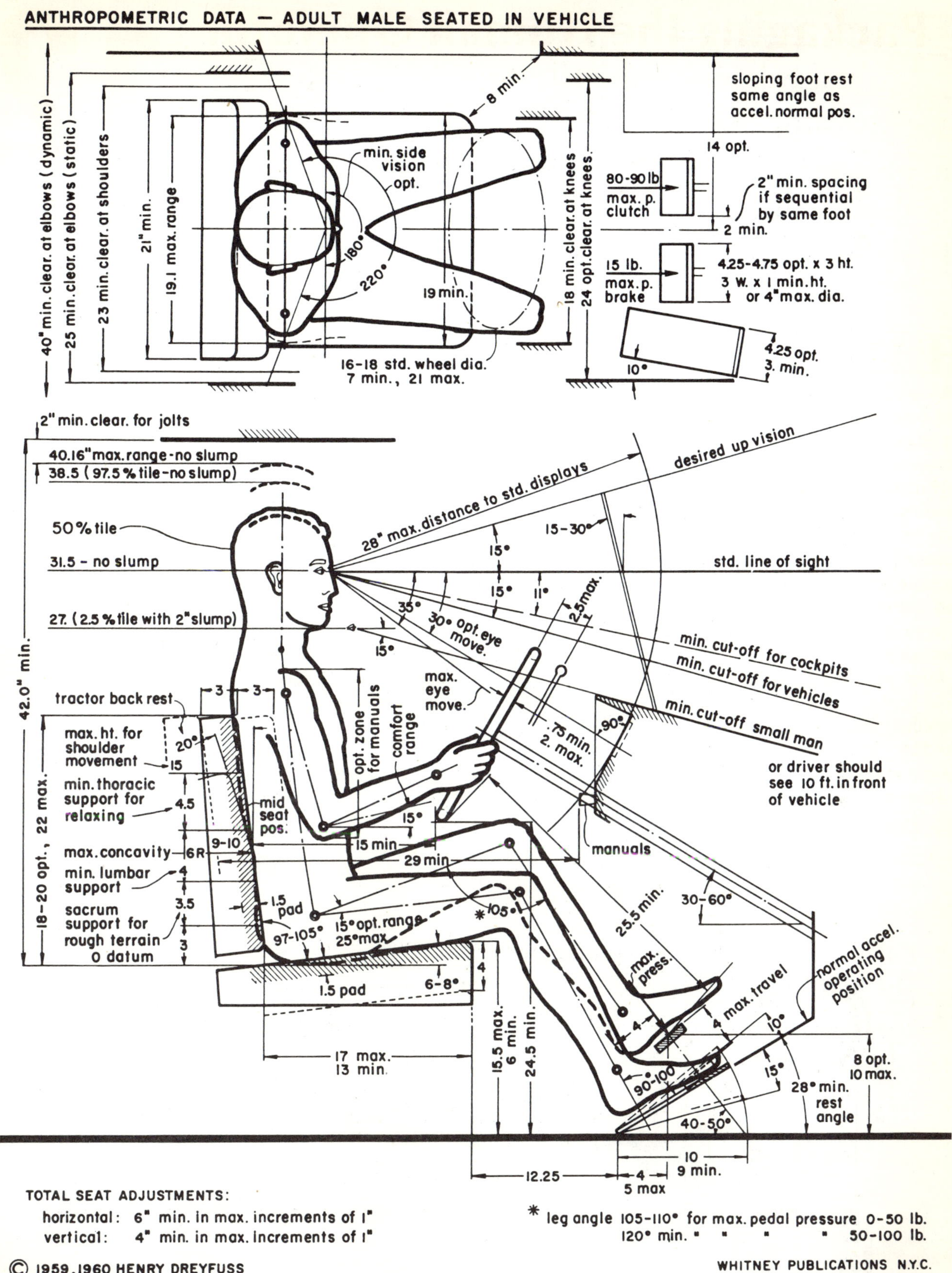

Fig. 12-2. Anthropometric data — seated adult male.

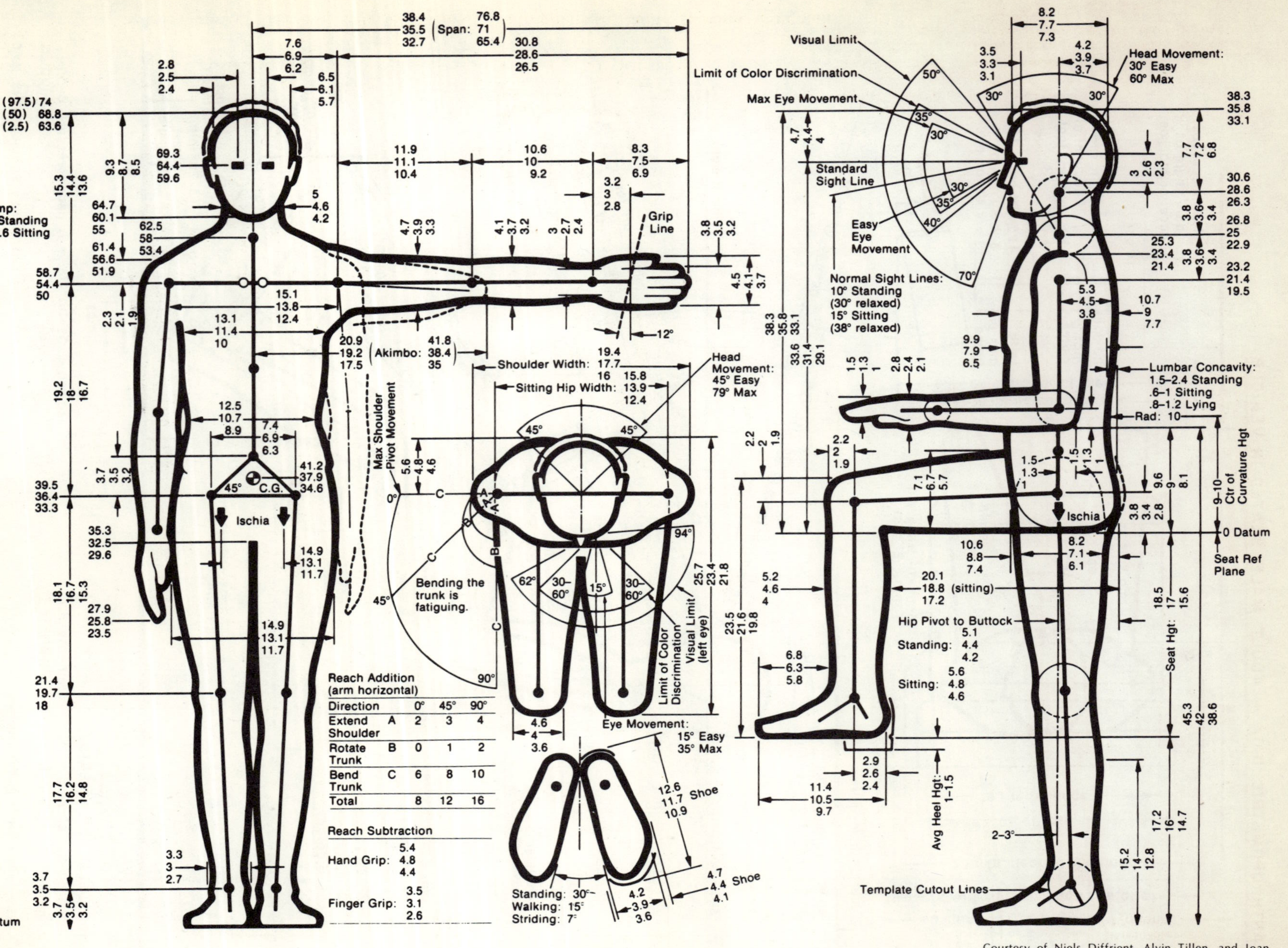

Fig. 12-3. Anthropometric data—standing adult male.

Courtesy of Niels Diffrient, Alvin Tillen, and Joan Bardagjy. *Humanscale 1/2/3*, The MIT Press, Cambridge, Mass.

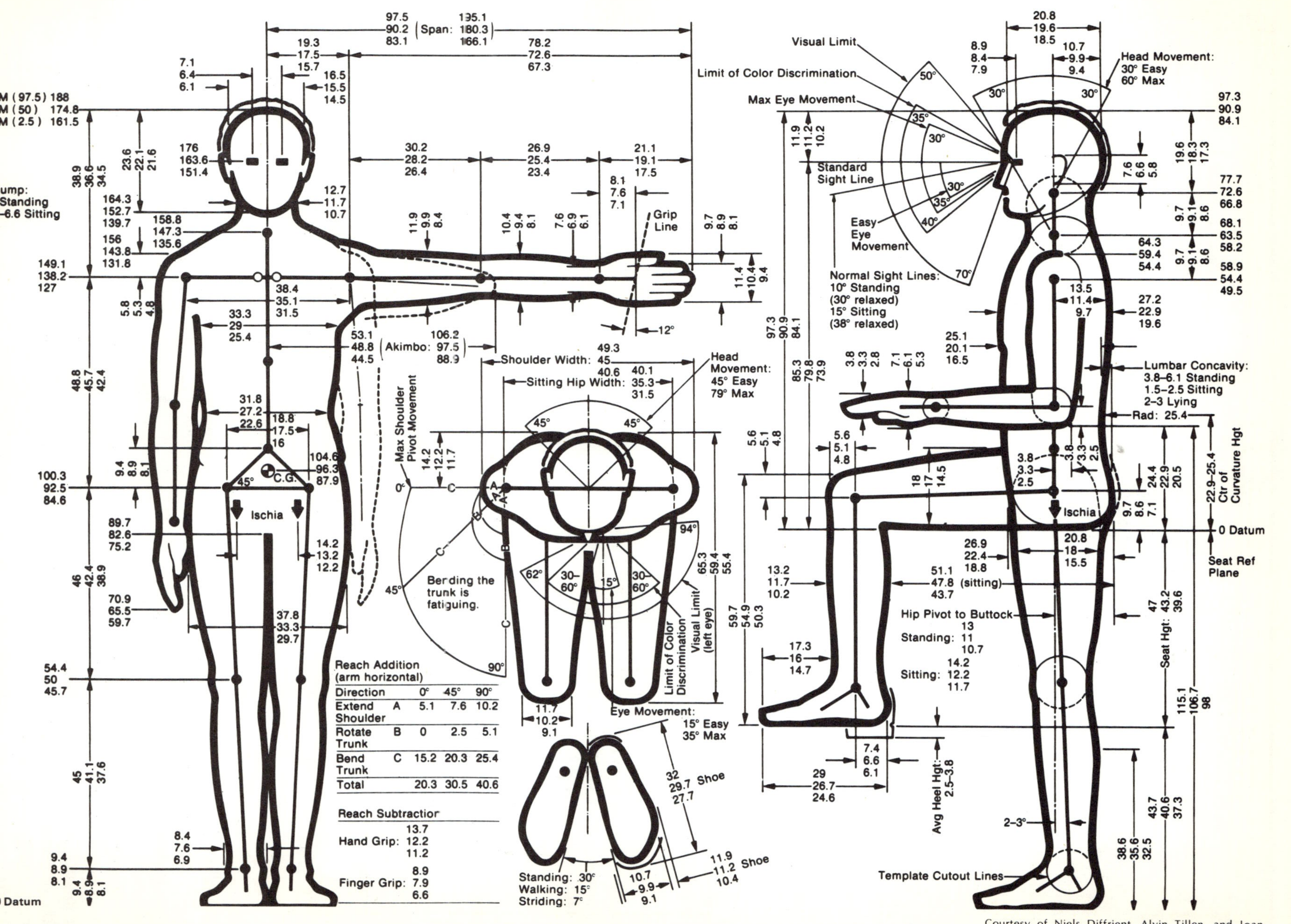

Reach Addition (arm horizontal)

Direction		0°	45°	90°
Extend Shoulder	A	5.1	7.6	10.2
Rotate Trunk	B	0	2.5	5.1
Bend Trunk	C	15.2	20.3	25.4
Total		20.3	30.5	40.6

Reach Subtraction

Hand Grip:	13.7 12.2 11.2
Finger Grip:	8.9 7.9 6.6

Fig. 12-3 (cont.). Anthropometric data—standing adult male.

Courtesy of Niels Diffrient, Alvin Tillen, and Joan Bardagjy. *Humanscale 1/2/3,* The MIT Press, Cambridge, Mass.

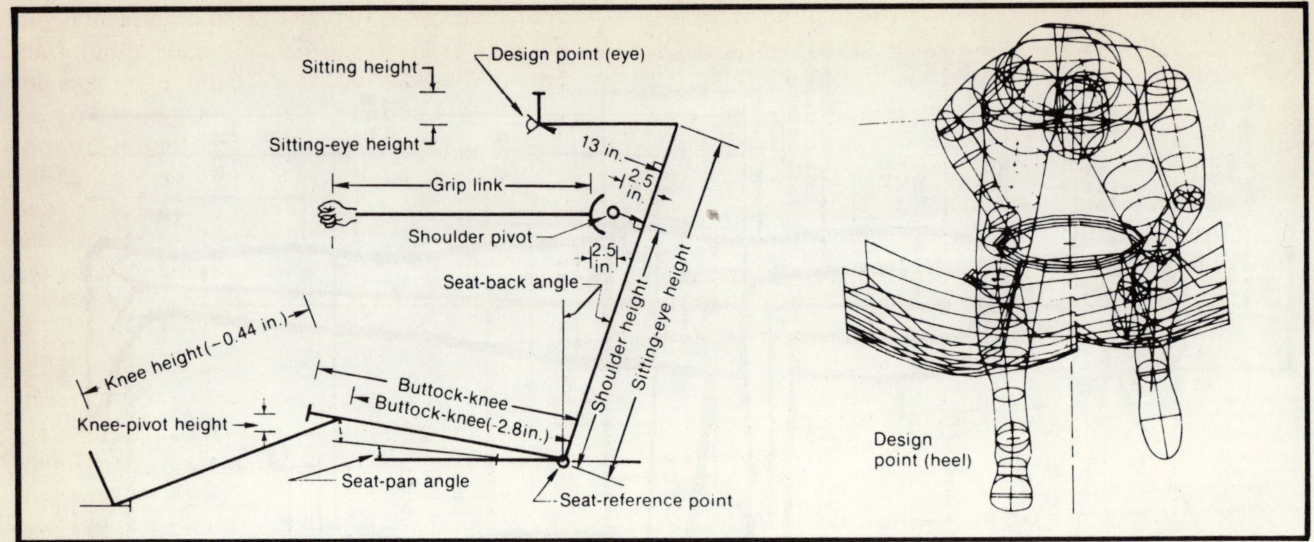

Courtesy of Penton/IPC, Inc., *Machine Design.*

Fig. 12-4. Human mechanism geometry.

A major automotive manufacturer utilizes a two-dimensional manikin (Fig. 12-4) constructed to represent in profile the physical torso-buttocks contour and weight (75.8 kg) of a 50th percentile adult male. The lower leg and thigh segments are constructed to represent the 90th percentile. The torso room template is used to determine equivalent head room at various seat-back angles. The H-point, or hip point, is the designated reference point on the human body. The 0-point for angular measurement is directly above the H-point on the template. (See Fig. 12-5.)

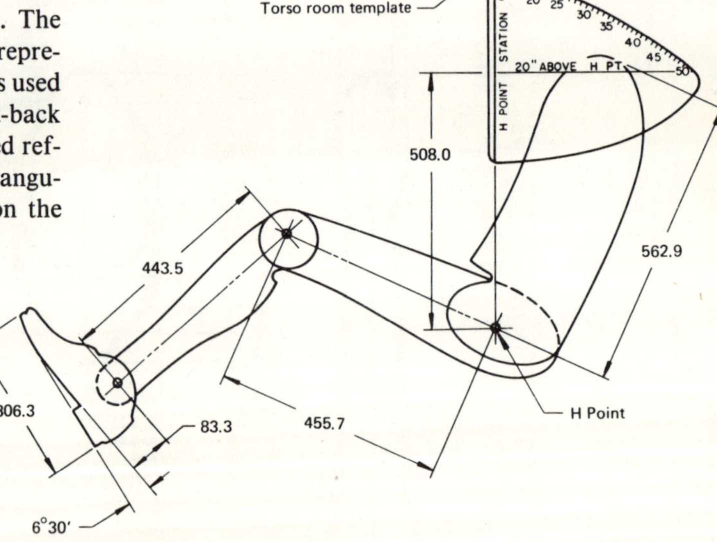

Fig. 12-5. Two-dimensional manikin.

Courtesy of General Motors Corp., *General Motors Drafting Standards.*

Problem Assignment

The following graphical problems and design questions are intended to provide you with experience in graphically analyzing and calculating basic mechanism constraints. Select, with instructor direction, the problem-sheet assignments suitable for you. Then complete the numbered problems in ordered sequence because each solution builds on preceding problems. Be sure to use the references suggested by your instructor (*list follows*) as sources for working through the assignment. Problem sheets are located at the back of the book.

Problem Sheet 12A

Problem 1. Utilizing measurements for a 97.5 percentile U.S. male, construct and analyze a standing kinematic representation. Assume that the figure is facing you as you view the drawing sheet. Use the C.G. as a starting point. Draw the representation by using metric values and show only the joints and connectors in your drawing. (Scale: 1.0 mm = 10.0 mm.) Represent the following conditions:

A. Raise the figure's left arm 30° (to your right) and

bend the elbow so that the wrist joint is directly above the elbow on a vertical line. Indicate the distance from the floor (0 datum) to the fingertips.

B. The C.G. has been positioned to be 65.0 cm above the floor (0 datum). The figure's back, arms, and fingers are pointed straight up in a vertical direction. Show a left-side view (figure's right side) indicating the position of the hands and knee joint. How high can the fingertips reach? How far in front of the vertical centerline does the skin surface of the knee extend?

Problem 2. Utilizing measurements for a 50.0 percentile U.S. male, construct and analyze a top view of a sitting figure. The shoulder centerline *s-s* positions the figure; it should be used as a starting reference. Draw the representation by using metric values, showing only the joints and connectors in your drawing. (Scale: 1.0 mm = 10.0 mm.) Represent the following conditions:

A. Move the figure's right leg out 20° from the vertical centerline about the right hip joint. Extend the right leg so that it sticks straight out on a horizontal plane referenced to the figure. Indicate the distance from the right hip joint to the bottom of the bare foot.

B. The grip line of the figure's two hands are clasped directly in front of the center of vision. If the grip line is 25.0 mm in front of the shoulder line *s-s*, determine the positions of the elbow joints. Assume the shoulders are kept in a straight line as shown.

Problem 3. Utilizing measurements for a 2.5 percentile U.S. male, construct and analyze the right kinematic representation. The side (left profile) of the figure is positioned so as to use the same floor surface (0 datum) as used in problem 1, above. Draw the representation by using metric values, showing only the joints and connectors in your drawing. (Scale: 1.0 mm = 10.0 mm.) Represent the following conditions:

A. Show the figure stepping up to a horizontal surface that is 45 cm above the floor (0 datum). Indicate the height from the floor to the top skin surface of the knee.

B. Move the hand and arm straight out in front of the figure on a horizontal plane. Rotate the fully extended arm and hand straight down and straight up on a vertical centerline. Indicate the height of the fingertips from the floor (0 datum) in all three positions.

Problem Sheet 12B

Problem 1. Utilizing measurements for a 50.0 percentile U.S. male, construct and analyze a sitting kinematic representation. Start by locating the H-point (points directly below the C.G.). Draw the representation by using metric values. Show the joints and connectors in your drawing but also show torso outline as needed. (Scale: 1.0 mm = 10.0 mm.) Represent the following conditions:

A. Locate the torso room template's horizontal centerline 50.80 cm above the H-point. (The 0-point for angular measurement is directly above the H-point on the template; see Fig. 12-5.) Lay out a race car seat with the following requirements: (1) the backrest angle is 28°; (2) the seat cushion makes an angle of 95° with the seat backrest; (3) the seat cushion depth is 43.20 cm; and (4) the car's floor is only 10.60 cm below the H-point. Determine the following by constructing a graphical layout: (1) the footrest angle when the foot is at 90° to the leg, (2) the seat backrest contour when the seat backrest is 55.90 cm high, and (3) the hand grip line when the arm is extended directly on a horizontal line. Use coordinate dimensions for all of the figure's joint positions.

B. Using a blank problem sheet from the back of the book, lay out your own joint dimensions for the same race car seat. Have a classmate assist you in making the necessary metric measurements.

References

The following source and page references can be used in developing the graphical solutions and calculations for the problem assignments.

Esposito, Anthony. *Kinematics for Technology.* Columbus, Ohio: Charles E. Merrill Publishing Co., 1973, pp. 1-27.

Hinkle, Rolland T. *Kinematics of Machines.* 2nd ed. Englewood Cliffs, N.J.: Prentice-Hall, 1960, pp. 1-10.

Hirschhorn, Jeremy. *Dynamics of Machinery.* New York: Barnes & Noble, 1967, pp. 1-44.

Kepler, Harold B. *Basic Graphical Kinematics.* 2nd ed. New York: McGraw-Hill Book Co., 1973, pp. 1-12, 29-71.

Lent, Deane. *Analysis and Design of Mechanisms.* 2nd ed. Englewood Cliffs, N.J.: Prentice-Hall, 1970, pp. 17-38.

Michels, Walter J., and Wilson, Charles E. *Mechanism: Design-Oriented Kinematics.* Chicago: American Technical Society, 1969, pp. 1-82.

Oberg, Erik; Jones, Franklin D.; and Horton, Holbrook L. *Machinery's Handbook.* 20th ed. New York: Industrial Press, 1976, pp. 291-311.

Patton, William J. *Kinematics.* Reston, Va.: Reston Publishing Co., 1979, pp. 1-29.

Tao, D.C. *Fundamentals of Applied Kinematics.* Reading, Mass.: Addison-Wesley Publishing Co., 1967, pp. 7-34.

Mechanical Handling Shuttle 13

hypo-cycloidal drive mechanism

THIS mechanical handling shuttle mechanism (Fig. 13-1) consists of a drive crank whose length is one-quarter of the desired index, a driven crank whose length is one-half of the index and which has a midpoint that is allowed to pivot about the outer end of the drive crank, and a guide for the driven crank. The guide assures the path of the driven crank when the center of the shuttle cart is directly over the center of the actuator. This unit uses a *hypo-cycloidal* drive mechanism that was developed (by the Mechanical Handling Section, Production Engineering Activity of Fisher Body Division, GM) for use in a parts transfer machine.

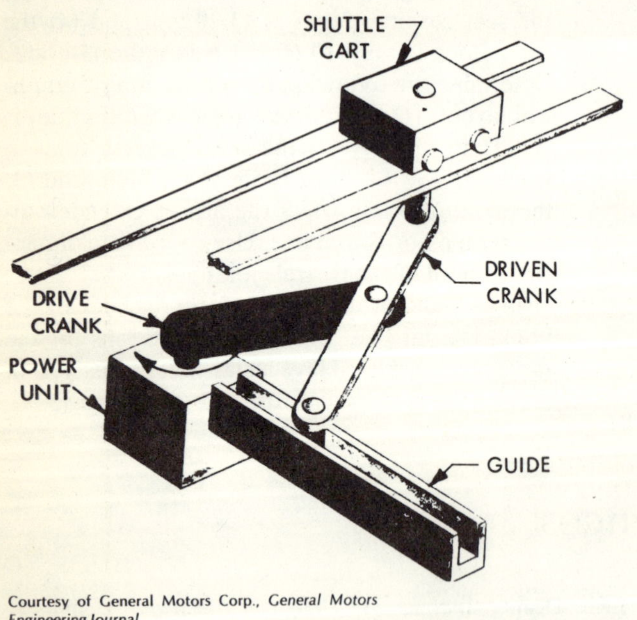

Courtesy of General Motors Corp., *General Motors Engineering Journal.*

Fig. 13-1. Hypo-cycloidal travel mechanism.

The drive unit shown in Fig. 13-2 uses a hypo-cycloidal travel mechanism to convert the constant rotation of the electrical motor into controlled linear motion. Through a series of drive belts, the motor rotates the flywheel. The shaft, which is secured to the flywheel, directs the power of the motor into a right-angle speed reducer that is connected to the drive crank of the travel mechanism. A clutch and brake are coupled to the power shaft to interrupt the power supply and to stop the shuttle cart at the end of the drive unit cycle.

The effectiveness of the design of a mechanical handling shuttle is based partially on the relative values of velocity that the unit develops. Velocity influences the behavior of parts being transferred by the shuttle. To determine values for these parameters, an equation for each must be derived. From these equations the design engineer can predict the performance of the unit (without going to the expense of construction) by assessing the parameters during the design stage. An example of this procedure follows:

> Given the design specifications listed in Fig. 13-3, what is the velocity of the shuttle cart and the angular velocity of the crank?

The solution for determining the velocity starts with the resolution of displacement S. The displacement is a function of the drive crank angle θ which, in turn, is the product of the drive crank's angular velocity ω and a point of time T. This is shown by the following equations:

$$\theta = \omega T \tag{13.1}$$
$$S = R_2 - R_3 \tag{13.2}$$

From the solution of right triangles, we know that:

$$R_3 = 2R_1 \cos\theta \tag{13.3}$$

Therefore,

$$S = R_2 - 2R_1 \cos\theta \tag{13.4}$$

And since

$$R_2 = 2R_1 \tag{13.5}$$

then

$$S = 2R_1 - 2R_1 \cos\theta \tag{13.6}$$
$$S = 2R_1(1 - \cos\theta) \tag{13.7}$$

Substituting ωT for θ in equation (13.7) and differentiating for S, equation (13.11) for the velocity V_s of the shuttle is obtained:

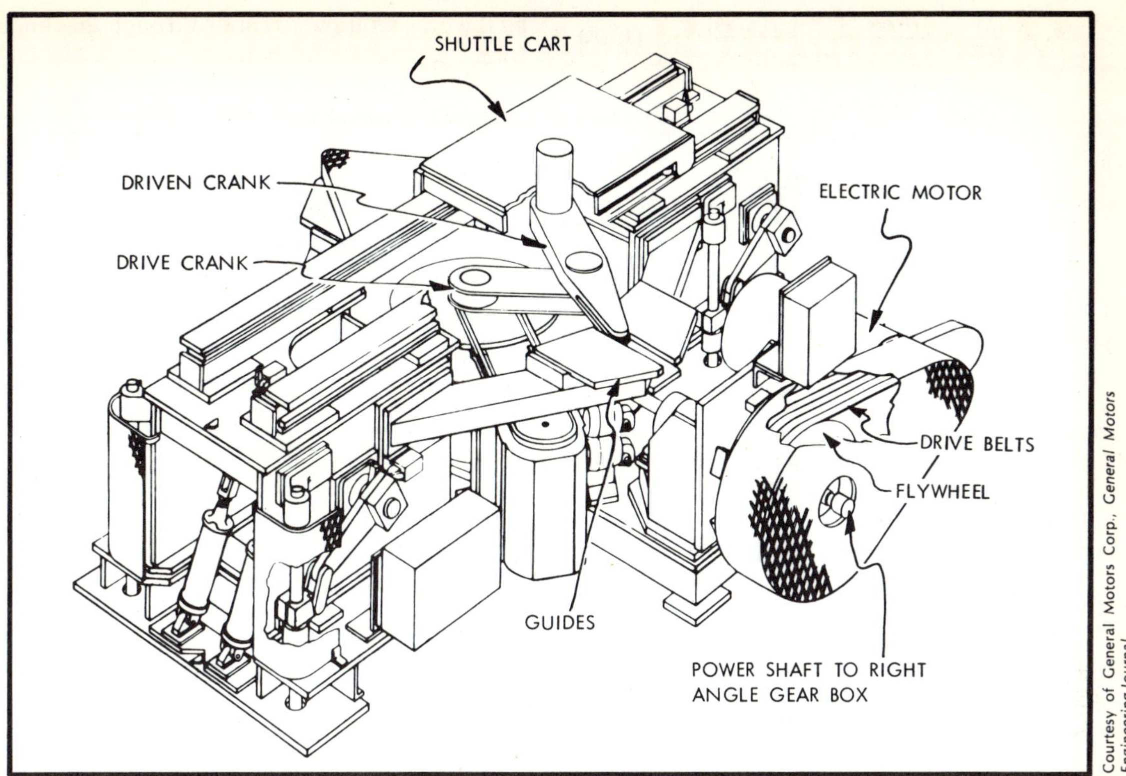

Fig. 13-2. Electrically powered mechanical handling shuttle.

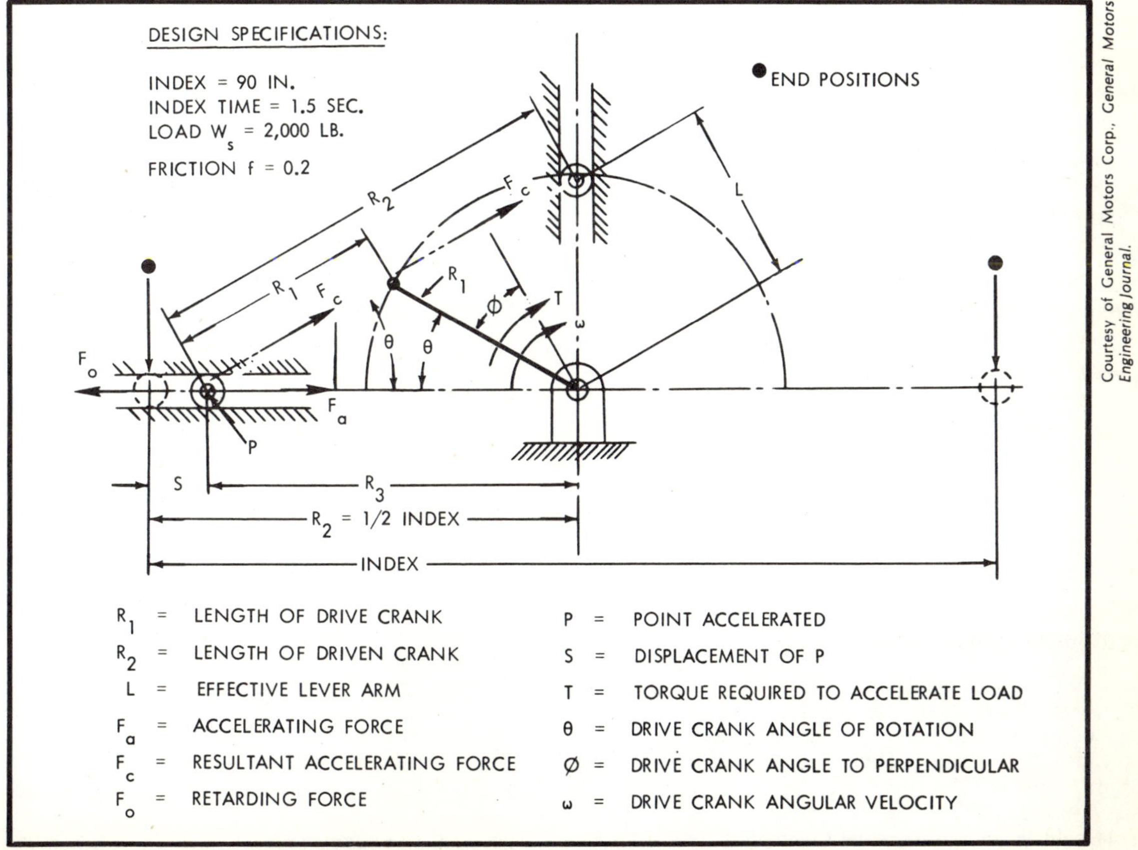

Fig. 13-3. Force system for hypo-cycloidal travel mechanism.

$$V_s = \frac{dS}{dT} = \frac{d[2R_1(1 - \cos \omega T)]}{dT} \quad (13.8)$$

$$V_s = \frac{d(2R_1)}{dT} - \frac{d(2R_1 \cos \omega T)}{dT} \quad (13.9)$$

$$V_s = 0 - 2R_1(-\sin \omega T)\frac{d\omega T}{dT} \quad (13.10)$$

$$V_s = 2R_1\omega \sin \omega T \quad (13.11)$$

Substituting back θ for ωT and R_2 for $2R_1$ in equation (13.11) gives:

$$V_s = R_2\omega \sin \theta \quad (13.12)$$

The angular velocity ω in equation (13.12) is equal to the total angular displacement of the crank arm divided by the time required to rotate through that angle. Therefore,

$$\text{angular displacement index time} = \left(\frac{180°}{1.5 \text{ sec.}}\right)\left(\frac{\pi \text{ rad}}{180°}\right) = 2.095 \text{ rad/sec.}$$

Length R_2 of the driven crank equals one-half of the 90-in. index. Hence R_2 equals 3.75 ft.

Substituting the calculated constants into equation (13.12) gives the following incremental values for velocity V:

$$V = (3.75)(2.095)(\sin \theta)$$
$$V = (7.86)(\sin \theta) \text{ ft./sec.}$$
$$V_{40°} = 5.05 \text{ ft./sec.}$$

Problem Assignment

The following graphical problems and design questions are intended to provide you with experience in graphically analyzing and calculating basic mechanism constraints. Select, with instructor direction, the problem-sheet assignments suitable for you. Then complete the numbered problems in ordered sequence because each solution builds on preceding problems. Be sure to use the references suggested by your instructor (*list follows*) as sources for working through the assignment. Problem sheets are located at the back of the book.

Problem Sheet 13A

Problem 1. Rotate crank link 2 clockwise every 30° starting on line *AD* as shown in the illustration. Determine the total index for the shuttle cart and for the guide pin. (Scale: .50 in. = 10.00 in.)

Problem 2. Plot the path that point *E* makes. What geometric shape is traced by its path?

Problem 3. Determine the angular velocity of crank link 2 when it rotates at 875 rpm.

Problem 4. Determine the maximum velocities for the shuttle cart and the guide pin. What position(s) does crank link 2 have when the maximum velocities occur?

Problem 5. Graphically determine the linear velocities for guide pin *D* and for shuttle cart *C* when crank link 2 rotates at 875 rpm. Determine the velocities when crank link 2 is at the 0°, 30°, and 60° positions.

Problem Sheet 13B

Problem 1. Plot the displacement graphs of shuttle cart *C* and guide pin *D*. The origin of the measurements should begin with point *D* in its furthermost position at the left. Measurements made to the right and up are plotted as positive values, while measurements plotted to the left and down are negative values. Rotate crank link 2 clockwise every 30° starting along line *AD*. Determine the total index for the shuttle cart and for the guide pin. (Scale: .50 in. = 10.00 in.)

Problem 2. Using a different line type, repeat problem 1 with point *B* relocated. Maintain the same link lengths, but locate *B* so that *BC* is twice the length of *BD*.

References

The following source and page references can be used in developing the graphical solutions and calculations for the problem assignments.

Esposito, Anthony. *Kinematics for Technology*. Columbus, Ohio: Charles E. Merrill Publishing Co., 1973, pp. 35, 36.

Hinkle, Rolland T. *Kinematics of Machines.* 2nd ed. Englewood Cliffs, N.J.: Prentice-Hall, 1960, pp. 334, 335.

Hirschhorn, Jeremy. *Dynamics of Machinery.* New York: Barnes & Noble, 1967, pp. 36, 37.

Kepler, Harold B. *Basic Graphical Kinematics.* 2nd ed. New York: McGraw-Hill Book Co., 1973, pp. 69-72.

Lent, Deane. *Analysis and Design of Mechanisms.* 2nd ed. Englewood Cliffs, N.J.: Prentice-Hall, 1970, pp. 25, 388.

Michels, Walter J., and Wilson, Charles E. *Mechanism: Design-Oriented Kinematics.* Chicago: American Technical Society, 1969, pp. 21-82.

Oberg, Erik; Jones, Franklin D.; and Horton, Holbrook L. *Machinery's Handbook*. 20th ed. New York: Industrial Press, 1976, pp. 295-305, 326-356.

Patton, William J. *Kinematics.* Reston, Va.: Reston Publishing Co., 1979, pp. 1-29.

Tao, D.C. *Fundamentals of Applied Kinematics.* Reading, Mass.: Addison-Wesley Publishing Co., 1967, pp. 152-183.

External Geneva Drives

14

intermittent mechanisms

THE external Geneva wheel is one of the most commonly used mechanisms for producing *intermittent rotary motion* from a uniform input speed. The driven member, or *star wheel*, contains a number of slots into which the roller of the driving crank fits. The number of slots determines the ratio between dwell and motion period of the driven shaft.

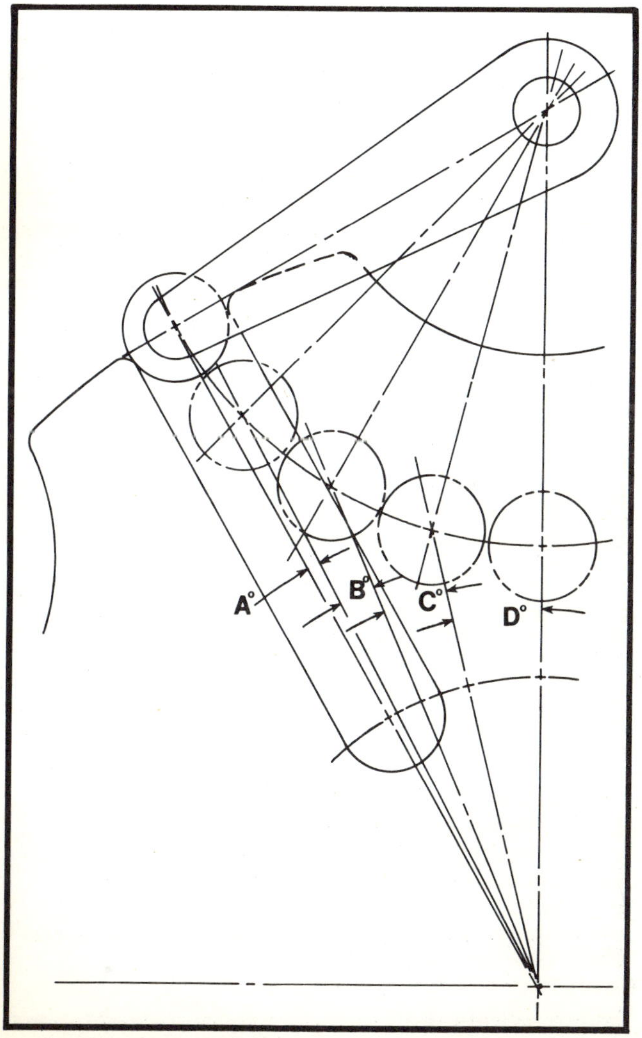

Fig. 14-1. Driving crank and star wheel.

As Fig. 14-1 illustrates, the driving member rotates at a constant speed but imparts an angular acceleration to the wheel. To start the wheel, the arm turns through 15°; the wheel turns A°. Using this angle as a unit, note that the wheel turns through an increasing number of degrees for every 15° increment on the driving arm. Maximum acceleration occurs when the pin is about one-quarter of the distance from the bottom of the slot; maximum wear also occurs here.

As the crank for a Geneva drive rotates at a constant speed, it engages a slotted wheel which rotates through the angle between the slots; the wheel then stops during the time the crank completes its rotation and engages the adjacent slot.

The slots must be at equal angles to each other, with the minimum number of slots being three and the practical maximum 18. The pin of the driving member strikes the slot, causing the wheel to start turning abruptly. The shock to the mechanism is at a minimum when the center of the arm is perpendicular to the slot, as shown in Fig. 14-2. The shock is greater if the angle is more than 90°, as in Fig. 14-3. Design conditions may require that shaft centers bc in the position shown; in this case, the rpm of the arm must be slow in order to limit the shock and subsequent wear.

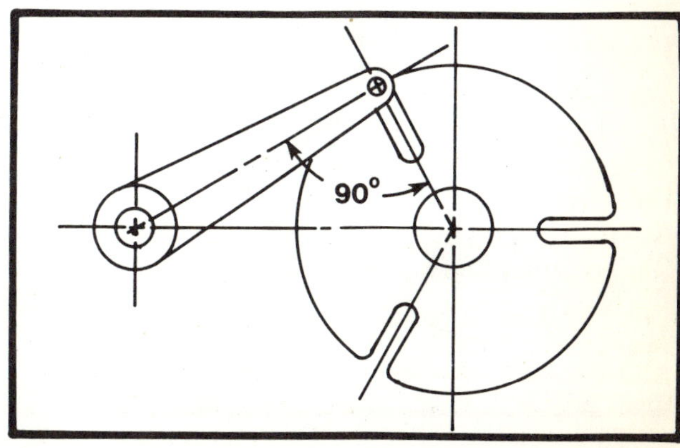

Fig. 14-2. Geneva drive's driving crank at the instant it engages the star wheel slot.

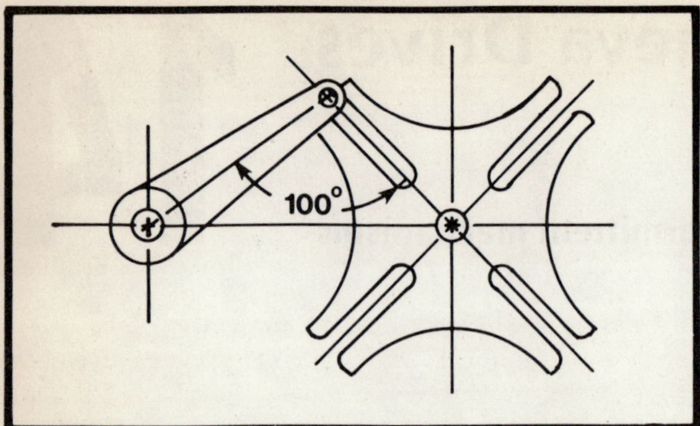

Fig. 14-3. Shock imparted to star wheel.

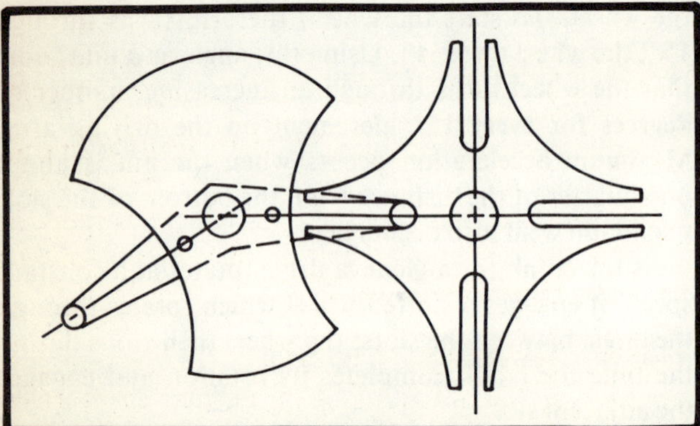

Fig. 14-4. Double arms at unequal angles for varying idling speeds.

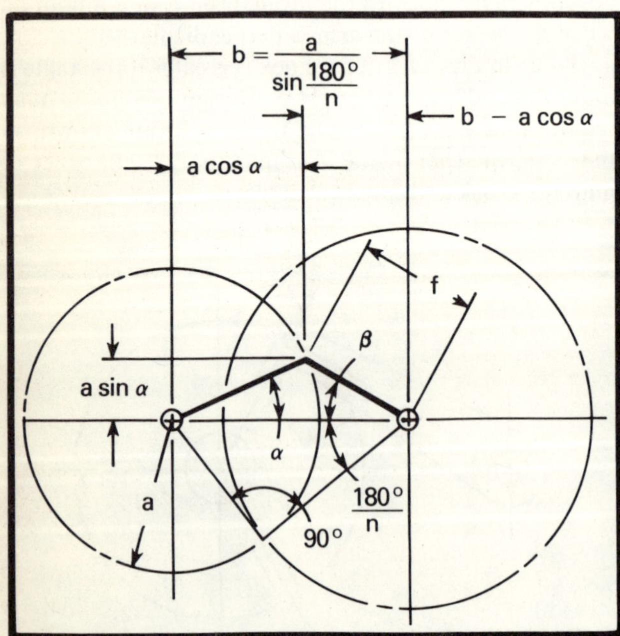

Fig. 14-5. Basic outline sketch for the external Geneva drive. (The symbols are identified for application in the basic equations.)

Fig. 14-4 shows a Geneva drive with two driving arms, at something less than 180° to each other. The time required to turn the wheel remains constant, but note that the dwell periods vary. As shown, the wheel is at its maximum velocity, but it has zero acceleration from this point.

Listed below are equivalents and equations which will be used in completing the problem assignments at the end of this unit. Figs. 14-5 and 14-6 illustrate the notation and corresponding formulas for external Geneva drives.

NOTATION

a = crank radius of driving member

b = center distance

D = diameter of driven member

d_r = roller diameter

n = number of slots

p = constant velocity of driving crank (in rpm)

α = angular position of driving crank at any time

β = angular displacement of driven member corresponding to crank angle α

ω = constant angular velocity of driving crank

FORMULAS

$$b = am \qquad m = \frac{1}{\sin \frac{180°}{n}}$$

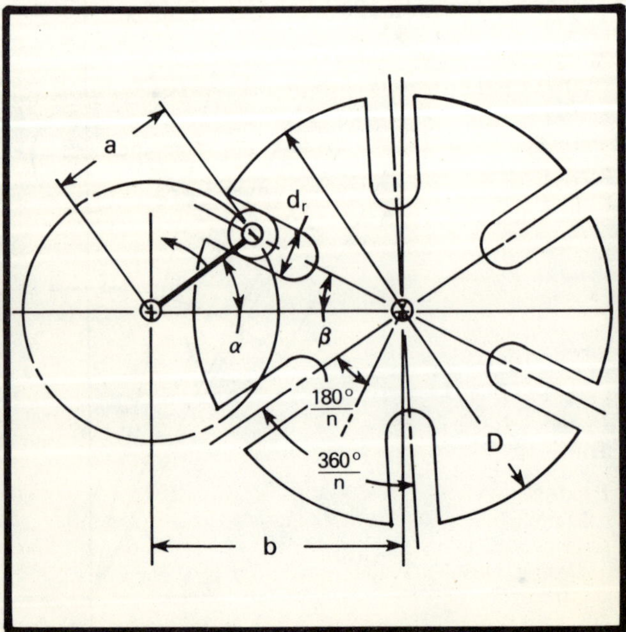

Fig. 14-6. A six-slot Geneva drive.

$$D = 2\sqrt{\frac{d_r^2}{4} + a^2\cot^2\frac{180°}{n}}$$

$$\omega = \frac{p\pi}{30}\text{ rad/sec.}$$

$$\cos\beta = \frac{m - \cos\alpha}{\sqrt{1 + m^2 - 2m\cos\alpha}}$$

Angular Velocity of driven member =

$$\omega\left[\frac{m\cos\alpha - 1}{1 + m^2 - 2m\cos\alpha}\right]$$

Angular Acceleration of driven member =

$$\omega^2\left[\frac{(m\sin\alpha)(1 - m^2)}{(1 + m^2 - 2m\cos\alpha)^2}\right]$$

Maximum Angular Velocity occurs at $\alpha = 0°$,

and equals $\frac{\omega}{m - 1}$ rad/sec.

Maximum Angular Acceleration occurs when

$$\cos\alpha = \sqrt{\left(\frac{1 + m^2}{4m}\right)^2 + 2} - \frac{1 + m^2}{4m}$$

Standardized Geneva drives are available as open or enclosed assemblies. All types of Geneva drives are available in any number of indexes from 3 to 24. Center distances less than .750 and greater than 6.000 exist upon special request. The engineering drawings and specifications on the next three pages (Figs. 14-7, 14-8, 14-9) show some of the choices that are commercially available.

Problem Assignment

The following graphical problems and design questions are intended to provide you with experience in graphically analyzing and calculating basic mechanism constraints. Select, with instructor direction, the problem-sheet assignments suitable for you. Be sure to use the references suggested by your instructor (*list follows*) as sources for working through the assignment. Problem sheets are located at the back of the book.

Problem Sheet 14A

Problem 1. The illustration shows a four-position Geneva mechanism, which is a well-known intermittent motion and indexing device. The Geneva wheel 1, shown, has four slots. It will make 1/4 revolution when driving crank wheel 2 makes one revolution. The driving crank wheel revolves about point B at a constant speed of rotation in a clockwise direction. Rotate the driving crankpin C about point B every 15° to point C'. Determine the lengths and configurations of the two slots, starting at C and C' and drawn toward point A. Show the lines connecting A to C and B to C in each 15° rotational position. Graphically lay out the design for the driving crank wheel; show it in position 2 and in position 2'. Determine the angular velocity of the Geneva wheel if the angular velocity of the driving crank wheel is a constant 20.0 rad/min. Show the velocity for each of the seven positions. (Scale: 1.00 in. = 1.00 in.)

Problem Sheet 14B

Problem 1. Make a graph of the angular displacement of the Geneva mechanism shown on Problem Sheet 14A. Use a protractor to measure the angles through which the Geneva wheel rotates for the corresponding angles through which the driving crank wheel rotates. Locate the point on the graph where the maximum velocity of the Geneva wheel occurs. (Scale: 1.00 in. = 1.00 in.)

Problem Sheet 14C

Problem 1. Lay out a six-position Geneva mechanism with the given center distance. The driving crank wheel revolves counterclockwise about point B at 30 rpm. The cam follower or roller diameter is .50 in. Determine the maximum angular velocity of the Geneva wheel. (Scale: 1.00 in. = 1.00 in.)

References

The following source and page references can be used in developing the graphical solutions and calculations for the problem assignments.

Esposito, Anthony. *Kinematics for Technology.* Columbus, Ohio: Charles E. Merrill Publishing Co., 1973, p. 16.

Greenwood, Douglas C., ed. *Engineering Data for Product Design.* New York: McGraw-Hill Book Co., 1961, pp. 240, 241.

———. *Product Engineering Design Manual.* New York: McGraw-Hill Book Co., 1959, pp. 254-274.

Hardison, Thomas B. *Introduction to Kinematics.* Reston, Va.: Reston Publishing Co., 1979, pp. 123-133.

Hinkle, Rolland T. *Kinematics of Machines.* 2nd ed. Englewood Cliffs, N.J.: Prentice-Hall, 1960, p. 102.

Kepler, Harold B. *Basic Graphical Kinematics.* 2nd ed. New York: McGraw-Hill Book Co., 1973, pp. 398-403.

Lent, Deane. *Analysis and Design of Mechanisms.* 2nd ed. Englewood Cliffs, N.J.: Prentice-Hall, 1970, pp. 338, 339, 392-394.

Patton, William J. *Kinematics.* Reston, Va.: Reston Publishing Co., 1979, pp. 110, 111, 149-153.

Tao, D.C. *Fundamentals of Applied Kinematics.* Reading, Mass.: Addison-Wesley Publishing Co., 1967, pp. 335-337.

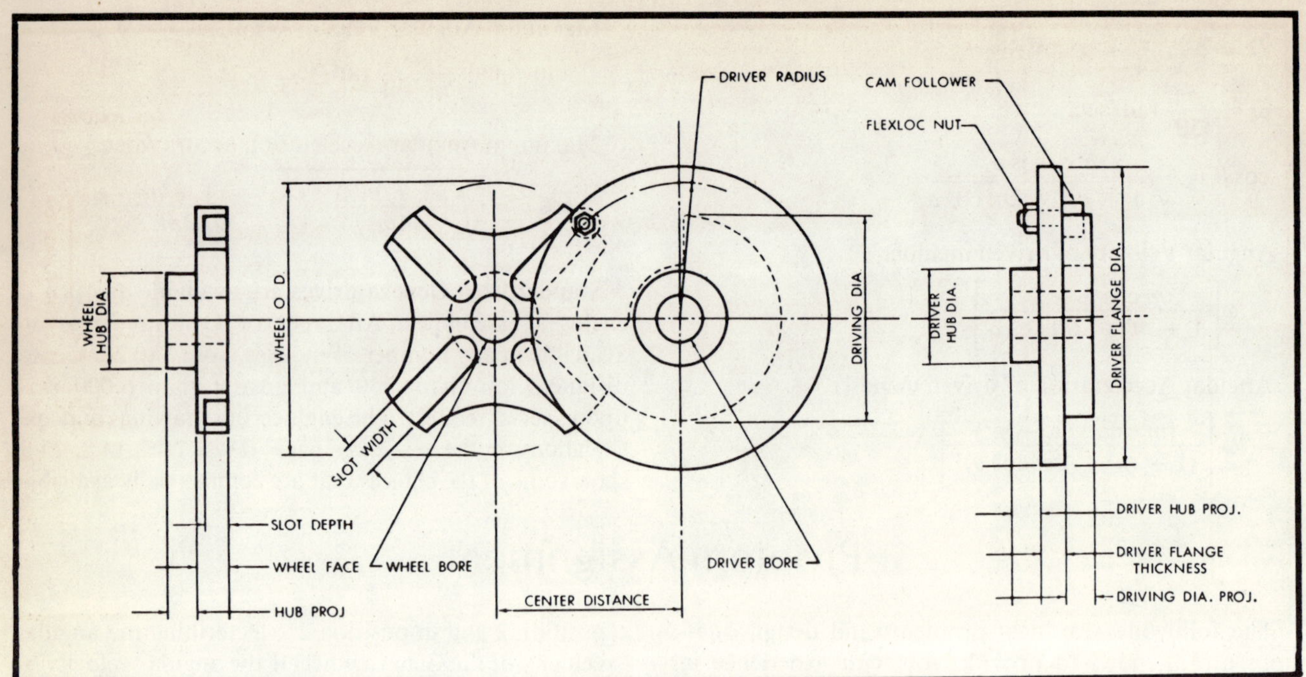

4 POINT MACRO GENEVA DRIVE

Intermediate Duty

ENGINEERING SPECIFICATIONS

PART NUMBER	4PI-21	4PI-22	4PI-23	4PI-25	4PI-26	4PI-27	4PI-28	4PI-30
Center Distance	2.125	2.250	2.375	2.500	2.625	2.750	2.875	3.000
Wheel O.D. Ref.	3.062	3.250	3.375	3.562	3.750	3.937	4.093	4.312
Wheel Face Max.	.625	.625	.625	.625	.625	.625	.625	.625
Wheel Slot Width	.501	.501	.501	.501	.626	.626	.626	.626
Wheel Slot Depth	.437	.437	.437	.437	.437	.437	.437	.437
Cam Follower Dia.	.500	.500	.500	.500	.625	.625	.625	.625
Wheel Bore Max.	.625	.625	.625	.625	.750	.750	.750	.750
Stock Wheel Bore	.500	.625	.625	.625	.625	.625	.750	.750
Wheel Hub Dia. Max.	1.250	1.250	1.250	1.250	1.500	1.500	1.500	1.500
Wheel Hub Proj. Max.	.500	.500	.500	.500	.500	.500	.500	.500
Driver Radius Ref.	1.500	1.593	1.671	1.765	1.859	1.937	2.031	2.125
Driving Dia. Ref.	2.250	2.437	2.609	2.781	2.828	3.015	3.187	3.375
Driving Dia. Proj. Max.	.625	.625	.625	.625	.625	.625	.625	.625
Driver Flange Dia. Ref.	3.500	3.687	3.859	4.031	4.328	4.515	4.687	4,875
Driver Flange Thick.	.375	.375	.375	.375	.437	.437	.437	.437
Driver Bore Max.	.625	.625	.625	.625	.750	.750	.750	.750
Stock Driver Bore	.500	.500	.500	.625	.625	.625	.750	.750
Driver Hub Dia. Max.	1.250	1.250	1.250	1.250	1.500	1.500	1.500	1.500
Driver Hub Proj. Max.	.500	.500	.500	.500	.500	.500	.500	.500
Output Torque in. lbs. 1 to 60 indexes per min.	485	500	535	560	900	940	980	1090
Output Torque in. lbs. 60 to 200 indexes per min.	240	255	265	280	450	470	490	545
Output Torque in. lbs. 200 to 2000 indexes per min.	**	**	**	**	**	**	**	**

Center distances less, intermediate, or greater than specified above are available on special order. Keyways and set screws may be specified as required at additional charge. Wheel material alloy steel, driver material—meehanite. Stainless Steel available on special order. Protective finishes available on request.

**Consult our Engineering Department for output torques for high speed indexing.

Courtesy of Geneva Mechanisms Corp.

Fig. 14-7. Four-point macro Geneva drive.

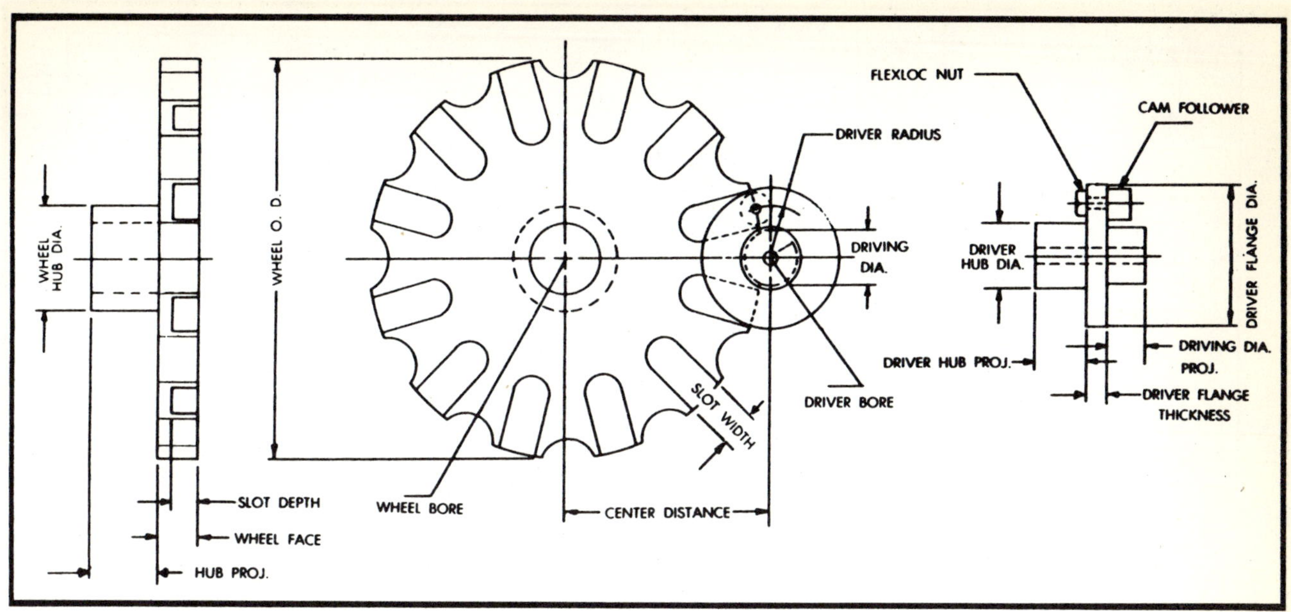

12 POINT GENEVAMATION DRIVE

Intermediate and Heavy Duty

ENGINEERING SPECIFICATIONS

PART NUMBER	12 PI-30	12 PI-32	12 PI-35	12 PI-37	12 PH-40	12 PH-42	12 PH-45	12 PH-47	12 PH-50	12 PH-52	12 PH-55	12 PH-57	12 PH-60
Center Distance	3.000	3.250	3.500	3.750	4.000	4.250	4.500	4.750	5.000	5.250	5.500	5.750	6.000
Wheel O.D. Ref.	5.843	6.312	6.812	7.281	7.765	8.250	8.718	9.218	9.687	10.187	10.656	11.125	11.625
Wheel Face Max.	.750	.750	.750	.750	1.000	1.000	1.000	1.000	1.125	1.125	1.125	1.125	1.125
Wheel Slot Width	.626	.626	.626	.626	.751	.751	.751	.751	1.001	1.001	1.001	1.001	1.001
Wheel Slot Depth	.500	.500	.500	.500	.562	.562	.562	.562	.687	.687	.687	.687	.687
Cam Follower Dia.	.625	.625	.625	.625	.750	.750	.750	.750	1.000	1.000	1.000	1.000	1.000
Wheel Bore Max.	2.000	2.000	2.000	2.000	2.000	2.000	2.000	2.000	2.000	2.000	2.000	2.000	2.000
Stock Wheel Bore	.750	.750	.750	.750	1.000	1.000	1.125	1.125	1.250	1.250	1.250	1.250	1.500
Wheel Hub Dia. Max.	3.000	3.000	3.000	3.000	3.000	3.000	3.000	3.000	3.000	3.000	3.000	3.000	3.000
Wheel Hub Proj. Max.	.500	.500	.500	.500	1.000	1.000	1.000	1.000	1.000	1.000	1.000	1.000	1.000
Driver Radius Ref.	.781	.843	.906	.968	1.031	1.093	1.156	1.234	1.296	1.359	1.421	1.484	1.546
Driving Dia. Ref.	.671	.812	.937	1.062	.937	1.078	1.203	1.328	1.093	1.218	1.343	1.468	1.609
Driving Dia. Proj. Max.	.750	.750	.750	.750	1.000	1.000	1.000	1.000	1.125	1.125	1.125	1.125	1.125
Driver Flange Dia. Ref.	2.187	2.312	2.437	2.562	2.812	2.937	3.062	3.203	3.562	3.718	3.843	3.968	4.109
Driver Flange Thick.	.500	.500	.500	.500	.562	.562	.562	.562	.625	.625	.625	.625	.625
Driver Bore Max., For Thru Round Shaft	.156	.156	.187	.187	.218	.218	.250	.250	.281	.281	.312	.312	.375
Stock Driver Bore, use Shaft with Clearance Flat	.750	.750	.750	.750	1.000	1.000	1.125	1.125	1.250	1.250	1.250	1.250	1.500
Driver Hub Dia. Max.	.750	.750	.750	.750	1.000	1.000	1.000	1.000	1.500	1.500	1.500	1.500	1.500
Driver Hub Proj. Max.	.500	.500	.500	.500	1.000	1.000	1.000	1.000	1.000	1.000	1.000	1.000	1.000
Output Torque in. lbs. 1 to 60 indexes per min.	1575	1675	1800	1975	4075	4275	4475	4675	10725	11125	11525	11925	12725
Output Torque in. lbs. 60 to 200 indexes per min.	800	850	925	975	2050	2150	2250	2350	5350	5550	5775	5950	6350
Output Torque in. lbs. 200 to 2000 indexes per min.	**	**	**	**	**	**	**	**	**	**	**	**	**

Center distances less, intermediate, or greater than specified above are available on special order. Keyways and set screws may be specified as required at additional charge. Wheel material alloy steel, driver material—meehanite. Stainless Steel available on special order. Protective finishes available on request.

**Consult our Engineering Department for output torques for high speed indexing.

Courtesy of Geneva Mechanisms Corp.

Fig. 14-8. Twelve-point Genevamation drive.

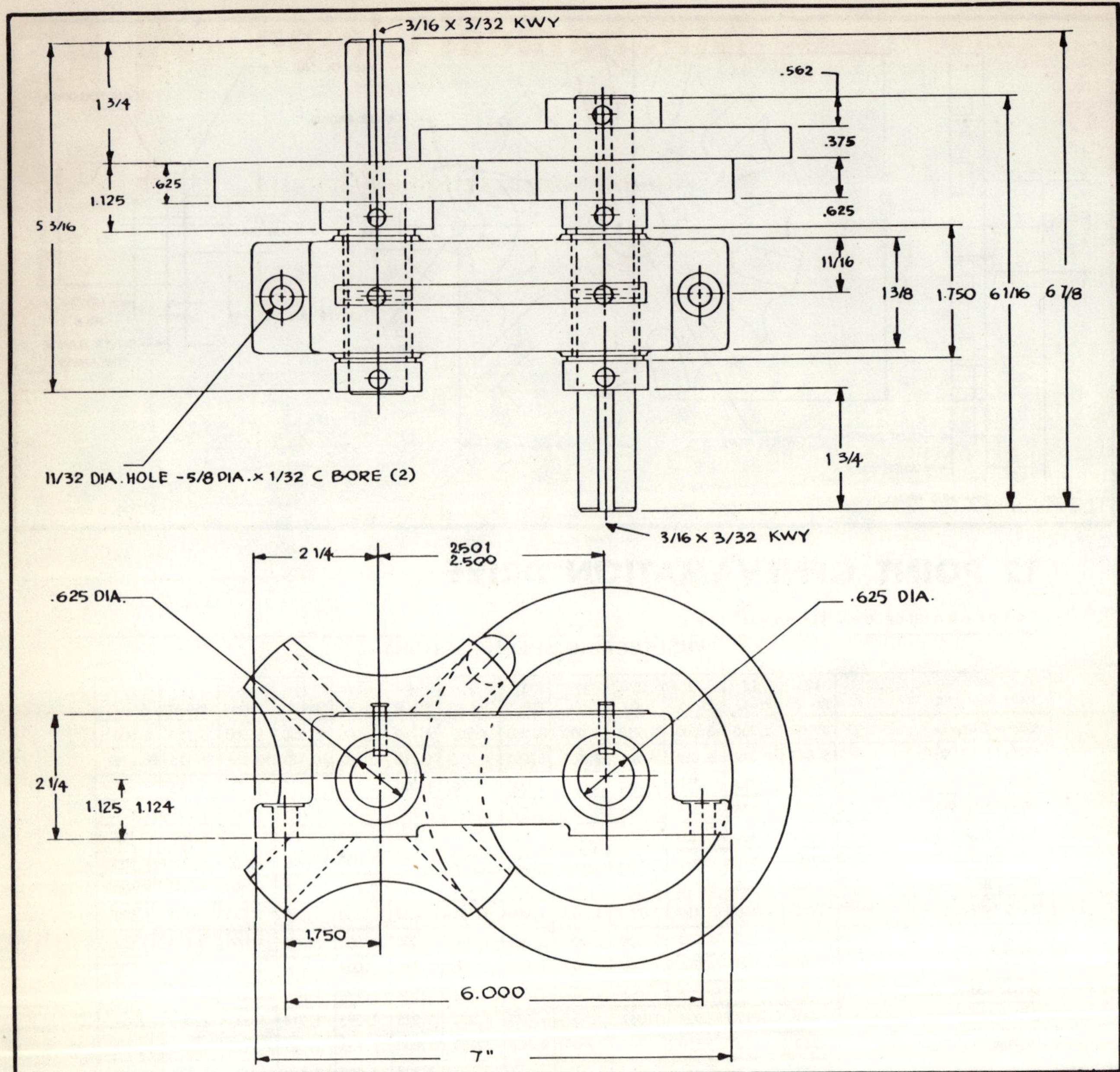

MACRO MOUNTED UNIT

2.500 Center Distance   No. 24 Mounting Base

FEATURES OF MOUNTED UNIT

Available with 4-6-8 Point Geneva Components. Cast Aluminum Housing. Oilite Bearings. Alloy Steel Shafts with Standard Keyways. Oil Reservoir with Oilers. Shaft to Shaft center distance precision bored to maintain alignment of Geneva Wheel and Driver. Proper running clearances built in. Mounting distances from base of Mounting Block to shaft center machined to close tolerance for ease of aligning in put and out put shafts with customer components. Readily disassembled for maintenance or modification. Available with 4 shaft arrangements.

Special shaft modifications or lengths other than standard are available on special order.

Mounted drives with center distances less, intermediate on greater than those listed above are also available on special order.

Courtesy of Geneva Mechanisms Corp.

Fig. 14-9. Macro-mounted Geneva drive unit.

Mechanical Clamp 15

double toggle action mechanism

A toggle mechanism is basically a four-bar mechanism. This kinematic device is used where a high mechanical advantage (M.A.) is desired. (M.A. = ratio of output force to input force = F_{output}/F_{input}.)

The maximum clamping or exerting force developed in any toggle action clamp is theoretically attained when the three pivot points of the mechanism lie on a straight line. In reality, however, vibration and intermittent load conditions found in industrial applications must be considered, since such conditions would soon unlock an im-

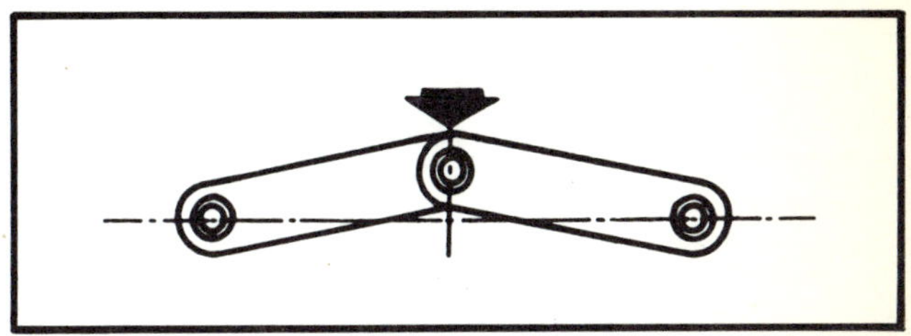

Courtesy of De-Sta-Co Division, Dover Corp.

Fig. 15-1. Toggle linkage.

MODEL DTV-400

TEMPLATE & PHOTO

¼ ACTUAL SIZE

Lapeer Mfg. Co.
KNU-VISE PRODUCTS
Lapeer, Michigan

RECOMMENDED LOAD

15/16″ from Bottom of Toggle Bar . . . . 600 Lbs.

1 5/8″ from Bottom of Toggle Bar . . . . 400 Lbs.

2 5/16″ from Bottom of Toggle Bar . . . . 300 Lbs.

Weight, 2.35 Lbs.

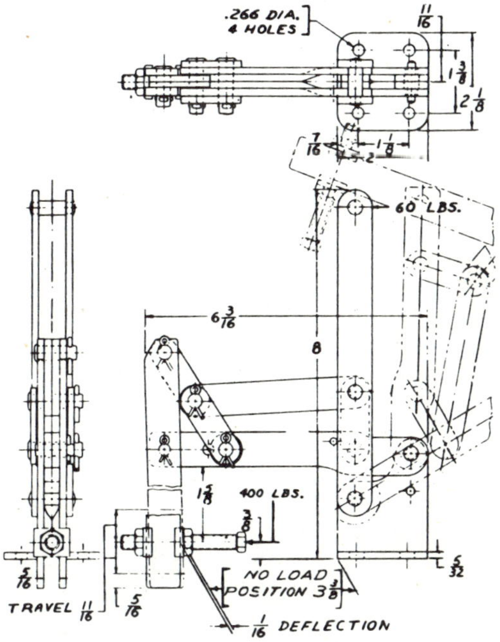

.266 DIA. 4 HOLES

60 LBS.

400 LBS.

NO LOAD POSITION 3 3/8

1/16 DEFLECTION

TRAVEL 11/16

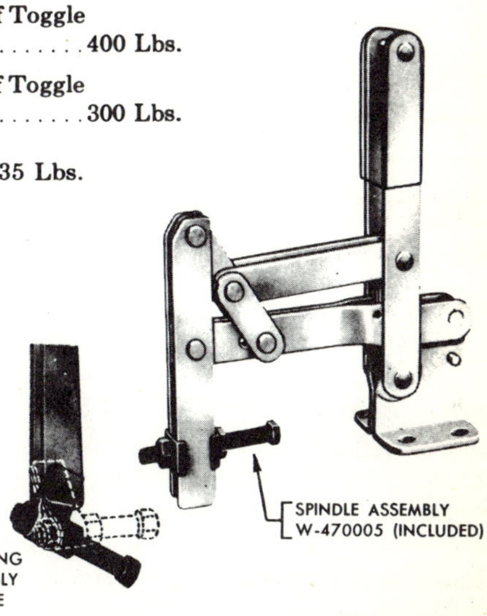

SPINDLE ASSEMBLY W-470005 (INCLUDED)

FOR *ANGULAR* CLAMPING BOLT RETAINER ASSEMBLY NR-470009 AVAILABLE

Courtesy of Lapeer Manufacturing Company.

Fig. 15-2. Toggle action swing clamp template.

properly designed clamp. The proper amount of over-center travel to produce maximum holding force and yet ensure positive locking is a carefully calculated and controlled dimension developed by years of experimentation and experience.

Most manufacturers rate a clamp by its "holding capacity." This is the maximum load or force that the clamp will sustain in the closed and locked position without permanent deflection. Exerting forces applied as the clamp closes are less than the holding capacity

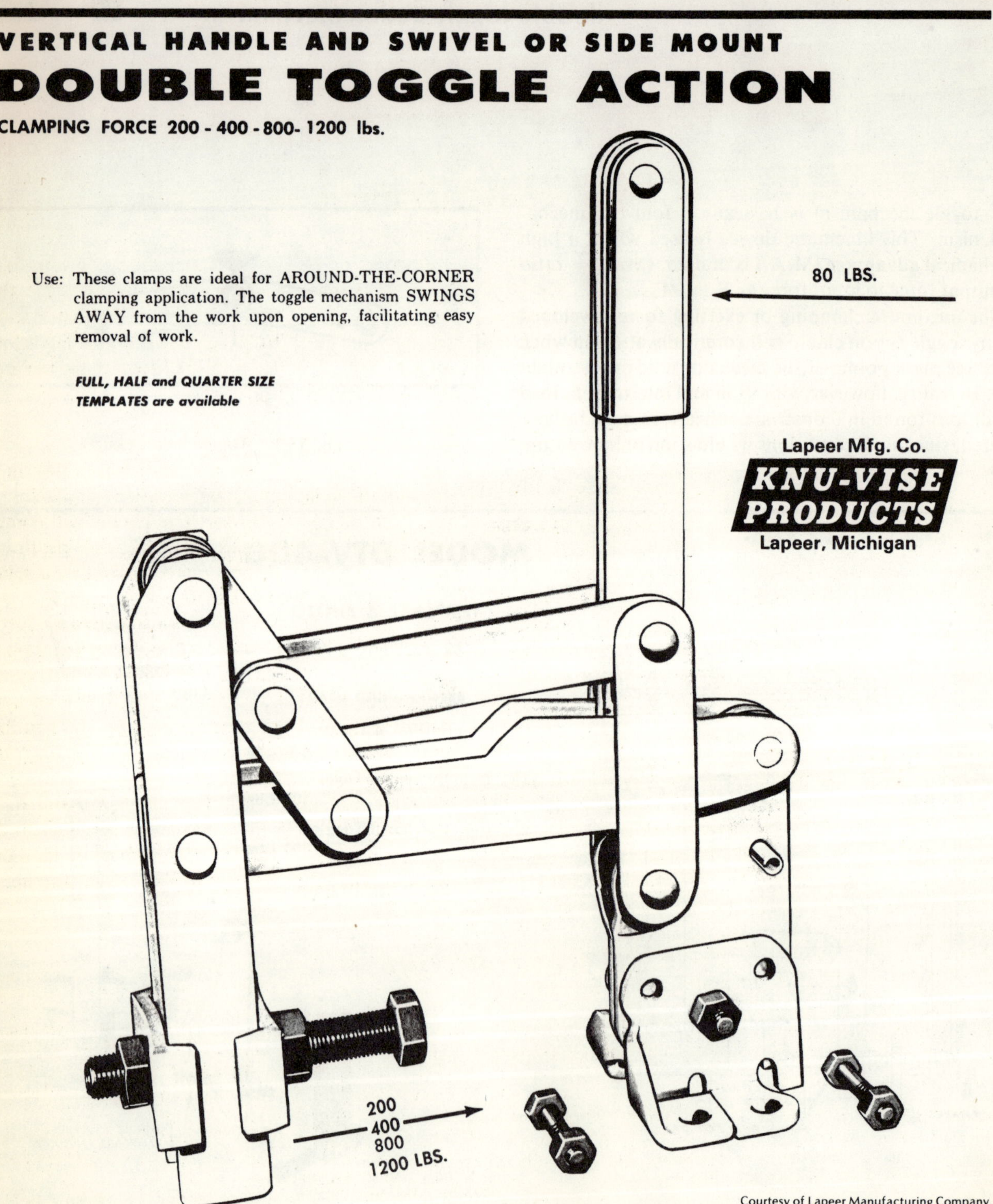

Courtesy of Lapeer Manufacturing Company.

Fig. 15-3. Toggle action swing clamp.

and are dependent on such variables as the position of the operator's hand on the handle, the amount of force applied, and the position of the spindle on the bar.

The toggle mechanism is made from an adaption of the basic slider-crank mechanism. It is often used in combination with another link or links, or with one link extended, as shown in Fig. 15-3. Practical uses include vise pliers, automotive door components, punch presses, and clamps. The clamp shown in Figs. 15-2 and 15-3 is a good example of a toggle mechanism. A force of 80 lb. on the handle can result in a 1,200-lb. clamping force.

Toggle clamps provide ready-made, compact devices for moving or holding components. A little imagination will often show the product designer a way to obtain motion or locking force with no tooling costs, and often at less cost than custom-designed components.

The same toggle action force multiplying principle can be applied to other tasks as well. Certain toggle clamps can be used to position parts, to power fixtures, or to perform mechanical functions such as piercing sheet metal or staking rivets.

Problem Assignment

The following graphical problems and design questions are intended to provide you with experience in graphically analyzing and calculating basic mechanism constraints. Select, with instructor direction, the problem-sheet assignments suitable for you. Then complete the numbered problems in ordered sequence because each solution builds on preceding problems. Be sure to use the references suggested by your instructor (*list follows*) as sources for working through the assignment. Problem sheets are located at the back of the book.

Problem Sheet 15A

Problem 1. The mechanism illustrated permits angular clamping by bar *AB*. Points *F* and *G* represent stop pins which limit the handle *CH*. Link *DE* is a hidden toggle link. Rotate the handle every 10° from its present position and record the positions of all the moveable links.

Problem Sheet 15B

Problem 1. Plot the angular displacements of the clamp handle and the clamping bar used in Problem Sheet 15A, problem 1. Use the same 10° increment.

Problem 2. Plot three new displacement curves when point *E* is moved .12, .25, and .38 in. along line *CH* toward point *H*. Use the space at the bottom of the sheet to develop a single chart showing the three curves. Which position provides for the fastest movement of the bar (assuming constant angular velocity of the handle)? Which position provides for the greatest clearance under the bar?

Problem Sheet 15C

Problem 1. Plot the open position of each link in the toggle action clamp mechanism when handle link *AB* has been rotated clockwise. The handle will come to rest against the stop pin *M*. Show the location of point *K*. Determine the exact straight-line displacement that point *K* travels to its new location K_1.

Problem 2. Develop resultant-vector diagrams for the forces acting on points *G* and *E*. Vector components and resultants should be directed along link centerlines when possible. Use the starting points of application illustrated on the drawing and the parallelogram method for development of the diagrams. (Scale: 1.00 in. = 500 lbs.)

Problem 3. Calculate the forces at *K*, K_2, and K_3 respectively. Indicate the Theoretical Mechanical Advantage (T.M.A.) for the mechanism in its closed position when the clamp point *K* is used in a setup.

References

The following source and page references can be used in developing the graphical solutions and calculations for the problem assignments.

Esposito, Anthony. *Kinematics for Technology.* Columbus, Ohio: Charles E. Merrill Publishing Co., 1973, pp. 84, 85.

Hinkle, Rolland T. *Kinematics of Machines.* 2nd ed. Englewood Cliffs, N.J.: Prentice-Hall, 1960, pp. 61-63.

Hirschhorn, Jeremy. *Dynamics of Machinery.* New York: Barnes & Noble, 1967, pp. 47-72.

Kepler, Harold B. *Basic Graphical Kinematics.* 2nd ed. New York: McGraw-Hill Book Co., 1973, pp. 8, 9, 64, 65.

Lent, Deane. *Analysis and Design of Mechanisms.* 2nd ed. Englewood Cliffs, N.J.: Prentice-Hall, 1970, pp. 381, 387.

Michels, Walter J., and Wilson, Charles E. *Mechanism: Design-Oriented Kinematics.* Chicago: American Technical Society, 1969, pp. 118-121.

Oberg, Erik; Jones, Franklin D.; and Horton, Holbrook L. *Machinery's Handbook.* 20th ed. New York: Industrial Press, 1976, pp. 295-311.

Patton, William J. *Kinematics.* Reston, Va.: Reston Publishing Co., 1979, pp. 1-29.

Tao, D.C. *Fundamentals of Applied Kinematics.* Reading, Mass.: Addison-Wesley Publishing Co., 1967, pp. 67, 75.

V-Belt Drives 16

angular and linear velocity

V-BELTS are used to economically transfer power from a motor to a shaft too far away to be driven by gears, or when the motor is at an rpm too high for roller chain. Many lengths of belts and sizes are available as stock items which can be selected for almost any application. V-belts operate without lubrication and need no dressings to keep them pliable. They operate quietly and transmit power highly efficiently, with no shock.

V-belts perform satisfactorily under severe dust conditions or in wet surroundings. Oil, of course, is very destructive to the rubber, as are pebbles or hard bits of foreign material falling between the belt and its groove. A simple guard, however, can be made rather cheaply to protect the belts and the hands of people working nearby while the machines are operating.

The speeds at which V-belts operate can be quite high, with the limit being in the range of 5,000 ft./min. Beyond this velocity, centrifugal force tends to throw the belt away from the pulley. High operating speed is an important factor in selecting a suitable belt drive for a machine. A high-speed electric motor or gas engine is cheaper, lighter in weight, and more available than the slow-speed motors of the same horsepower rating. V-belts operate at low velocities, but power transmission is correspondingly low.

Speed ratios for V-belt drives can be as high as 10 to 1 and still operate satisfactorily at close centers without undue tension on the belts, as would be required for a flat-belt drive. Fig. 16-1 illustrates the limited area in contact with the small pulley of a flat-belt drive operating at close centers. This condition requires either high tension on the belt in order to create enough friction to drive the large pulley, or else the use of an idler to get more wrap around the small pulley.

For a V-belt drive, this is less of a problem. The wedging action of the V-belt prevents slippage and lessens the pressure on the bearings, thus providing more horsepower. As the belt enters the groove, it bends and expands against the sloping sides of the grooves on the sheave.

Belts operate in either direction and in any position. The slack or loose side of the belt can be on the top or bottom of a V-belt drive where the shafts are at the same level.

As Fig. 16-2 illustrates, open drives can be positioned at any angle. These arrangements are ideal for high capacity drives; they allow a wider range of speed ratios than any other type. There are no minimum or maximum center-distance limits other than those set by available belt lengths, but provisions must be made for take-up.

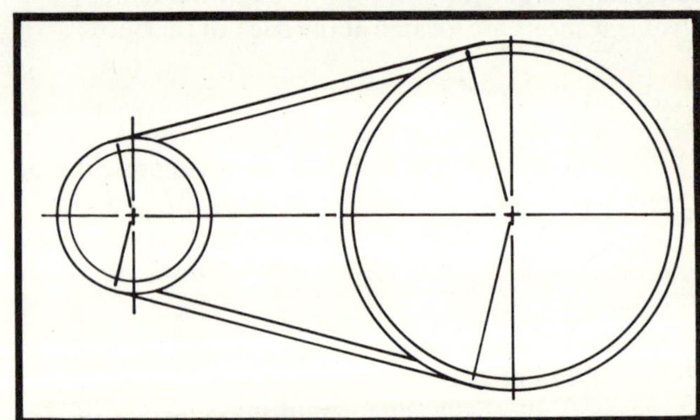

Fig. 16-1. Limited area of flat belt in contact with small pulley limits horsepower output.

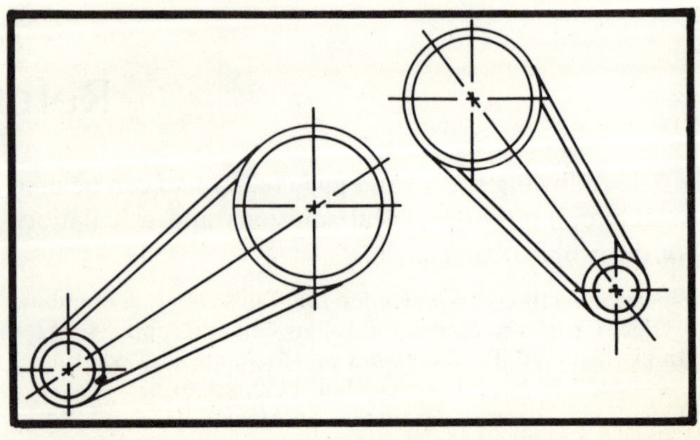

Fig. 16-2. Open drive for regular V, Poly-V, or flat-belt operation.

Three-pulley drive (Fig. 16-3) is usually limited in capacity to 8-10 horsepower because of small high-speed pulleys and limited provisions for take-up. This arrangement is typically used with one unit hinge-mounted. Large belts can stretch beyond the capacity of a hinged unit to take up slack.

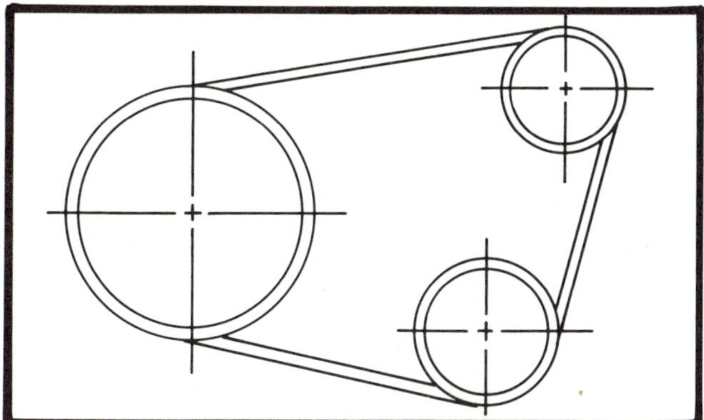

Fig. 16-3. Three-pulley drive for regular V, Poly-V, or flat-belt operation.

Fig. 16-4 illustrates a reverse bend idler, which is used when driver and driven sheaves are fixed and there is no provision for take-up. The idler is placed on the slack side of the belt near the point where the belt leaves the driver sheave. The idler gives increased wrap and increased arc of contact.

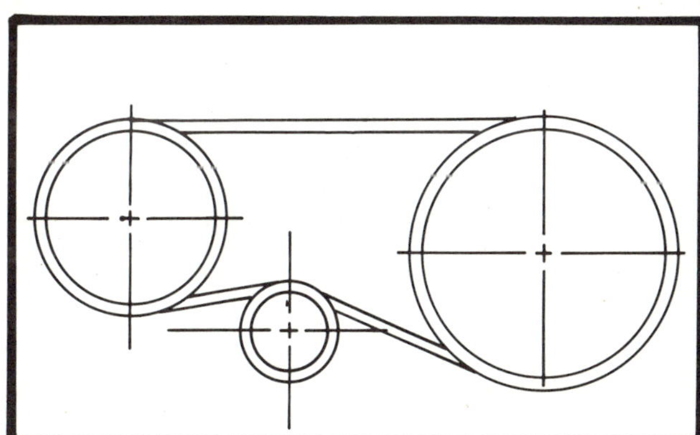

Fig. 16-4. Reverse bend idler for V, Poly-V, or flat-belt operation.

Serpentine drives (Fig. 16-5) are useful where several units are driven from a central shaft. The belt that was developed especially for this drive is shown in Fig. 16-6.

V-belt manufacturers offer five standard sizes of belts for industrial use; these are readily available from local suppliers of power transmission equipment. The five sizes are shown in proportionate scale in Fig. 16-7, but the concavity of the sides is slightly exaggerated.

As the belts bend around the sheave grooves, the sides

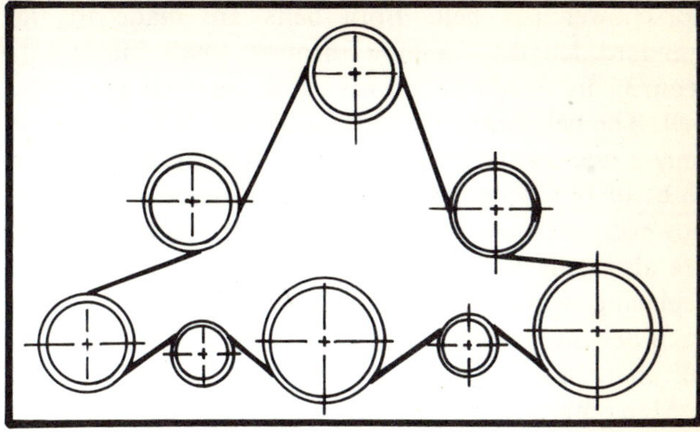

Fig. 16-5. Serpentine drive for V, Poly-V, or flat-belt operation.

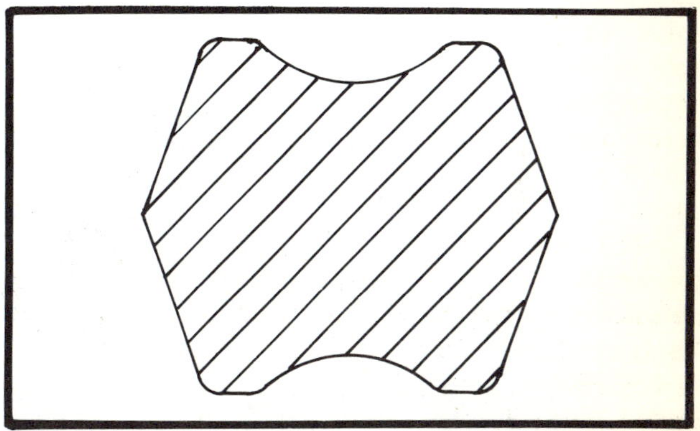

Fig. 16-6. Section profile of Poly-V belt.

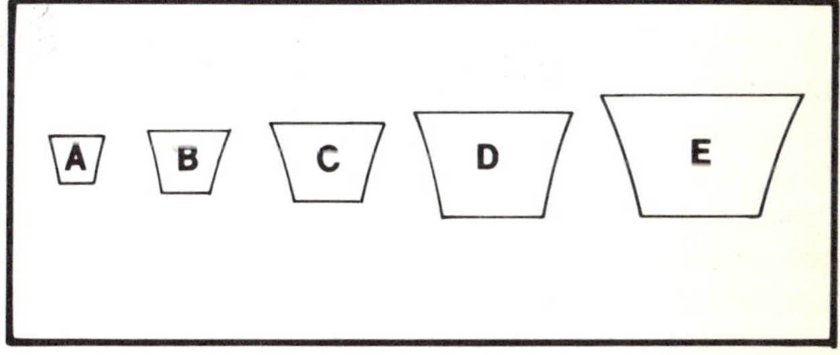

Fig. 16-7. Section profiles of five standard V-belts.

bulge and exert large forces against the sides of the grooves. The concave surfaces are an attempt to make the forces equal at all points along the side. The belts must be used with sheaves large enough to prevent excessive bulging on the inside surface of the belt and excessive stretching of the outside edge. Using sheaves smaller than those suggested shortens the belt life. Catalogs suggest a minimum pitch diameter for "A" belts of 3 in. and a minimum for "E" belts of 21 in.

Special, heavy-duty V-belts offered by several companies have the same general shape and use the same grooves as standard V-belts, but they transmit greater horsepower per belt. Both belts are made to the standard lengths, ranging through small increments from 35 in. to 660 in. measured on the pitch line of the belt. The belt number is close to the pitch length but is only a nominal length. Slight variations in length seem to be unavoidable in the manufacturing process; belts, however, are matched for length at the supplier. They are also delivered in sets, which suggests the need for replacing worn belts with a complete new set so that each belt sits properly in the grooves and transmits its full share of the required horsepower.

Many manufacturers offer stocks of standard V-belts in rolls, which can be cut to any desired length and fixed together with a flexible fastener. Another V-belt of standard cross section is made of short links that are riveted together through slots, permitting easy removal of links to make any belt length desired.

An excellent example of the many applications for a large number of V-belt drives can be found in a self-propelled combine (Figs. 16-8 and 16-9). The manufacturer provides some interesting hints to the operator for care of the V-belts:

> Belts should be kept tight enough to avoid slippage when the combine is operating under normal load. If the belt is allowed to slip unnecessarily, excessive heat is generated which causes early deterioration. Belts should be washed occasionally with mild soap and water. This will remove some of the embedded grime and increase belt efficiency. If the machine is to be stored, remove belts and store them in a cool, dry environment.

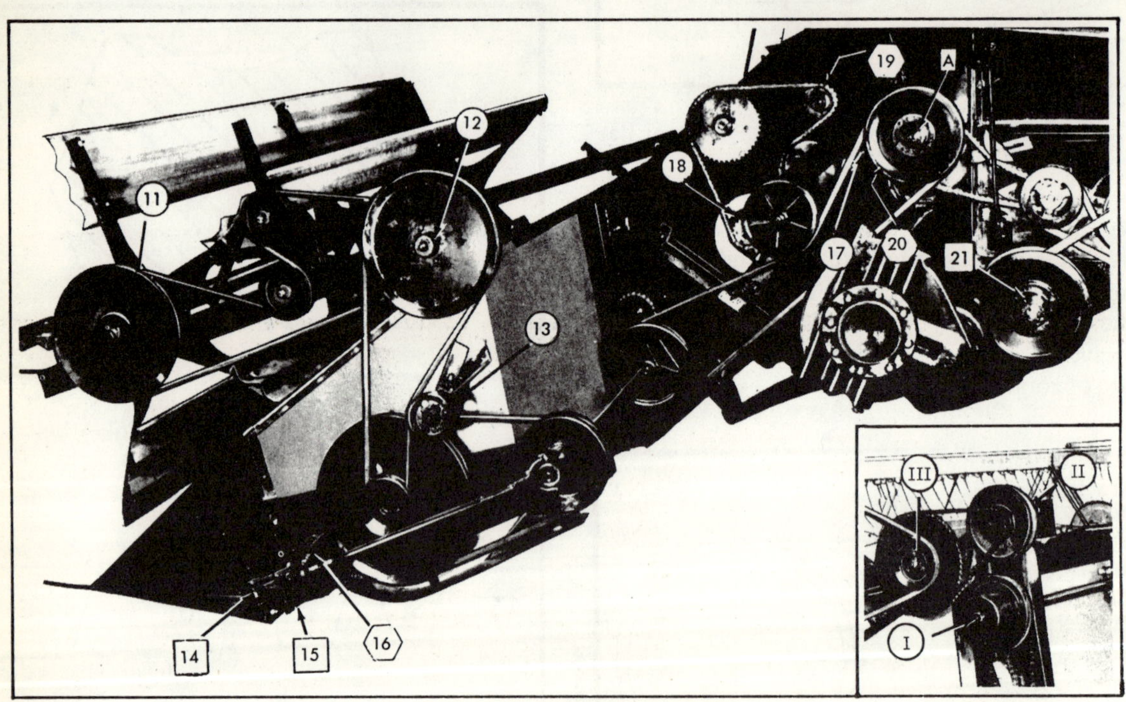

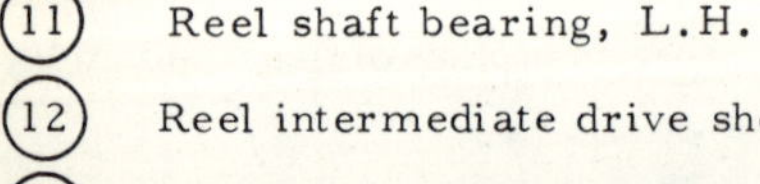

(11) Reel shaft bearing, L.H.

(12) Reel intermediate drive sheave bearing

(13) Reel intermediate drive belt idler

[14] Bellcrank ball, front

[15] Bellcrank ball, rear

(16) Sickle bellcrank bearing

(17) Idler arm bearing, rear

(18) Cylinder bearing, L.H.

(19) Slip clutch hub

(20) Pivot shaft bearing, L.H.

[21] Input sheave hub, inner, put in low range to lubricate.

(I) Variable driver sheave*

(II) Reel variable driven sheave*

(III) Reel pivot hub*

A Repack bearings once a season.

* Combines with variable speed reel drive.

Courtesy of Allis-Chalmers Corp.

Fig. 16-8. Belt drives for a self-propelled combine.

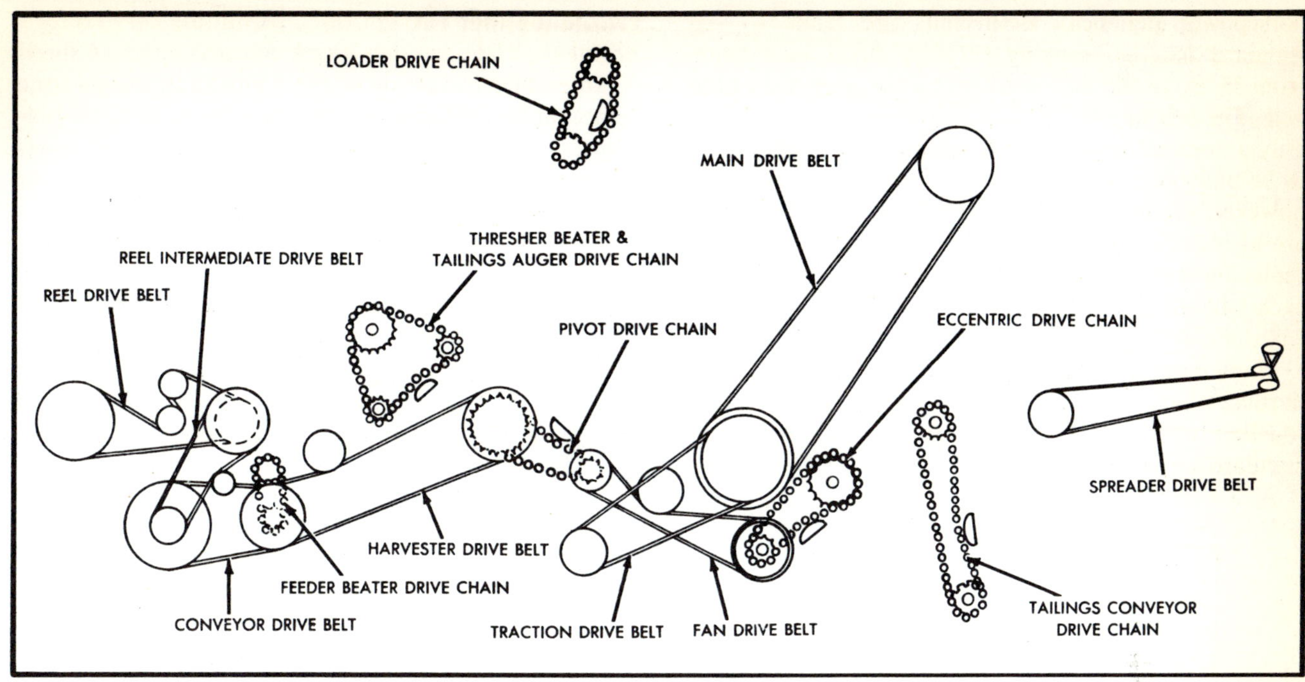

Courtesy of Allis-Chalmers Corp.

Fig. 16-9. Belt diagram for left-hand drive.

Problem Assignment

The following graphical problems and design questions are intended to provide you with experience in graphically analyzing and calculating basic mechanism constraints. Select, with instructor direction, the problem-sheet assignments suitable for you. Then complete the numbered problems in ordered sequence because each solution builds on preceding problems. Be sure to use the references suggested by your instructor (*list follows*) as sources for working through the assignment. Problem sheets are located at the back of the book.

Problem Sheet 16A

Problem 1. The serpentine V-belt drive illustrated uses a Poly-V belt in its operation. The lines represent the pitch line of the belt and the pitch circles of each pulley. The speed of the belt is 100 fpm, and drive pulley *A* rotates in a clockwise direction. (Scale: .25 in. = 1.00 in.) Determine and indicate the following characteristics of the system:

A. The rpm for each pulley.
B. The rad/min. (ω) for each pulley.
C. The wrap angle (θ) for each pulley.
D. The direction of rotation for each pulley.

Problem Sheet 16B

Problem 1. This V-belt drive shows the pitch circle for each pulley and the pitch line for the belt. Belt length L can be calculated by the formula:

$$L = 2C + 1.57(D + d) + \frac{(D - d)^2}{4C}$$

Center distance C is determined by:

$$C = \frac{b + \sqrt{b^2 - 32(D - d)^2}}{16}$$

where $b = 4L - 6.28(D + d)$
D = diameter of large pulley
d = diameter of small pulley
L = length of belt
C = center distance

The pitch line velocities (linear) of the two pulleys are equal. The velocity in each case is:

$$V_b = \omega R$$

where ω and R are the angular velocity and radius of the pulley, respectively. Or,

$$V_b = \omega_L R = 1/2 \omega_L D$$

and

$$V_b = \omega_s r = 1/2 \omega_s d$$

Equating these two quantities gives

$$1/2 \omega_L D = 1/2 \omega_s d$$

$$\frac{\omega_L}{\omega_s} = \frac{d}{D}$$

The wrap angles can be determined by

$$\theta_L = \pi + 2 \arcsin \frac{D - d}{2C}$$

and

$$\theta_s = \pi - 2 \arcsin \frac{D - d}{2C}$$

Using the equations above, and given that C = 52.0 in., D = 12.0 in., ω_L = 120 rad/min., and d = 8.0 in., calculate the following:

A. Belt length L.

B. Wrap angles θ_L and θ_s.

C. Belt velocity V_b.

Problem 2. Given the drive data that L = 120.0 in., V_b = 1,200 in./min., D = 10.0 in., d = 8.0 in., calculate the following:

A. Center distance C.

B. Angular velocities of the pulleys, ω_L and ω_s.

C. Wrap angles θ_L and θ_s.

Problem Sheet 16C

Problem 1. Special sheaves permit a change of speed ratios on a V-belt drive. The illustration shows the mechanism component (B and C) which can be adjusted while the sheave is turning; this provides an infinite range of speeds through quite a wide ratio. Drive pulley A is rotating at 1,750 rpm in a clockwise direction. (Scale: .25 in. = 1.00 in.) Determine the following characteristics of the system:

A. Indicate the rpm of pulleys B and C and pulley D in the condition shown.

B. Move pulleys B and C to the position shown in the new top sectional view. Complete the illustration in section, showing the changes in the double pulley B and C. Also, show the new pitch lines and circles in the front view. Indicate the rpm of pulleys B and C and pulley D in the new condition.

References

The following source and page references can be used in developing the graphical solutions and calculations for the problem assignments.

Esposito, Anthony. *Kinematics for Technology.* Columbus, Ohio: Charles E. Merrill Publishing Co., 1973, pp. 331-333.

Greenwood, Douglas C., ed. *Product Engineering Design Manual.* New York: McGraw-Hill Book Co., 1959, pp. 194-200.

Hardison, Thomas B. *Introduction to Kinematics.* Reston, Va.: Reston Publishing Co., 1979, pp. 253-260.

Hinkle, Rolland T. *Kinematics of Machines.* 2nd ed. Englewood Cliffs, N.J.: Prentice-Hall, 1960, pp. 222-225.

Hirschhorn, Jeremy. *Dynamics of Machinery.* New York: Barnes & Noble, 1967, pp. 430, 431.

Kepler, Harold B. *Basic Graphical Kinematics.* 2nd ed. New York: McGraw-Hill Book Co., 1973, pp. 231, 322-376.

Lent, Deane. *Analysis and Design of Mechanisms.* 2nd ed. Englewood Cliffs, N.J.: Prentice-Hall, 1970, pp. 402-405.

Michels, Walter J., and Wilson, Charles E. *Mechanism: Design-Oriented Kinematics.* Chicago: American Technical Society, 1969, pp. 417-423.

Oberg, Erik; Jones, Franklin D.; and Horton, Holbrook L. *Machinery's Handbook.* 20th ed. New York: Industrial Press, 1976, pp. 1,046-1,075.

Patton, William J. *Kinematics.* Reston, Va.: Reston Publishing Co., 1979, pp. 193-228.

Tao, D.C. *Fundamentals of Applied Kinematics.* Reading, Mass.: Addison-Wesley Publishing Co., 1967, pp. 321-332.

Universal Joints and Control Shafts

17

Hooke's joint

THE universal joint is a form of coupling used for connecting two rotating shafts, the axes of which intersect at some angle. There are many different designs of universal couplings based on the original "Hooke's Joint." Theoretically, the universal joint will operate at any angle between 0° and 90°, but because of physical limitations and friction forces, the joint does not work well if the angle exceeds 45°. If practicable, the angle should be limited to about 25° unless the speed is slow and little power is transmitted.

The simple universal joint (Fig. 17-1) is basically comprised of two Y-shaped yokes connected by a crossmember consisting of a center block and two pins. In automotive applications this crossmember is called a spider. The spider is shaped like an X and the arms that extend from it are called trunnions. (See Fig. 17-2.)

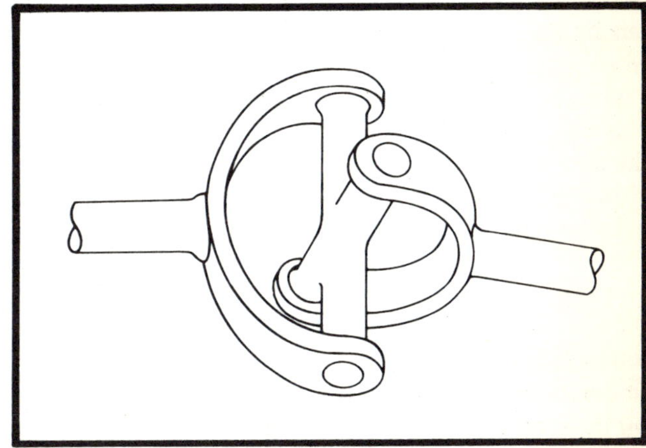

Courtesy of Chevrolet Division, General Motors Corp.

Fig. 17-2. Simple universal joint.

7 reasons why Curtis sets the standard

1 2 3 4 5 6 7

Figs. 1 and 2 FORKS —
heat-treated to obtain maximum wear and strength; inside ear surfaces accurately ground for close, smooth-working operation; pin holes broached and honed for good bearing surface and precise alignment; the O. D. cylindrically ground on centers for perfect concentricity.

Fig. 3 CENTER BLOCK —
holes intersecting accurately in the center, with the faces parallel; hardened and precision ground for long wear; with recess for the "Telltale" Lock ring.

Figs. 4 and 5 PINS —
hardened and centerless ground to close tolerances to reduce radial play.

Fig. 6 OILER —
Curtis pioneered proper and essential lubrication, and simplified it by providing patented oilers in the large pin on all joints of 1" O. D. and larger. Smaller joints are easily lubricated through holes in large pins.

Fig. 7 "TELLTALE" LOCK RING —
patented by Curtis. Snaps around either pin groove and fits into recess of center block. This insures firm, positive lock-up of entire assembly — makes disassembly and reassembly quick and easy. The "Telltale" end shows at a glance that the ring is in proper locking position after final assembly.

Courtesy of Curtis Universal Joint Co., Inc.

Fig. 17-1. Elements of the basic universal joint.

The crossmember (spider) allows the two yoke shafts to operate at an angle to one another. When torque is transmitted at an angle through this type of joint, the driving yoke rotates at a constant speed while the driven yoke speeds up and slows down twice per revolution. This change of velocity (acceleration) of the driven yoke increases as the angle between the two yoke shafts increases. This is the prime reason why single high-speed universal joints are not used for angles greater than 3° to 4°. At 4°, for example, the change of velocity is .5 percent. At 10° it is 3 percent. If the universal joint were set at 30° and the driving yoke were turning at 1,000 rpm, the velocity of the driven yoke would change from 866 rpm to 1,155 rpm in one-quarter of a revolution. In the remaining quarter revolution, the velocity would change from 1,155 rpm to 866 rpm.

On a one-piece drive shaft this problem can be eliminated by arranging two simple universal joints so that the two driving yokes are rotated 90° to each other. However, the angle between the drive and driven yokes must be very nearly the same on both joints in order for this to work. (See Fig. 17-3.) This allows the alternate acceleration and deceleration of one joint to be offset by the alternate deceleration and acceleration of the second joint. When the two joints do not run at approximately the same angle, operation can be rough and an objectionable vibration produced.

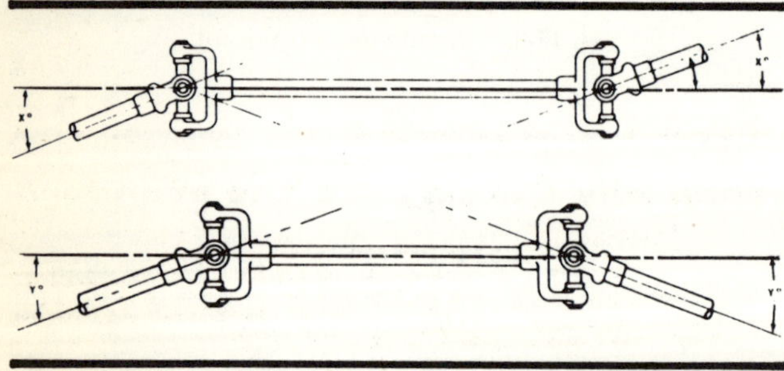

Courtesy of Chevrolet Division, General Motors Corp.

Fig. 17-3. Drive shafts with connecting yokes.

As mentioned previously, the simple universal joint will operate efficiently through small angles only. Also, two simple universal joints phased properly and operating through the same angle will transmit constant velocity. When a large angle is encountered in a driveline, a simple universal joint will introduce two vibrations in each revolution. It is in this situation that a constant velocity joint is used.

Essentially, a constant velocity joint consists of two simple universal joints closely coupled by a coupling yoke, and phased properly for constant velocity. A centering ball socket between the joints maintains the relative position of the two units. This centering device causes each of the two units to operate through one-half of the complete angle between the input and output shafts. Fig. 17-4 illustrates the theory and construction of the constant velocity joint.

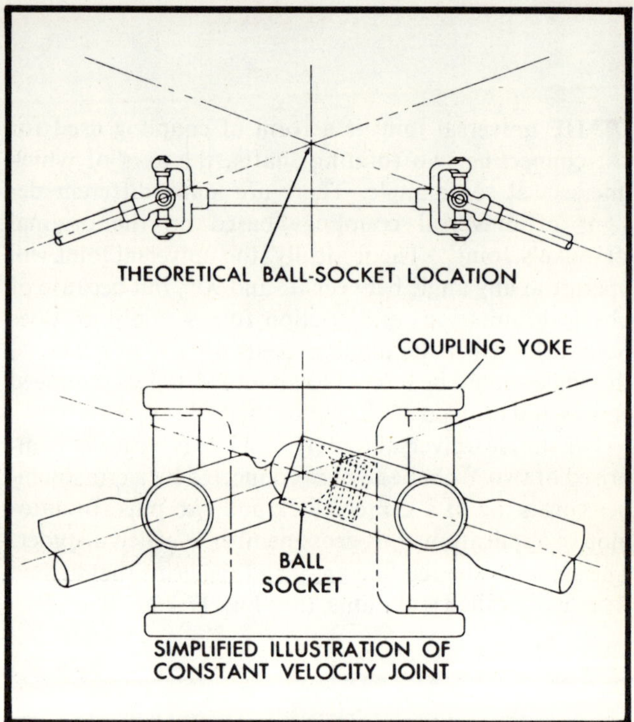

Courtesy of Chevrolet Division, General Motors Corp.

Fig. 17-4. Constant velocity joint.

General rules to be followed when designing conventional universal joints and control shafts are:

1. Use two universal joints at opposite ends of a universal shaft.

2. Control shafts, which are to be joined by a universal shaft, must be co-planar.

3. Angles of bend at the ends of the universal shaft must be the same for each joint.

4. The individual parts of a universal joint do not rotate; they oscillate or slide on each other, and should therefore be designed for lubrication at least every 40 or 50 hours of operation for normal use.

Problem Assignment

The following graphical problems and design questions are intended to provide you with experience in graphically analyzing and calculating basic mechanism constraints. Select, with instructor direction, the problem-sheet assignments suitable for you. Then complete the numbered problems in ordered sequence because each solution builds on preceding problems. Be sure to use the references suggested by your instructor (*list follows*) as sources for working through the assignment. Problem sheets are located at the back of the book.

Problem Sheet 17A

Problem 1. The input velocity ω_1 of shaft 1 is 1,000 rpm. Angle α is 10°. When ω_1 is constant, the output velocity varies instantaneously from a minimum value below ω_1, given by

$$\omega_{2(\min)} = \omega_1 \cos \alpha$$

up to a maximum value above ω_1, given by

$$\omega_{2(\max)} = \frac{\omega_1}{\cos \alpha}$$

The output velocity ω_2 can be determined at any angular position of shaft 1 by the formula:

$$\omega_2 = \omega_1 \left(\frac{\cos \alpha}{1 - \sin^2 \alpha \sin^2 \emptyset} \right)$$

where $\emptyset$ = the input rotational angle of shaft 1 (0°, 45°, 90°, 180°, or any angle from 0° through 360°).

Calculate and graph the values of output velocity ω_2 for each 30° angle of input angle $\emptyset$ for shaft 1. Use the vertical axis to show the values for ω_2.

Problem 2. Given that output velocity ω_2 is to be a constant 1,750 rpm, determine and graph the necessary variable input velocity ω_1 when α is 5°.

Problem Sheet 17B

Problem 1. The double universal joint is often used to produce a constant velocity output. When the input velocity ω_1 from shaft 1 is held constant, the velocity of center shackle 2 will vary as in the case of a single universal joint. But when the motion is transmitted through the second shaft (3), the output velocity becomes constant again. This only happens when $\alpha = \beta$. If the two angles α and β are not equal, minimum and maximum output velocities can be determined by:

$$\omega_{3(\min)} = \omega_1 \cos (\beta - \alpha)$$

and

$$\omega_{3(\max)} = \frac{\omega_1}{\cos (\beta - \alpha)}$$

Calculate and graph the output velocities for the shackle (ω_2) and shaft 3 (ω_3). Shaft 1 has an input velocity of 1,000 rpm. Angle α = 5° and angle β = 10°. Plot curves for both the shackle and shaft 3 on the same chart. What is the final output velocity of shaft 3?

Problem 2. Given that output velocity ω_3 is to be a constant 200 rpm, determine and graph the necessary variable input velocity ω_1 when $\alpha = \beta = 8°$.

References

The following source and page references can be used in developing the graphical solutions and calculations for the problem assignments.

Esposito, Anthony. *Kinematics for Technology.* Columbus, Ohio: Charles E. Merrill Publishing Co., 1973, pp. 333, 334.

Hinkle, Rolland T. *Kinematics of Machines.* 2nd ed. Englewood Cliffs, N.J.: Prentice-Hall, 1960, pp. 329-331.

Kepler, Harold B. *Basic Graphical Kinematics.* 2nd ed. New York: McGraw-Hill Book Co., 1973, pp. 384, 385.

Michels, Walter J., and Wilson, Charles E. *Mechanism: Design-Oriented Kinematics.* Chicago: American Technical Society, 1969, pp. 463-465.

Oberg, Erik; Jones, Franklin D.; and Horton, Holbrook L. *Machinery's Handbook.* 20th ed. New York: Industrial Press, 1976, pp. 707-709.

Tao, D.C. *Fundamentals of Applied Kinematics.* Reading, Mass.: Addison-Wesley Publishing Co., 1967, pp. 327, 328.

Computing Mechanisms 18

analog and digital computers, controllers, and counters

LINKAGES, cams, and gears used as computing mechanisms have been generally replaced by electronic devices, controllers, and counters. These electronic devices provide more than adequate durability, accuracy, and speed. However, in harsh, hot, corrosive, explosive, or humid environments, mechanical systems have the advantage of being rugged and easily maintainable.

Computing mechanisms consist of two types: analog and digital. Evaluating when to use each type should take into consideration the following comparisons:

1. Analog computers deal with analogous (equivalent) systems which parallel mathematical operations, thereby allowing the computer to handle functions directly rather than through discrete steps. They may be mechanical, electrical, pneumatic, hydraulic, or a combination of these. Advantages include:
    a. Programming is not as complex as with the digital computer.
    b. More complex mathematical or vector solutions can be handled.
    c. Operations are performed instantaneously.

    Disadvantages include:
    a. The computing unit cannot have as high a degree of accuracy built into it as the digital can.
    b. It is not flexible enough to handle different situations (such as certain types of input).
2. Digital computers deal with numerical quantities for which a series of mathematical operations are performed in discrete steps. These discrete steps, no matter how complex the input equation, must be performed through a process of addition. Advantages include:
    a. The system may be mechanical, electrical, pneumatic, hydraulic, or a combination of these.
    b. Programming may be done through punched cards, magnetic tape, disks, or drums.
    c. A digital computer can be constructed to give a high degree of accuracy.
    d. It has a large memory storage unit.

    Disadvantages include:
    a. Because of the discrete steps in computing through addition, the machine is relatively slow in operation.
    b. Programming for calculations is often complicated and time consuming.
    c. It lacks the flexibility to handle certain types of calculations.

The most common analog mechanism is the differential. Linkages, racks and pinions, and gear trains can be used to add or subtract shaft speeds and linear measurements.

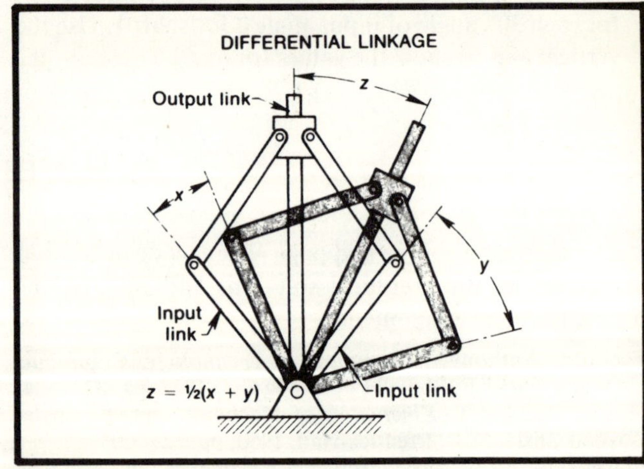

Courtesy of Penton/IPC, Inc., *Machine Design*.

Fig. 18-1. Analog mechanism adds or subtracts angular measurements.

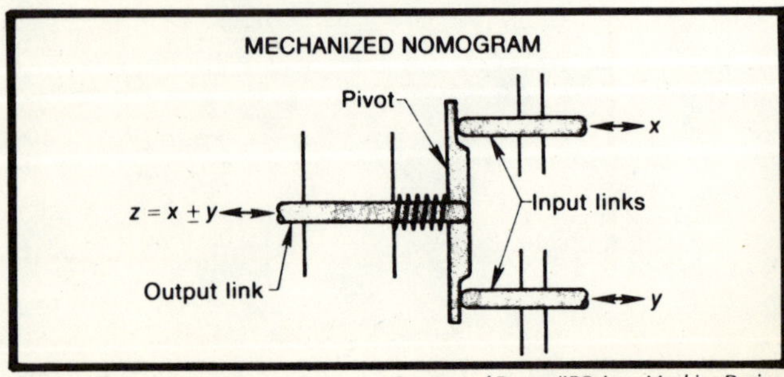

Courtesy of Penton/IPC, Inc., *Machine Design*.

Fig. 18-2. Analog mechanism adds or subtracts linear measurements.

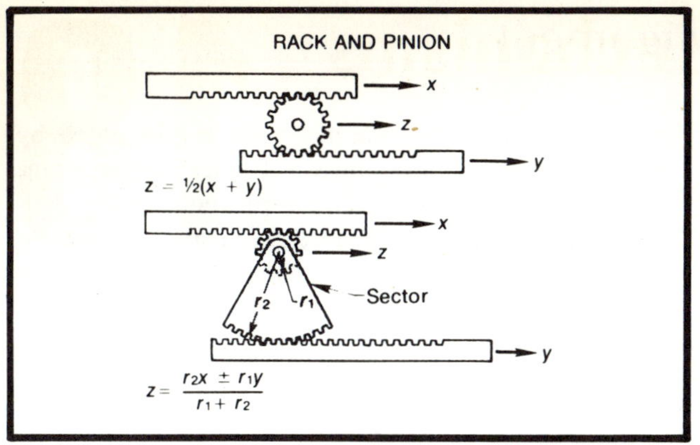

Courtesy of Penton/IPC, Inc., *Machine Design*.

Fig. 18-3. Analog mechanism consisting of a rack and pinion to provide an adjustable scale factor.

DIFFERENTIAL GEARBOX
Input
Central bevel gear
2:1 ratio
Output

Courtesy of Penton/IPC, Inc., *Machine Design*.

Fig. 18-4. Analog mechanism consisting of a differential gearbox which adds or subtracts shaft speeds.

Digital mechanisms are comprised of linkages or gear trains. The most widely used examples are counters and calculators. Inputs to these devices can come from electrical, mechanical, or fluid components.

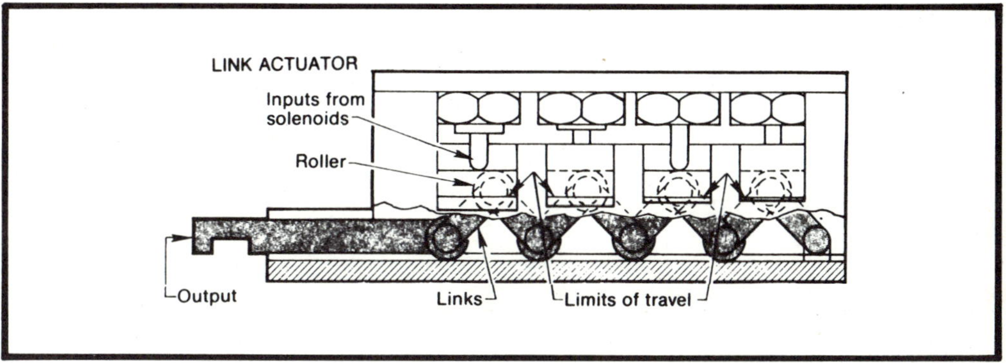

Courtesy of Penton/IPC, Inc., *Machine Design*.

Fig. 18-5. Digital mechanism uses the downward movement of solenoid plungers to extend the length of the link chain.

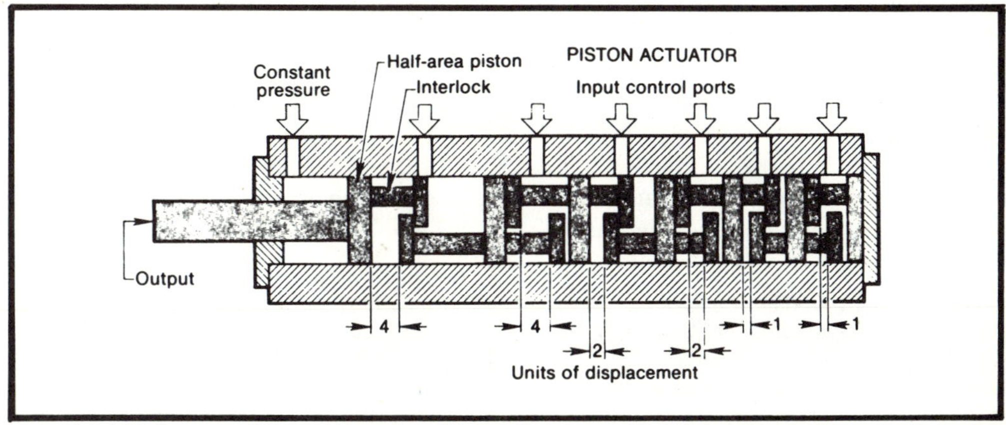

Courtesy of Penton/IPC, Inc., *Machine Design*.

Fig. 18-6. Digital mechanism similar to link actuator. The position of the output rod is controlled by mechanical interlocks that determine the distance between adjacent pistons.

Problem Assignment

The following graphical problems and design questions are intended to provide you with experience in graphically analyzing and calculating basic mechanism constraints. Select, with instructor direction, the problem-sheet assignments suitable for you. Then complete the numbered problems in ordered sequence because each solution builds on preceding problems. Be sure to use the references suggested by your instructor (*list follows*) as sources for working through the assignment. Problem sheets are located at the back of the book.

Problem Sheet 18A

Problem 1. The illustration depicts a simple slide and link adder. The displacements of slides 1 and 3 represent inputs x and y in the equation

$$W = \frac{x + y}{2}$$

and the movement of slide 2 represents the sum of $x + y$. The 2 in the denominator is eliminated by making the scale of sum $x + y$ twice that of the inputs x and y. Subtraction can be performed by extending the scales to the left. Determine the scale W and lay it out accurately, directly above the existing scale of 0 to 100. Carefully perform and illustrate the following operations: (Scale: 1.00 in. = 1.00 in.)

A. 5,000 + 2,500 = ______.
B. 325 + 55 = ______.
C. 25.2 − 12.5 = ______.

Problem 2. The differential adder functions in a manner very similar to the slide and link adder. The displacements of bars 1 and 2 represent the inputs x and y; the displacement of bar 4 represents the sum of $x + y$. Again,

$$W = \frac{x + y}{2}$$

and the 2 in the denominator can be eliminated by making the scale of the sum $x + y$ twice that of the inputs x and y. Slipping can be prevented by substituting racks and pinions. Determine and lay out the three scales, x, y, and W, starting at the pointers on the zeros. Carefully perform and illustrate the following operations: (Scale: 1.00 in. = 1.00 in.)

A. 25 + 35 = ______.
B. 325 + 475 = ______.
C. 15.50 + 25.75 = ______.

Problem Sheet 18B

Problem 1. Trigonometric mechanisms produce outputs that are some trigonometric function of the input. The Scotch yoke is the basic mechanism for producing sine-wave linear motion from angular inputs. The double Scotch yoke (illustrated) provides both sine and cosine output. Link 4 is a crank with length R which rotates about a fixed point A. The radius R of the crank determines the scale factor of the mechanism. The continuous trigonometric functions are:

$$y = R \sin \theta$$
$$y = R \cos \theta'$$

and

$$x = R \cos \theta$$
$$x = R \sin \theta'$$

Choosing a scale factor for R of 1.0 mm = 100.0 mm, when R represents the length of the hypotenuse of a right-angle triangle, determine and plot sides x and y of the triangle. Start your graph by rotating crank link 4 counterclockwise every 30°. Plot two curves, one for each side of the triangle. Start your graphing with point B in the B' position.

References

The following source and page references can be used in developing the graphical solutions and calculations for the problem assignments.

Greenwood, Douglas C., ed. *Product Engineering Design Manual.* New York: McGraw-Hill Book Co., 1959, pp. 250-253.

Hinkle, Rolland T. *Kinematics of Machines.* 2nd ed. Englewood Cliffs, N.J.: Prentice-Hall, 1960, pp. 293-305.

Michels, Walter J., and Wilson, Charles E. *Mechanism: Design-Oriented Kinematics.* Chicago: American Technical Society, 1969, pp. 163-167.

Patton, William J. *Kinematics.* Reston, Va.: Reston Publishing Co., 1979, pp. 231-256.

Special Mechanism Design Problems 19

contemporary mechanisms

IN this unit several problems are presented for which a number of different solutions are equally acceptable. These permit you to exercise ingenuity in design with considerable latitude.

The development of a particular design may be aided by combining a number of ideas. It is best to record these individual ideas as they occur.

A visual record of a designer's thought pattern acts as a very important self-aid. For each problem assigned to you, some half-dozen or more thumbnail sketches or ideas should be drawn. After setting down several of these ideas, or combinations of them, proceed to the final development. Specific details for each problem are given, but your instructor may add additional limitations and considerations.

Problem 1

Develop a mechanism layout necessary to design a popcorn machine that pops corn continuously (rather than one batch at a time).

Limitations and Considerations

1. Physical properties of popcorn are:
   A. 60 lb./bu.
   B. Will pop in range of 250°-500°F.
   C. A popped kernel occupies 1/8 in^3.
2. The popper will get its power supply from a 120 V circuit.
3. Determine the number of "person units" the popper will supply (allotting 1/2 person unit for ages 0-14, 1-1/2 person units for ages 14-18, and 1 person unit for ages 18+). One person unit will eat up to one popped kernel per second.
4. A butterer and/or salter may be included either as part of the machine or designed as an accessory.

Problem 2

Develop a mechanism layout necessary to design a power hacksaw that would be suitable to sell to home-workshop users.

Limitations and Considerations

The hacksaw should be made to retail for approximately $200-$210, including motor. A national hardware chain will buy 8,000 of these annually for the next five years. A complete initial design layout should be made of the power hacksaw. Some features the saw should include are:

1. Pressure relief on reverse stroke—lift finger.
2. 10-12 in. high-speed steel blade.
3. Hinged motor mount.
4. Adjustable bronze gibs (easily replaced).
5. Hookup bar.
6. 1/3 h.p. motor—3-phase, 110 V.
7. Two speeds (120 and 160 strokes/min.).
8. Vise jaws swivel 45°.
9. Gauge bar for cutting uniform lengths.
10. Bench sizes approximately 11-1/4 × 34-1/4 × 15-1/2 in. high.
11. Dry cut.
12. Automatic stop switch.
13. 5 × 4 in. capacity.

Problem 3

Develop a mechanism layout necessary to design an automatic feeder for an aquarium.

Limitations and Considerations

1. Quantity of food per feeding should accommodate a normal fish population in a 20-gallon tank.
2. Dry flake food will be used.
3. Feeder must be fully automatic to operate seven days, two feedings daily, 7 a.m. and 3 p.m.

Problem 4

Develop a mechanism layout necessary to design a portable pump capable of moving one quart per minute of light oil in an intermittent continuous flow.

Limitations and Considerations

1. Fluid is to be carried in an unbroken continuous plastic hose which has a .5 in. inside diameter

and a .625 in. outside diameter.

2. The drive for the pump will be an electric motor which revolves at 1,750 rpm.

Problem 5

Develop a mechanism layout necessary to design an automatic machine to cut miters of picture framing stock or similar moldings made of either wood, plastic, or soft metals.

Limitations and Considerations

1. Width of moldings to be cut is limited to 4-1/2 in. or less.
2. Angle of any cut is limited to 45°.
3. Provision should be made to clamp corners, with a total frame limitation to be determined.

Problem 6

Develop a mechanism layout necessary to design a coin packaging machine.

Limitations and Considerations

1. The design must be able to handle all coins (1, 5, 10, 25, and 50 cent pieces as well as $1 coins).
2. Different coins do not have to be packaged simultaneously.
3. The machine will be hand-operated.
4. Coins are to be packaged in standard paper container tubes.
5. *Suggestion*: The system could consist of a coin hopper, counter, package filler, and package sealer. It should discharge the package at the end of the cycle. The counter, filler, and sealer, if necessary, must be so designed that only a slight, simple modification is required to change from one coin to another.

Problem 7

Develop a mechanism layout necessary to design a traveling sprinkler.

Limitations and Considerations

1. It will be used on athletic fields and large lawns.
2. It should water an area up to 140 × 650 ft. in one setting.
3. The device should pull itself and be water-powered.
4. Water shut-off is automatic.

Problem 8

Develop a mechanism layout necessary to design a safe recumbent bicycle.

Limitations and Considerations

1. Steering (handlebars, etc.) should be located below the seat level.
2. Pedals should be located in front of the front wheel; pedal forces should be directed in a near-horizontal direction (instead of the usual vertical direction).
3. The back of the driver should be supported in a chair-like manner, with the body and leg relationship not unlike a position they would take in a small urban vehicle.

Problem 9

Develop a mechanism layout necessary to design a device to store and dispense granulated sugar at a small office coffee bar.

Limitations and Considerations

1. The dispenser is to stand on a table top and dispense a fixed amount (one level teaspoon) into a cup placed on the table.
2. The dispenser should hold a week's supply of sugar; it will be used by 50 persons, five days per week, twice each day.
3. The device must be lightweight (not more than 15 lbs.), moisture proof, and easily cleaned.

Problem 10

Develop a mechanism layout necessary to design an exoskeleton mechanical hand.

Limitations and Considerations

1. During an average work task, the thumb applies 45 percent of the total effort, the forefinger 20 percent, the middle and ring fingers 10 percent each, and the little finger 15 percent.
2. Control and power systems could be gear, linkage, hydraulic and/or pneumatic.
3. Motion inputs should closely resemble an operator's hand.

Problem 11

Develop a mechanism layout necessary to design an automatic change dispenser.

Limitations and Considerations

1. The dispenser must accept one, two, five, and ten dollar bills.
2. Change must contain two quarters, four dimes, two nickels, and remaining bills.
3. It should not be able to accept counterfeit currency.

Problem 12

Develop a mechanism layout necessary to design a portable lifting tower.

Limitations and Considerations

1. The tower must be compact and also extendable to 40 ft. in height.
2. The lifting requirement is 2,000 lbs.
3. The device should be designed so that it can be moved by a small truck.
4. Uses include building repair, window washing, TV camera usage, etc.

Problem 13

Develop a mechanism layout necessary to design a cement block laying machine.

Limitations and Considerations

1. Machine must be capable of laying 4, 8, 10, and 12 in. cement blocks.
2. The foundation footing or cement floor exists before the machine is utilized.
3. The machine should accept blocks and previously mixed mortar.
4. The machine should be able to lay the wall to a maximum of 12 blocks high, point the wall, and grout.

Problem 14

Develop a mechanism layout necessary to design a self-basting rotisserie for use in any kitchen oven (range).

Limitations and Considerations

1. The rotisserie should continuously baste a piece of meat (such as a turkey, ham, or roast) that has a maximum weight of 25 lbs.
2. The basting device should be capable of supplying 1-2 lbs. of basting liquid to the piece of meat over a maximum of six hours, without any attention needed by the cook.

Problem 15

Develop a mechanism layout necessary to design a device which will function as a muscle-building exerciser.

Limitations and Considerations

1. The device should be permanently, or semi-permanently, placed on the floor of a workout room.
2. A number of appropriate exercises should be possible by the manipulation of various linkage ratios or body-positioning mechanisms.

Problem 16

Develop a mechanism layout necessary to design an automatic or semi-automatic trap and skeet shooting device.

Limitations and Considerations

1. The device is to be remote-controlled by the shooter.
2. It should be able to throw a series of clay targets without reloading.

Problem 17

Develop a mechanism layout necessary to design a storage device for a kitchen.

Limitations and Considerations

1. The device is to operate as part of the upper unit of a kitchen cabinet.
2. The mechanism should be able to raise and lower the shelves to heights easily accessible to a person 4 ft. 8 in. tall.
3. The person must be able to reach the top shelves without standing on a chair or footstool.

Problem 18

Develop a mechanism layout necessary to design a device that will raise a door 3/4 in. when open and will return the door to zero height when closed.

Limitations and Considerations

1. The raising device should work with any door thickness, width, or height.
2. The mechanism can, but is not required to, be integrated into the door hinge design.

Problem 19

Develop a mechanism layout necessary to design a machine that will climb a vertical right-angled corner of a room from floor to ceiling.

Limitations and Considerations

1. The mechanism can use only the flat wall surfaces to produce lifting reactions.
2. No practical use needs to be identified for the machine.

Problem 20

Develop a mechanism layout necessary to design an analog computer that will automatically calculate the north-south velocity and east-west velocity from an input speed (V) and a directional deviation (θ).

Limitations and Considerations

1. Consider the two statements: $V_{n\text{-}s} = V \sin \theta$ and $V_{e\text{-}w} = V \cos \theta$ (where θ is the angle between the velocity vector V and the east-west velocity vector $V_{e\text{-}w}$).
2. Consider using a Scotch yoke mechanism to generate sine and cosine functions.

Appendix A

notation and abbreviations

NOTATION

(Points on a link will be designated by letters. Links will be designated by numbers.)

a	Linear acceleration
D	Driver
I	Instant center of rotation
S	Linear displacement
T	Time
V_B	Absolute linear velocity of point B
V_{BA}	Linear velocity of point B relative to point A
A_B	Absolute linear acceleration of point B
A_{BA}	Linear acceleration of point B relative to point A
ω_2	Angular velocity of link 2 (omega)
α_2	Angular acceleration of link 2 (alpha)
β	Label for angle (beta)
Δ	Small amount, small change (delta)
θ	Angular displacement (theta)
π	3.1416 constant (pi)

ABBREVIATIONS

cm	Centimeter
F	Force
fpm	Feet per minute
fps	Feet per second
ft.	Feet
in.	Inch
m	Meter
M.A.	Mechanical advantage
mm	Millimeter
mph	Miles per hour
rad/min.	Radians per minute
rad/sec.	Radians per second
rpm	Revolutions per minute
rps	Revolutions per second

Appendix B

equivalency tables

DECIMAL EQUIVALENTS — INCH-MILLIMETER CONVERSION TABLE

1/2	1/4	1/8	1/16	1/32	1/64	Decimals	Millimeters
					1	.015625	.396875
				1		.031250	.793750
					3	.046875	1.190625
			1			.062500	1.587500
					5	.078125	1.984375
				3		.093750	2.381250
					7	.109375	2.778125
		1				.125000	3.175000
					9	.140625	3.571875
				5		.156250	3.968750
					11	.171875	4.365625
			3			.187500	4.762500
					13	.203125	5.159375
				7		.218750	5.556250
					15	.234375	5.953125
	1					.250000	6.350000
					17	.265625	6.746875
				9		.281250	7.143750
					19	.296875	7.540625
			5			.312500	7.937500
					21	.328125	8.334375
				11		.343750	8.731250
					23	.359375	9.128125
		3				.375000	9.525000
					25	.390625	9.921875
				13		.406250	10.318750
					27	.421875	10.715625
			7			.437500	11.112500
					29	.453125	11.509375
				15		.468750	11.906250
					31	.484375	12.303125
1						.500000	12.700000
					33	.515625	13.096875
				17		.531250	13.493750
					35	.546875	13.890625
			9			.562500	14.287500
					37	.578125	14.684375
				19		.593750	15.081250
					39	.609375	15.478125
		5				.625000	15.875000
					41	.640625	16.271875
				21		.656250	16.668750
					43	.671875	17.065625
			11			.687500	17.462500
					45	.703125	17.859375
				23		.718750	18.256250
					47	.734375	18.653125
	3					.750000	19.050000
					49	.765625	19.446875
				25		.781250	19.843750
					51	.796875	20.240625
			13			.812500	20.637500
					53	.828125	21.034375
				27		.843750	21.431250
					55	.859375	21.828125
		7				.875000	22.225000
					57	.890625	22.621875
				29		.906250	23.018750
					59	.921875	23.415625
			15			.937500	23.812500
					61	.953125	24.209375
				31		.968750	24.606250
					63	.984375	25.003125
2	4	8	16	32	64	1.000000	25.400000

Courtesy of New Departure—Hyatt Division, General Motors Corp.

INCHES TO MILLIMETERS

in.	mm	in.	mm	in.	mm	in.	mm
1	25.4	26	660.4	51	1295.4	76	1930.4
2	50.8	27	685.8	52	1320.8	77	1955.8
3	76.2	28	711.2	53	1346.2	78	1981.2
4	101.6	29	736.6	54	1371.6	79	2006.6
5	127.0	30	762.0	55	1397.0	80	2032.0
6	152.4	31	787.4	56	1422.4	81	2057.4
7	177.8	32	812.8	57	1447.8	82	2082.8
8	203.2	33	838.2	58	1473.2	83	2108.2
9	228.6	34	863.6	59	1498.6	84	2133.6
10	254.0	35	889.0	60	1524.0	85	2159.0
11	279.4	36	914.4	61	1549.4	86	2184.4
12	304.8	37	939.8	62	1574.8	87	2209.8
13	330.2	38	965.2	63	1600.2	88	2235.2
14	355.6	39	990.6	64	1625.6	89	2260.6
15	381.0	40	1016.0	65	1651.0	90	2286.0
16	406.4	41	1041.4	66	1676.4	91	2311.4
17	431.8	42	1066.8	67	1701.8	92	2336.8
18	457.2	43	1092.2	68	1727.2	93	2362.2
19	482.6	44	1117.6	69	1752.6	94	2387.6
20	508.0	45	1143.0	70	1778.0	95	2413.0
21	533.4	46	1168.4	71	1803.4	96	2438.4
22	558.8	47	1193.8	72	1828.8	97	2463.8
23	584.2	48	1219.2	73	1854.2	98	2489.2
24	609.6	49	1244.6	74	1879.6	99	2514.6
25	635.0	50	1270.0	75	1905.0	100	2540.0

The above table is exact on the basis: 1 in. = 25.4 mm

MILLIMETERS TO INCHES

mm	in.	mm	in.	mm	in.	mm	in.
1	0.039370	26	1.023622	51	2.007874	76	2.992126
2	0.078740	27	1.062992	52	2.047244	77	3.031496
3	0.118110	28	1.102362	53	2.086614	78	3.070866
4	0.157480	29	1.141732	54	2.125984	79	3.110236
5	0.196850	30	1.181102	55	2.165354	80	3.149606
6	0.236220	31	1.220472	56	2.204724	81	3.188976
7	0.275591	32	1.259843	57	2.244094	82	3.228346
8	0.314961	33	1.299213	58	2.283465	83	3.267717
9	0.354331	34	1.338583	59	2.322835	84	3.307087
10	0.393701	35	1.377953	60	2.362205	85	3.346457
11	0.433071	36	1.417323	61	2.401575	86	3.385827
12	0.472441	37	1.456693	62	2.440945	87	3.425197
13	0.511811	38	1.496063	63	2.480315	88	3.464567
14	0.551181	39	1.535433	64	2.519685	89	3.503937
15	0.590551	40	1.574803	65	2.559055	90	3.543307
16	0.629921	41	1.614173	66	2.598425	91	3.582677
17	0.669291	42	1.653543	67	2.637795	92	3.622047
18	0.708661	43	1.692913	68	2.677165	93	3.661417
19	0.748031	44	1.732283	69	2.716535	94	3.700787
20	0.787402	45	1.771654	70	2.755906	95	3.740157
21	0.826772	46	1.811024	71	2.795276	96	3.779528
22	0.866142	47	1.850394	72	2.834646	97	3.818898
23	0.905512	48	1.889764	73	2.874016	98	3.858268
24	0.944882	49	1.929134	74	2.913386	99	3.897638
25	0.984252	50	1.968504	75	2.952756	100	3.937008

The above table is approximate on the basis: 1 in. = 25.4 mm, 1/25.4 = 0.039370078740 +

Courtesy of New Departure—Hyatt Division, General Motors Corp.

Appendix C

circular measurement

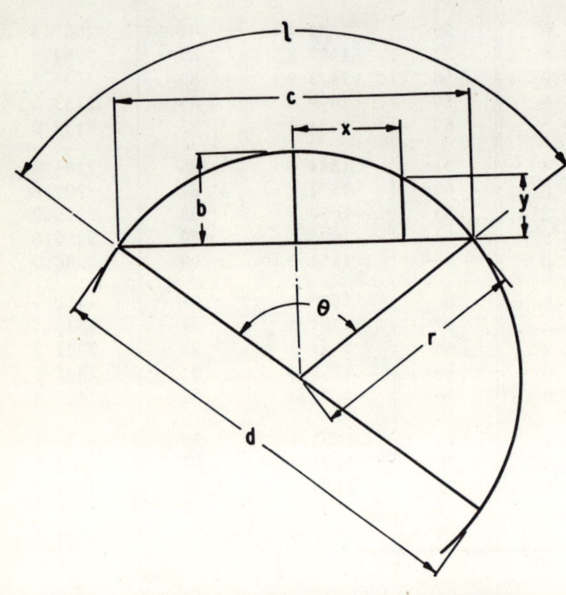

PROPERTIES OF THE CIRCLE

Circumference of Circle of Diameter $1 = \pi = 3.14159265$
Circumference of Circle $= 2\pi r = \pi d$
Diameter of Circle = Circumference × 0.31831
Diameter of Circle of equal periphery as square = side × 1.27324
Side of Square of equal periphery as circle = diameter × 0.78540
Diameter of Circle circumscribed about square = side × 1.41421
Side of Square inscribed in Circle = diameter × 0.70711

Arc, $l = \dfrac{\pi r \theta^\circ}{180} = 0.017453\, r\, \theta^\circ$

Angle, $\theta = \dfrac{180^\circ\, l}{\pi r} = 57.29578\, \dfrac{l}{r}$

Radius, $r = \dfrac{4b^2 + c^2}{8b}$ Diameter, $d = \dfrac{4b^2 + c^2}{4b}$

Chord, $c = 2\sqrt{2br - b^2} = 2r \sin\dfrac{\theta}{2} = d \sin\dfrac{\theta}{2}$

Rise, $b = r - \dfrac{1}{2}\sqrt{4r^2 - c^2} = \dfrac{c}{2}\tan\dfrac{\theta}{4} = 2r\sin^2\dfrac{\theta}{4}$

Rise, $b = r + y - \sqrt{r^2 - x^2}$ $y = b - r + \sqrt{r^2 - x^2}$ $x = \sqrt{r^2 - (r + y - b)^2}$

$\pi = 3.14159265$	$\log = 0.4971499$	$\pi^2 = 9.869604$	$\log = 0.994300$
		$\pi^3 = 31.006277$	$\log = 1.491450$
$\dfrac{1}{\pi} = 0.318310$	$\log = 9.502850 - 10$	$\dfrac{1}{\pi^2} = 0.101321$	$\log = 9.005700 - 10$
$\dfrac{2}{\pi} = 0.636620$	$\log = 9.803880 - 10$	$\dfrac{1}{\pi^3} = 0.032252$	$\log = 8.508557 - 10$
$\dfrac{180}{\pi} = 57.295780$	$\log = 1.758123$	$\sqrt{\pi} = 1.772454$	$\log = 0.248575$
		$1/\sqrt{\pi} = 0.564190$	$\log = 9.751425 - 10$
$\dfrac{\pi}{180} = 0.017453$	$\log = 8.241870 - 10$	$\sqrt[3]{\pi} = 1.464592$	$\log = 0.165717$
		$1/\sqrt[3]{\pi} = 0.682784$	$\log = 9.834283 - 10$

Courtesy of New Departure—Hyatt Division, General Motors Corp.

RADIAN MEASURE OF ANGLES

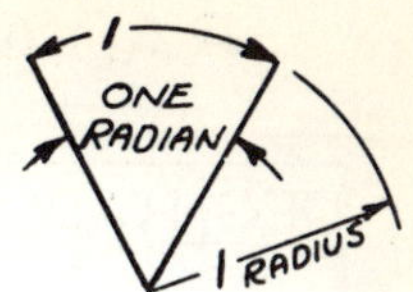

THE SOLUTION OF MANY GEAR PROBLEMS IS GREATLY SIMPLIFIED BY THE USE OF THE RADIAN MEASURE OF ANGLES.

THE RADIAN MEASURE OF AN ANGLE IS EQUAL TO THE LENGTH OF THE ARC OF A CIRCLE, WHEN THE RADIUS IS UNITY.

REFERRING TO FIGURE 1, A CIRCLE OF UNIT RADIUS IS SHOWN.

CIRCUMFERENCE OF A CIRCLE $= 2\pi R$

" " " " $= 2\pi$ WHEN RADIUS IS UNITY.

THEREFORE $360° = 2\pi$ RADIANS

AND $1° = 2\pi \div 360$

$1° = .0174533$ RADIANS

ONE RADIAN $= 360° \div 2\pi = 57.295780°$

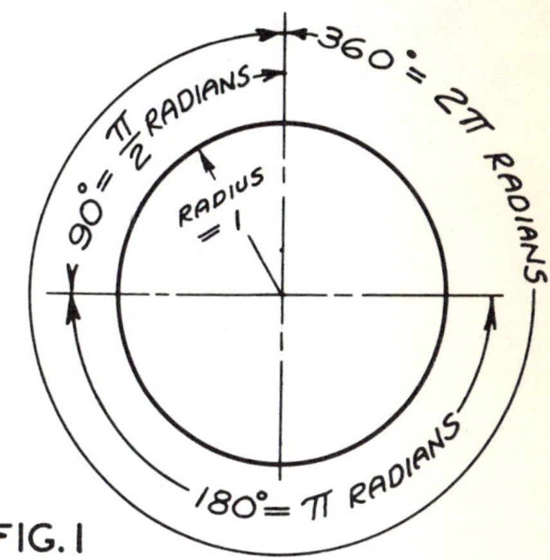

FIG. 1

PROBLEM NO. 1

DETERMINE THE ANGLE β IN RADIANS WHEN ARC AND RADIUS ARE GIVEN.

REFERRING TO FIGURE 2.

ARC FOR 2.5 RADIUS $= 1.5$

ARC FOR 1. RADIUS $= 1.5 \div 2.5 = .6$

THEREFORE ANGLE $\beta = .6$ RADIANS

ANGLE IN RADIANS = ARC ÷ RADIUS

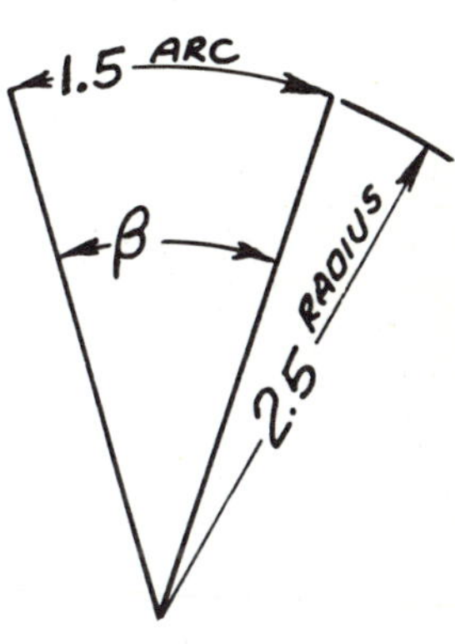

FIG. 2

PROBLEM NO. 2

DETERMINE THE ARC LENGTH "A" WHEN ANGLE AND RADIUS ARE GIVEN.

REFERRING TO FIGURE 3

SINCE $1° = .0174533$ RADIANS

$30° = .523599$ RADIANS

ARC FOR UNIT RADIUS $= .523599$

ARC FOR 2.5 RADIUS $= .523599 \times 2.5$

ARC "A" $= 1.3089975$

ARC = ANGLE IN RADIANS × RADIUS

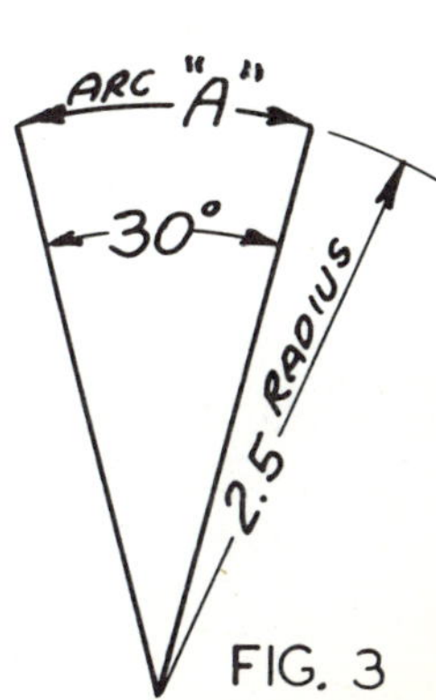

FIG. 3

Courtesy of INVO Spline, Inc.

RADIANS FOR • DEGREES • MINUTES • SECONDS •

DEG.	RADIANS	DEG.	RADIANS	MIN.	RADIANS	SEC.	RADIANS
1	.0174533	61	1.0646508	1	.0002909	1	.0000048
2	.0349066	62	1.0821041	2	.0005818	2	.0000097
3	.0523599	63	1.0995574	3	.0008727	3	.0000145
4	.0698132	64	1.1170107	4	.0011636	4	.0000194
5	.0872665	65	1.1344640	5	.0014544	5	.0000242
6	.1047198	66	1.1519173	6	.0017453	6	.0000291
7	.1221730	67	1.1693706	7	.0020362	7	.0000339
8	.1396263	68	1.1868239	8	.0023271	8	.0000388
9	.1570796	69	1.2042772	9	.0026180	9	.0000436
10	.1745329	70	1.2217305	10	.0029089	10	.0000485
11	.1919862	71	1.2391838	11	.0031998	11	.0000533
12	.2094395	72	1.2566371	12	.0034907	12	.0000582
13	.2268928	73	1.2740904	13	.0037815	13	.0000630
14	.2443461	74	1.2915436	14	.0040724	14	.0000679
15	.2617994	75	1.3089969	15	.0043633	15	.0000727
16	.2792527	76	1.3264502	16	.0046542	16	.0000776
17	.2967060	77	1.3439035	17	.0049451	17	.0000824
18	.3141593	78	1.3613568	18	.0052360	18	.0000873
19	.3316126	79	1.3788101	19	.0055269	19	.0000921
20	.3490659	80	1.3962634	20	.0058178	20	.0000970
21	.3665191	81	1.4137167	21	.0061087	21	.0001018
22	.3839724	82	1.4311700	22	.0063995	22	.0001067
23	.4014257	83	1.4486233	23	.0066904	23	.0001115
24	.4188790	84	1.4660766	24	.0069813	24	.0001164
25	.4363323	85	1.4835299	25	.0072722	25	.0001212
26	.4537856	86	1.5009832	26	.0075631	26	.0001261
27	.4712389	87	1.5184364	27	.0078540	27	.0001309
28	.4886922	88	1.5358897	28	.0081449	28	.0001357
29	.5061455	89	1.5533430	29	.0084358	29	.0001406
30	.5235988	90	1.5707963	30	.0087266	30	.0001454
31	.5410521	91	1.5882496	31	.0090175	31	.0001503
32	.5585054	92	1.6057029	32	.0093084	32	.0001551
33	.5759587	93	1.6231562	33	.0095993	33	.0001600
34	.5934119	94	1.6406095	34	.0098902	34	.0001648
35	.6108652	95	1.6580628	35	.0101811	35	.0001697
36	.6283185	96	1.6755161	36	.0104720	36	.0001745
37	.6457718	97	1.6929694	37	.0107629	37	.0001794
38	.6632251	98	1.7104227	38	.0110538	38	.0001842
39	.6806784	99	1.7278760	39	.0113446	39	.0001891
40	.6981317	100	1.7453293	40	.0116355	40	.0001939
41	.7155850	101	1.7627825	41	.0119264	41	.0001988
42	.7330383	102	1.7802358	42	.0122173	42	.0002036
43	.7504916	103	1.7976891	43	.0125082	43	.0002085
44	.7679449	104	1.8151424	44	.0127991	44	.0002133
45	.7853982	105	1.8325957	45	.0130900	45	.0002182
46	.8028515	106	1.8500490	46	.0133809	46	.0002230
47	.8203047	107	1.8675023	47	.0136717	47	.0002279
48	.8377580	108	1.8849556	48	.0139626	48	.0002327
49	.8552113	109	1.9024089	49	.0142535	49	.0002376
50	.8726646	110	1.9198622	50	.0145444	50	.0002424
51	.8901179	111	1.9373155	51	.0148353	51	.0002473
52	.9075712	112	1.9547688	52	.0151262	52	.0002521
53	.9250245	113	1.9722221	53	.0154171	53	.0002570
54	.9424778	114	1.9896753	54	.0157080	54	.0002618
55	.9599311	115	2.0071286	55	.0159989	55	.0002666
56	.9773844	116	2.0245819	56	.0162897	56	.0002715
57	.9948377	117	2.0420352	57	.0165806	57	.0002763
58	1.0122910	118	2.0594885	58	.0168715	58	.0002812
59	1.0297443	119	2.0769418	59	.0171624	59	.0002860
60	1.0471976	120	2.0943951	60	.0174533	60	.0002909

Courtesy of INVO Spline, Inc.

MINUTES AND SECONDS
IN DECIMALS OF ONE DEGREE

Min.	Seconds 0	10	20	30	40	50	Min.	Seconds 0	10	20	30	40	50
0	.0000	.0028	.0056	.0083	.0111	.0139	30	.5000	.5028	.5056	.5083	.5111	.5139
1	.0167	.0195	.0223	.0250	.0278	.0306	31	.5167	.5195	.5223	.5250	.5278	.5306
2	.0333	.0361	.0389	.0417	.0444	.0472	32	.5333	.5361	.5389	.5417	.5444	.5472
3	.0500	.0528	.0556	.0583	.0611	.0639	33	.5500	.5528	.5556	.5583	.5611	.5639
4	.0667	.0695	.0723	.0750	.0778	.0806	34	.5667	.5695	.5723	.5750	.5778	.5806
5	.0833	.0861	.0889	.0917	.0944	.0972	35	.5833	.5861	.5889	.5917	.5944	.5972
6	.1000	.1028	.1056	.1083	.1111	.1139	36	.6000	.6028	.6056	.6083	.6111	.6139
7	.1167	.1195	.1223	.1250	.1278	.1306	37	.6167	.6195	.6223	.6250	.6278	.6306
8	.1333	.1361	.1389	.1417	.1444	.1472	38	.6333	.6361	.6389	.6417	.6444	.6472
9	.1500	.1528	.1556	.1583	.1611	.1639	39	.6500	.6528	.6556	.6583	.6611	.6639
10	.1667	.1695	.1723	.1750	.1778	.1806	40	.6667	.6695	.6723	.6750	.6778	.6806
11	.1833	.1861	.1889	.1917	.1944	.1972	41	.6833	.6861	.6889	.6917	.6944	.6972
12	.2000	.2028	.2056	.2083	.2111	.2139	42	.7000	.7028	.7056	.7083	.7111	.7139
13	.2167	.2195	.2223	.2250	.2278	.2306	43	.7167	.7195	.7223	.7250	.7278	.7306
14	.2333	.2361	.2389	.2417	.2444	.2472	44	.7333	.7361	.7389	.7417	.7444	.7472
15	.2500	.2528	.2556	.2583	.2611	.2639	45	.7500	.7528	.7556	.7583	.7611	.7639
16	.2667	.2695	.2723	.2750	.2778	.2806	46	.7667	.7695	.7723	.7750	.7778	.7806
17	.2833	.2861	.2889	.2917	.2944	.2972	47	.7833	.7861	.7889	.7917	.7944	.7972
18	.3000	.3028	.3056	.3083	.3111	.3139	48	.8000	.8028	.8056	.8083	.8111	.8139
19	.3167	.3195	.3223	.3250	.3278	.3306	49	.8167	.8195	.8223	.8250	.8278	.8306
20	.3333	.3361	.3389	.3417	.3444	.3472	50	.8333	.8361	.8389	.8417	.8444	.8472
21	.3500	.3528	.3556	.3583	.3611	.3639	51	.8500	.8528	.8556	.8583	.8611	.8639
22	.3667	.3695	.3723	.3750	.3778	.3806	52	.8667	.8695	.8723	.8750	.8778	.8806
23	.3833	.3861	.3889	.3917	.3944	.3972	53	.8833	.8861	.8889	.8917	.8944	.8972
24	.4000	.4028	.4056	.4083	.4111	.4139	54	.9000	.9028	.9056	.9083	.9111	.9139
25	.4167	.4195	.4223	.4250	.4278	.4306	55	.9167	.9195	.9223	.9250	.9278	.9306
26	.4333	.4361	.4389	.4417	.4444	.4472	56	.9333	.9361	.9389	.9417	.9444	.9472
27	.4500	.4528	.4556	.4583	.4611	.4639	57	.9500	.9528	.9555	.9583	.9611	.9639
28	.4667	.4695	.4723	.4750	.4778	.4806	58	.9667	.9695	.9723	.9750	.9778	.9806
29	.4833	.4861	.4889	.4917	.4944	.4972	59	.9833	.9861	.9889	.9917	.9944	.9972
30	.5000	.5028	.5056	.5083	.5111	.5139	60	1.0000	1.0028	1.0055	1.0083	1.0111	1.0139

DECIMALS OF ONE DEGREE
IN MINUTES AND SECONDS

Degrees	.00	.01	.02	.03	.04	.05	.06	.07	.08	.09
0.0	0'00"	0'36"	1'12"	1'48"	2'24"	3'00"	3'36"	4'12"	4'48"	5'24"
0.1	6'00"	6'36"	7'12"	7'48"	8'24"	9'00"	9'36"	10'12"	10'48"	11'24"
0.2	12'00"	12'36"	13'12"	13'48"	14'24"	15'00"	15'36"	16'12"	16'48"	17'24"
0.3	18'00"	18'36"	19'12"	19'48"	20'24"	21'00"	21'36"	22'12"	22'48"	23'24"
0.4	24'00"	24'36"	25'12"	25'48"	26'24"	27'00"	27'36"	28'12"	28'48"	29'24"
0.5	30'00"	30'36"	31'12"	31'48"	32'24"	33'00"	33'36"	34'12"	34'48"	35'24"
0.6	36'00"	36'36"	37'12"	37'48"	38'24"	39'00"	39'36"	40'12"	40'48"	41'24"
0.7	42'00"	42'36"	43'12"	43'48"	44'24"	45'00"	45'36"	46'12"	46'48"	47'24"
0.8	48'00"	48'36"	49'12"	49'48"	50'24"	51'00"	51'36"	52'12"	52'48"	53'24"
0.9	54'00"	54'36"	55'12"	55'48"	56'24"	57'00"	57'36"	58'12"	58'48"	59'24"

VALUES OF RADIANS
IN DEGREES

Radians	.00	.01	.02	.03	.04	.05	.06	.07	.08	.09
	Deg.	Deg.	Deg.	Deg.	Deg.	Deg.	Deg.	Deg.	Deg.	Deg.
0.0	0.0000	0.5730	1.1459	1.7189	2.2918	2.8648	3.4377	4.0107	4.5837	5.1566
0.1	5.7296	6.3025	6.8755	7.4485	8.0214	8.5944	9.1673	9.7403	10.3132	10.8862
0.2	11.4592	12.0321	12.6051	13.1780	13.7510	14.3239	14.8969	15.4699	16.0428	16.6158
0.3	17.1887	17.7617	18.3346	18.9076	19.4806	20.0535	20.6265	21.1994	21.7724	22.3454
0.4	22.9183	23.4913	24.0642	24.6372	25.2101	25.7831	26.3561	26.9290	27.5020	28.0749
0.5	28.6479	29.2208	29.7938	30.3668	30.9397	31.1527	32.0856	32.6586	33.2316	33.8045
0.6	34.3775	34.9504	35.5234	36.0963	36.6693	37.2423	37.8152	38.3882	38.9611	39.5341
0.7	40.1070	40.6800	41.2530	41.8259	42.3989	42.9718	43.5448	44.1178	44.6907	45.2637
0.8	45.8366	46.4096	46.9825	47.5555	48.1285	48.7014	49.2744	49.8473	50.4203	50.9932
0.9	51.5662	52.1392	52.7121	53.2851	53.8580	54.4310	55.0039	55.5769	56.1499	56.7228

1 Radian = 57.29578 deg   2 Radians = 114.59156 deg   3 Radians = 171.88734 deg

Courtesy of New Departure—Hyatt Division, General Motors Corp.

LENGTHS OF CIRCULAR ARCS • MULTIPLICATION FACTOR

DEGREES

1	.017453	19	.331613	37	.645772	55	.959931	73	1.274090	91	1.588250	109	1.902409	127	2.216568	145	2.530728	163	2.844887
2	.034907	20	.349066	38	.663225	56	.977384	74	1.291544	92	1.605703	110	1.919862	128	2.234022	146	2.548181	164	2.862340
3	.052360	21	.366519	39	.680678	57	.994838	75	1.308997	93	1.623156	111	1.937316	129	2.251475	147	2.565634	165	2.879794
4	.069813	22	.383972	40	.698132	58	1.012291	76	1.326450	94	1.640610	112	1.954769	130	2.268928	148	2.583088	166	2.897247
5	.087266	23	.401426	41	.715585	59	1.029744	77	1.343904	95	1.658063	113	1.972222	131	2.286382	149	2.600541	167	2.914700
6	.104720	24	.418879	42	.733038	60	1.047198	78	1.361357	96	1.675516	114	1.989676	132	2.303835	150	2.617994	168	2.932153
7	.122173	25	.436332	43	.750492	61	1.064651	79	1.378810	97	1.692970	115	2.007129	133	2.321288	151	2.635447	169	2.949607
8	.139626	26	.453786	44	.767945	62	1.082104	80	1.396264	98	1.710423	116	2.024582	134	2.338741	152	2.652901	170	2.967060
9	.157080	27	.471239	45	.785398	63	1.099558	81	1.413717	99	1.727876	117	2.042035	135	2.356195	153	2.670354	171	2.984513
10	.174533	28	.488692	46	.802852	64	1.117011	82	1.431170	100	1.745329	118	2.059489	136	2.373648	154	2.687807	172	3.001967
11	.191986	29	.506146	47	.820305	65	1.134464	83	1.448623	101	1.762783	119	2.076942	137	2.391101	155	2.705261	173	3.019420
12	.209440	30	.523599	48	.837758	66	1.151917	84	1.466077	102	1.780236	120	2.094395	138	2.408555	156	2.722714	174	3.036873
13	.226893	31	.541052	49	.855211	67	1.169371	85	1.483530	103	1.797689	121	2.111849	139	2.426008	157	2.740167	175	3.054326
14	.244346	32	.558505	50	.872665	68	1.186824	86	1.500983	104	1.815143	122	2.129302	140	2.443461	158	2.757620	176	3.071780
15	.261799	33	.575959	51	.890118	69	1.204277	87	1.518437	105	1.832596	123	2.146755	141	2.460914	159	2.775074	177	3.089233
16	.279253	34	.593412	52	.907571	70	1.221731	88	1.535890	106	1.850049	124	2.164208	142	2.478368	160	2.792527	178	3.106686
17	.296706	35	.610865	53	.925025	71	1.239184	89	1.553343	107	1.867502	125	2.181662	143	2.495821	161	2.809980	179	3.124140
18	.314159	36	.628319	54	.942478	72	1.256637	90	1.570796	108	1.884956	126	2.199115	144	2.513274	162	2.827434	180	3.141593

MINUTES

1	.000291	11	.003200	21	.006109	31	.009018	41	.011926	51	.014835
2	.000582	12	.003491	22	.006400	32	.009308	42	.012217	52	.015126
3	.000873	13	.003782	23	.006690	33	.009599	43	.012508	53	.015417
4	.001164	14	.004072	24	.006981	34	.009890	44	.012799	54	.015708
5	.001454	15	.004363	25	.007272	35	.010181	45	.013090	55	.015999
6	.001745	16	.004654	26	.007563	36	.010472	46	.013381	56	.016290
7	.002036	17	.004945	27	.007854	37	.010763	47	.013672	57	.016581
8	.002327	18	.005236	28	.008145	38	.011054	48	.013963	58	.016872
9	.002618	19	.005527	29	.008436	39	.011345	49	.014254	59	.017162
10	.002909	20	.005818	30	.008727	40	.011636	50	.014544	60	.017453

$$\text{Multiplication Factor} = \frac{3.141593}{180} \times \text{number of degrees}$$

Length of Arc = Radius × Multiplication Factor

Example: To find the length of arc of a circular sector of 33° 15′ having a 20 inch radius:

Factor for 33° = .575959
Factor for 15′ = .004363
.580322

.580322 × 20 = 11.6064 inches.

Courtesy of New Departure—Hyatt Division, General Motors Corp.

Appendix D

trigonometric measurement

TRIGONOMETRIC FORMULAE

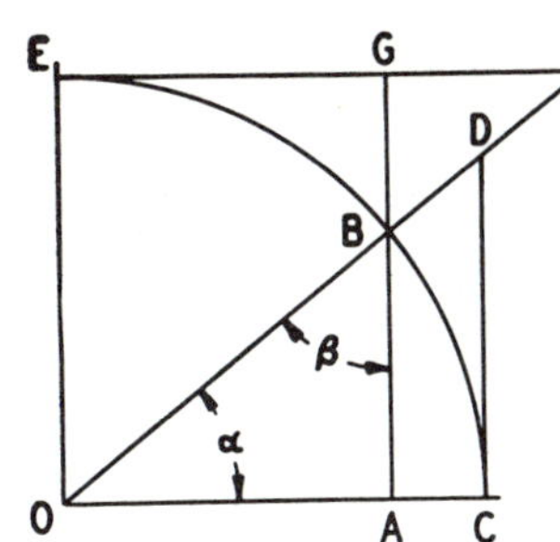

$OC = OB = OE = 1$
$AB = \text{Sin}\,\alpha$
$OA = \text{Cos}\,\alpha$
$CD = \text{Tan}\,\alpha$
$EF = \text{Cot}\,\alpha$
$OD = \text{Sec}\,\alpha$
$OF = \text{Cosec}\,\alpha$
$AC = \text{Vers}\,\alpha = 1 - \text{Cos}\,\alpha$
$BG = \text{Covers}\,\alpha = 1 - \text{Sin}\,\alpha$

$$\text{Radius } 1 = \sin^2\alpha + \cos^2\alpha = \sin\alpha \operatorname{cosec}\alpha = \cos\alpha \sec\alpha = \tan\alpha \cot\alpha$$

$$\sin\alpha = \frac{\cos\alpha}{\cot\alpha} = \frac{1}{\operatorname{cosec}\alpha} = \cos\alpha \tan\alpha = \sqrt{1-\cos^2\alpha}$$

$$\cos\alpha = \frac{\sin\alpha}{\tan\alpha} = \frac{1}{\sec\alpha} = \sin\alpha \cot\alpha = \sqrt{1-\sin^2\alpha}$$

$$\tan\alpha = \frac{\sin\alpha}{\cos\alpha} = \frac{1}{\cot\alpha} = \sin\alpha \sec\alpha \qquad \sec\alpha = \frac{\tan\alpha}{\sin\alpha} = \frac{1}{\cos\alpha}$$

$$\cot\alpha = \frac{\cos\alpha}{\sin\alpha} = \frac{1}{\tan\alpha} = \cos\alpha \operatorname{cosec}\alpha \qquad \operatorname{cosec}\alpha = \frac{\cot\alpha}{\cos\alpha} = \frac{1}{\sin\alpha}$$

$$\sin(\alpha \pm \beta) = \sin\alpha\cos\beta \pm \cos\alpha\sin\beta \qquad \tan(\alpha \pm \beta) = \frac{\tan\alpha \pm \tan\beta}{1 \mp \tan\alpha\tan\beta}$$

$$\cos(\alpha \pm \beta) = \cos\alpha\cos\beta \mp \sin\alpha\sin\beta \qquad \cot(\alpha \pm \beta) = \frac{\cot\alpha\cot\beta \mp 1}{\cot\beta \pm \cot\alpha}$$

$$\sin\alpha + \sin\beta = 2\sin\tfrac{1}{2}(\alpha+\beta)\cos\tfrac{1}{2}(\alpha-\beta) \qquad \tan\alpha + \tan\beta = \frac{\sin(\alpha+\beta)}{\cos\alpha\cos\beta}$$

$$\sin\alpha - \sin\beta = 2\cos\tfrac{1}{2}(\alpha+\beta)\sin\tfrac{1}{2}(\alpha-\beta) \qquad \tan\alpha - \tan\beta = \frac{\sin(\alpha-\beta)}{\cos\alpha\cos\beta}$$

$$\cos\alpha + \cos\beta = 2\cos\tfrac{1}{2}(\alpha+\beta)\cos\tfrac{1}{2}(\alpha-\beta) \qquad \cot\alpha + \cot\beta = \frac{\sin(\beta+\alpha)}{\sin\alpha\sin\beta}$$

$$\cos\beta - \cos\alpha = 2\sin\tfrac{1}{2}(\alpha+\beta)\sin\tfrac{1}{2}(\alpha-\beta) \qquad \cot\alpha - \cot\beta = \frac{\sin(\beta-\alpha)}{\sin\alpha\sin\beta}$$

$$\sin 2\alpha = 2\sin\alpha\cos\alpha \qquad \sin\tfrac{1}{2}\alpha = \sqrt{\frac{1-\cos\alpha}{2}} \qquad \sin^2\alpha = \frac{1-\cos 2\alpha}{2}$$

$$\cos 2\alpha = \cos^2\alpha - \sin^2\alpha \qquad \cos\tfrac{1}{2}\alpha = \sqrt{\frac{1+\cos\alpha}{2}} \qquad \cos^2\alpha = \frac{1+\cos 2\alpha}{2}$$

$$\tan 2\alpha = \frac{2\tan\alpha}{1-\tan^2\alpha} \qquad \tan\tfrac{1}{2}\alpha = \frac{\sin\alpha}{1+\cos\alpha} \qquad \tan^2\alpha = \frac{1-\cos 2\alpha}{1+\cos 2\alpha}$$

$$\cot 2\alpha = \frac{\cot^2\alpha - 1}{2\cot\alpha} \qquad \cot\tfrac{1}{2}\alpha = \frac{\sin\alpha}{1-\cos\alpha} \qquad \cot^2\alpha = \frac{1+\cos 2\alpha}{1-\cos 2\alpha}$$

$$\sin^2\alpha - \sin^2\beta = \sin(\alpha+\beta)\sin(\alpha-\beta) \qquad \cos^2\alpha - \sin^2\beta = \cos(\alpha+\beta)\cos(\alpha-\beta)$$

$$\frac{\sin\alpha \pm \sin\beta}{\cos\alpha + \cos\beta} = \tan\tfrac{1}{2}(\alpha \pm \beta), \qquad \frac{\sin\alpha \pm \sin\beta}{\cos\beta - \cos\alpha} = \cot\tfrac{1}{2}(\alpha \mp \beta)$$

Courtesy of New Departure—Hyatt Division, General Motors Corp.

TRIGONOMETRIC SOLUTIONS OF TRIANGLES

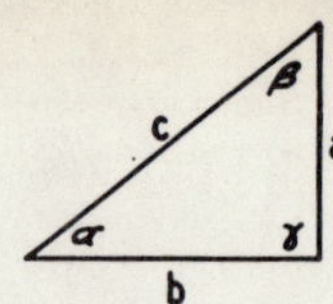

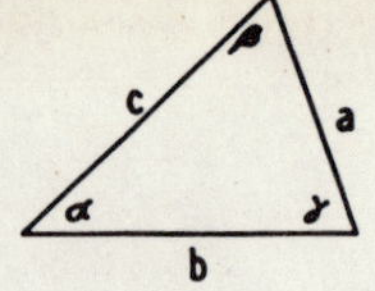

$$s = \frac{a+b+c}{2}$$

Given	Sought	Formulae		
RIGHT-ANGLE TRIANGLES				
a, c	α, β, b	$\sin\alpha = \frac{a}{c}$	$\cos\beta = \frac{a}{c}$	$b = \sqrt{c^2 - a^2}$
	area	$\text{area} = \frac{a}{2}\sqrt{c^2 - a^2}$		
a, b	α, β, c	$\tan\alpha = \frac{a}{b}$	$\tan\beta = \frac{b}{a}$	$c = \sqrt{a^2 + b^2}$
	area	$\text{area} = \frac{ab}{2}$		
α, a	β, b, c	$\beta = 90° - \alpha$	$b = a \cot\alpha$	$c = \frac{a}{\sin\alpha}$
	area	$\text{area} = \frac{a^2}{2}\cot\alpha$		
α, b	β, a, c	$\beta = 90° - \alpha$	$a = b\tan\alpha$	$c = \frac{b}{\cos\alpha}$
	area	$\text{area} = \frac{b^2}{2}\tan\alpha$		
α, c	β, a, b	$\beta = 90° - \alpha$	$a = c\sin\alpha$	$b = c\cos\alpha$
	area	$\text{area} = \frac{c^2}{2}\sin\alpha\cos\alpha = \frac{c^2}{4}\sin 2\alpha$		
OBLIQUE-ANGLE TRIANGLES				
a, b, c	α	$\sin\frac{\alpha}{2} = \sqrt{\frac{(s-b)(s-c)}{bc}}$	$\cos\frac{\alpha}{2} = \sqrt{\frac{s(s-a)}{bc}}$	$\tan\frac{\alpha}{2} = \sqrt{\frac{(s-b)(s-c)}{s(s-a)}}$
	β	$\sin\frac{\beta}{2} = \sqrt{\frac{(s-a)(s-c)}{ac}}$	$\cos\frac{\beta}{2} = \sqrt{\frac{s(s-b)}{ac}}$	$\tan\frac{\beta}{2} = \sqrt{\frac{(s-a)(s-c)}{s(s-b)}}$
	γ	$\sin\frac{\gamma}{2} = \sqrt{\frac{(s-a)(s-b)}{ab}}$	$\cos\frac{\gamma}{2} = \sqrt{\frac{s(s-c)}{ab}}$	$\tan\frac{\gamma}{2} = \sqrt{\frac{(s-a)(s-b)}{s(s-c)}}$
	area	$\text{area} = \sqrt{s(s-a)(s-b)(s-c)}$		
a, α, β	b, γ, c	$b = a\frac{\sin\beta}{\sin\alpha}$	$\gamma = 180° - \alpha - \beta$	$c = a\frac{\sin\gamma}{\sin\alpha} = a\frac{\sin(\alpha+\beta)}{\sin\alpha}$
	area	$\text{area} = \frac{ab}{2}\sin\gamma = \frac{a^2}{2} \times \frac{\sin\beta\sin\gamma}{\sin\alpha}$		
a, b, α	β, γ, c	$\sin\beta = \frac{b}{a}\sin\alpha$	$\gamma = 180° - \alpha - \beta$	$c = a\frac{\sin\gamma}{\sin\alpha} = b\frac{\sin\gamma}{\sin\beta}$
	area	$\text{area} = \frac{ab}{2}\sin\gamma$		$c = \sqrt{a^2 + b^2 - 2ab\cos\gamma}$
a, b, γ	α, β	$\tan\alpha = \frac{a\sin\gamma}{b - a\cos\gamma}$,	$\tan\frac{\alpha-\beta}{2} = \frac{a-b}{a+b}\cot\frac{\gamma}{2}$	
	c	$c = \sqrt{a^2 + b^2 - 2ab\cos\gamma} = a\frac{\sin\gamma}{\sin\alpha}$		
	area	$\text{area} = \frac{ab}{2}\sin\gamma$		

$a^2 = b^2 + c^2 - 2bc\cos\alpha$, $\quad b^2 = a^2 + c^2 - 2ac\cos\beta$, $\quad c^2 = a^2 + b^2 - 2ab\cos\gamma$

Courtesy of New Departure—Hyat Division, General Motors Corp.

TRIGONOMETRIC CONSTANTS AND RELATIONSHIPS

VARIATION OF NUMERICAL VALUE OF ANGULAR FUNCTIONS

Angular Function	In Quadrant I $0°—90°$	In Quadrant II $90°—180°$	In Quadrant III $180°—270°$	In Quadrant IV $270°—360°$
sin	0 to $+1$	$+1$ to 0	0 to -1	-1 to 0
cos	$+1$ to 0	0 to -1	-1 to 0	0 to $+1$
tan	0 to $+\infty$	$-\infty$ to 0	0 to $+\infty$	$-\infty$ to 0
cot	$+\infty$ to 0	0 to $-\infty$	$+\infty$ to 0	0 to $-\infty$
sec	$+1$ to $+\infty$	$-\infty$ to -1	-1 to $-\infty$	$+\infty$ to $+1$
cosec	$+\infty$ to $+1$	$+1$ to $+\infty$	$-\infty$ to -1	-1 to $-\infty$

NUMERICAL VALUE OF ANGULAR FUNCTIONS OF COMMON ANGLES

Angular Function	Angle				
	0°	30°	45°	60°	90°
sin	0	$\frac{1}{2} = .500$	$\frac{1}{2}\sqrt{2} = .707$	$\frac{1}{2}\sqrt{3} = .866$	1
cos	1	$\frac{1}{2}\sqrt{3} = .866$	$\frac{1}{2}\sqrt{2} = .707$	$\frac{1}{2} = .500$	0
tan	0	$\frac{1}{3}\sqrt{3} = .577$	$1 = 1.000$	$\sqrt{3} = 1.732$	∞
cot	∞	$\sqrt{3} = 1.732$	$1 = 1.000$	$\frac{1}{3}\sqrt{3} = .577$	0
sec	1	$\frac{2}{3}\sqrt{3} = 1.155$	$\sqrt{2} = 1.414$	$2 = 2.000$	∞
cosec	∞	$2 = 2.000$	$\sqrt{2} = 1.414$	$\frac{2}{3}\sqrt{3} = 1.155$	1

RELATIONSHIP OF ANGULAR FUNCTIONS OF FIRST-QUADRANT ANGLES AND RELATED ANGLES OF OTHER QUADRANTS

Angular Function	Related Angles				
	$-a$	$90° \pm a$	$180° \pm a$	$270° \pm a$	$360° \pm a$
$\sin a$	$-\sin a$	$+\cos a$	$\mp \sin a$	$-\cos a$	$\pm \sin a$
$\cos a$	$+\cos a$	$\mp \sin a$	$-\cos a$	$\pm \sin a$	$+\cos a$
$\tan a$	$-\tan a$	$\mp \cot a$	$\pm \tan a$	$\mp \cot a$	$\pm \tan a$
$\cot a$	$-\cot a$	$\mp \tan a$	$\pm \cot a$	$\mp \tan a$	$\pm \cot a$
$\sec a$	$+\sec a$	$\mp \csc a$	$-\sec a$	$\pm \csc a$	$+\sec a$
$\csc a$	$-\csc a$	$+\sec a$	$\mp \csc a$	$-\sec a$	$\pm \csc a$

Courtesy of New Departure—Hyatt Division, General Motors Corp.

NATURAL TRIGONOMETRIC FUNCTIONS

Degrees	SINES 0′	10′	20′	30′	40′	50′	60′	Cosines
0	0.00000	0.00291	0.00582	0.00873	0.01164	0.01454	0.01745	89
1	0.01745	0.02036	0.02327	0.02618	0.02908	0.03199	0.03490	88
2	0.03490	0.03781	0.04071	0.04362	0.04653	0.04943	0.05234	87
3	0.05234	0.05524	0.05814	0.06105	0.06395	0.06685	0.06976	86
4	0.06976	0.07266	0.07556	0.07846	0.08136	0.08426	0.08716	85
5	0.08716	0.09005	0.09295	0.09585	0.09874	0.10164	0.10453	84
6	0.10453	0.10742	0.11031	0.11320	0.11609	0.11898	0.12187	83
7	0.12187	0.12476	0.12764	0.13053	0.13341	0.13629	0.13917	82
8	0.13917	0.14205	0.14493	0.14781	0.15069	0.15356	0.15643	81
9	0.15643	0.15931	0.16218	0.16505	0.16792	0.17078	0.17365	80
10	0.17365	0.17651	0.17937	0.18224	0.18509	0.18795	0.19081	79
11	0.19081	0.19366	0.19652	0.19937	0.20222	0.20507	0.20791	78
12	0.20791	0.21076	0.21360	0.21644	0.21928	0.22212	0.22495	77
13	0.22495	0.22778	0.23062	0.23345	0.23627	0.23910	0.24192	76
14	0.24192	0.24474	0.24756	0.25038	0.25320	0.25601	0.25882	75
15	0.25882	0.26163	0.26443	0.26724	0.27004	0.27284	0.27564	74
16	0.27564	0.27843	0.28123	0.28402	0.28680	0.28959	0.29237	73
17	0.29237	0.29515	0.29793	0.30071	0.30348	0.30625	0.30902	72
18	0.30902	0.31178	0.31454	0.31730	0.32006	0.32282	0.32557	71
19	0.32557	0.32832	0.33106	0.33381	0.33655	0.33929	0.34202	70
20	0.34202	0.34475	0.34748	0.35021	0.35293	0.35565	0.35837	69
21	0.35837	0.36108	0.36379	0.36650	0.36921	0.37191	0.37461	68
22	0.37461	0.37730	0.37999	0.38268	0.38537	0.38805	0.39073	67
23	0.39073	0.39341	0.39608	0.39875	0.40142	0.40408	0.40674	66
24	0.40674	0.40939	0.41204	0.41469	0.41734	0.41998	0.42262	65
25	0.42262	0.42525	0.42788	0.43051	0.43313	0.43575	0.43837	64
26	0.43837	0.44098	0.44359	0.44620	0.44880	0.45140	0.45399	63
27	0.45399	0.45658	0.45917	0.46175	0.46433	0.46690	0.46947	62
28	0.46947	0.47204	0.47460	0.47716	0.47971	0.48226	0.48481	61
29	0.48481	0.48735	0.48989	0.49242	0.49495	0.49748	0.50000	60
30	0.50000	0.50252	0.50503	0.50754	0.51004	0.51254	0.51504	59
31	0.51504	0.51753	0.52002	0.52250	0.52498	0.52745	0.52992	58
32	0.52992	0.53238	0.53484	0.53730	0.53975	0.54220	0.54464	57
33	0.54464	0.54708	0.54951	0.55194	0.55436	0.55678	0.55919	56
34	0.55919	0.56160	0.56401	0.56641	0.56880	0.57119	0.57358	55
35	0.57358	0.57596	0.57833	0.58070	0.58307	0.58543	0.58779	54
36	0.58779	0.59014	0.59248	0.59482	0.59716	0.59949	0.60182	53
37	0.60182	0.60414	0.60645	0.60876	0.61107	0.61337	0.61566	52
38	0.61566	0.61795	0.62024	0.62251	0.62479	0.62706	0.62932	51
39	0.62932	0.63158	0.63383	0.63608	0.63832	0.64056	0.64279	50
40	0.64279	0.64501	0.64723	0.64945	0.65166	0.65386	0.65606	49
41	0.65606	0.65825	0.66044	0.66262	0.66480	0.66697	0.66913	48
42	0.66913	0.67129	0.67344	0.67559	0.67773	0.67987	0.68200	47
43	0.68200	0.68412	0.68624	0.68835	0.69046	0.69256	0.69466	46
44	0.69466	0.69675	0.69883	0.70091	0.70298	0.70505	0.70711	45
Sines	60′	50′	40′	30′	20′	10′	0′	Degrees
	COSINES							

Degrees	COSINES 0′	10′	20′	30′	40′	50′	60′	Sines
0	1.00000	1.00000	0.99998	0.99996	0.99993	0.99989	0.99985	89
1	0.99985	0.99979	0.99973	0.99966	0.99958	0.99949	0.99939	88
2	0.99939	0.99929	0.99917	0.99905	0.99892	0.99878	0.99863	87
3	0.99863	0.99847	0.99831	0.99813	0.99795	0.99776	0.99756	86
4	0.99756	0.99736	0.99714	0.99692	0.99668	0.99644	0.99619	85
5	0.99619	0.99594	0.99567	0.99540	0.99511	0.99482	0.99452	84
6	0.99452	0.99421	0.99390	0.99357	0.99324	0.99290	0.99255	83
7	0.99255	0.99219	0.99182	0.99144	0.99106	0.99067	0.99027	82
8	0.99027	0.98986	0.98944	0.98902	0.98858	0.98814	0.98769	81
9	0.98769	0.98723	0.98676	0.98629	0.98580	0.98531	0.98481	80
10	0.98481	0.98430	0.98378	0.98325	0.98272	0.98218	0.98163	79
11	0.98163	0.98107	0.98050	0.97992	0.97934	0.97875	0.97815	78
12	0.97815	0.97754	0.97692	0.97630	0.97566	0.97502	0.97437	77
13	0.97437	0.97371	0.97304	0.97237	0.97169	0.97100	0.97030	76
14	0.97030	0.96959	0.96887	0.96815	0.96742	0.96667	0.96593	75
15	0.96593	0.96517	0.96440	0.96363	0.96285	0.96206	0.96126	74
16	0.96126	0.96046	0.95964	0.95882	0.95799	0.95715	0.95630	73
17	0.95630	0.95545	0.95459	0.95372	0.95284	0.95195	0.95106	72
18	0.95106	0.95015	0.94924	0.94832	0.94740	0.94646	0.94552	71
19	0.94552	0.94457	0.94361	0.94264	0.94167	0.94068	0.93969	70
20	0.93969	0.93869	0.93769	0.93667	0.93565	0.93462	0.93358	69
21	0.93358	0.93253	0.93148	0.93042	0.92935	0.92827	0.92718	68
22	0.92718	0.92609	0.92499	0.92388	0.92276	0.92164	0.92050	67
23	0.92050	0.91936	0.91822	0.91706	0.91590	0.91472	0.91355	66
24	0.91355	0.91236	0.91116	0.90996	0.90875	0.90753	0.90631	65
25	0.90631	0.90507	0.90383	0.90259	0.90133	0.90007	0.89879	64
26	0.89879	0.89752	0.89623	0.89493	0.89363	0.89232	0.89101	63
27	0.89101	0.88968	0.88835	0.88701	0.88566	0.88431	0.88295	62
28	0.88295	0.88158	0.88020	0.87882	0.87743	0.87603	0.87462	61
29	0.87462	0.87321	0.87178	0.87036	0.86892	0.86748	0.86603	60
30	0.86603	0.86457	0.86310	0.86163	0.86015	0.85866	0.85717	59
31	0.85717	0.85567	0.85416	0.85264	0.85112	0.84959	0.84805	58
32	0.84805	0.84650	0.84495	0.84339	0.84182	0.84025	0.83867	57
33	0.83867	0.83708	0.83549	0.83389	0.83228	0.83066	0.82904	56
34	0.82904	0.82741	0.82577	0.82413	0.82248	0.82082	0.81915	55
35	0.81915	0.81748	0.81580	0.81412	0.81242	0.81072	0.80902	54
36	0.80902	0.80730	0.80558	0.80386	0.80212	0.80038	0.79864	53
37	0.79864	0.79688	0.79512	0.79335	0.79158	0.78980	0.78801	52
38	0.78801	0.78622	0.78442	0.78261	0.78079	0.77897	0.77715	51
39	0.77715	0.77531	0.77347	0.77162	0.76977	0.76791	0.76604	50
40	0.76604	0.76417	0.76229	0.76041	0.75851	0.75661	0.75471	49
41	0.75471	0.75280	0.75088	0.74896	0.74703	0.74509	0.74314	48
42	0.74314	0.74120	0.73924	0.73728	0.73531	0.73333	0.73135	47
43	0.73135	0.72937	0.72737	0.72537	0.72337	0.72136	0.71934	46
44	0.71934	0.71732	0.71529	0.71325	0.71121	0.70916	0.70711	45
Cosines	60′	50′	40′	30′	20′	10′	0′	Degrees
	SINES							

Courtesy of New Departure—Hyatt Division, General Motors Corp.

NATURAL TRIGONOMETRIC FUNCTIONS—Continued

Degrees	TANGENTS 0′	10′	20′	30′	40′	50′	60′	Cotangents
0	0.00000	0.00291	0.00582	0.00873	0.01164	0.01455	0.01746	89
1	0.01746	0.02036	0.02328	0.02619	0.02910	0.03201	0.03492	88
2	0.03492	0.03783	0.04075	0.04366	0.04658	0.04949	0.05241	87
3	0.05241	0.05533	0.05824	0 06116	0.06408	0.06700	0.06993	86
4	0.06993	0 07285	0.07578	0.07870	0.08163	0.08456	0.08749	85
5	0.08749	0.09042	0.09335	0.09629	0.09923	0.10216	0.10510	84
6	0.10510	0.10805	0.11099	0.11394	0.11688	0.11983	0.12278	83
7	0.12278	0.12574	0.12869	0.13165	0.13461	0.13758	0.14054	82
8	0.14054	0.14351	0.14648	0.14945	0.15243	0.15540	0.15838	81
9	0.15838	0.16137	0.16435	0.16734	0.17033	0.17333	0.17633	80
10	0.17633	0.17933	0.18233	0.18534	0.18835	0.19136	0.19438	79
11	0.19438	0.19740	0.20042	0.20345	0.20648	0.20952	0.21256	78
12	0.21256	0.21560	0.21864	0.22169	0.22475	0.22781	0.23087	77
13	0.23087	0.23393	0.23700	0.24008	0.24316	0.24624	0.24933	76
14	0.24933	0.25242	0.25552	0.25862	0.26172	0.26483	0.26795	75
15	0.26795	0.27107	0.27419	0.27732	0.28046	0.28360	0.28675	74
16	0.28675	0.28990	0.29305	0.29621	0.29938	0.30255	0.30573	73
17	0.30573	0.30891	0.31210	0.31530	0.31850	0.32171	0.32492	72
18	0.32492	0.32814	0.33136	0.33460	0.33783	0.34108	0.34433	71
19	0.34433	0.34758	0.35085	0.35412	0.35740	0.36068	0.36397	70
20	0.36397	0.36727	0.37057	0.37388	0.37720	0.38053	0.38386	69
21	0.38386	0.38721	0.39055	0.39391	0.39727	0.40065	0.40403	68
22	0.40403	0.40741	0.41081	0.41421	0.41763	0.42105	0.42447	67
23	0.42447	0.42791	0.43136	0.43481	0.43828	0.44175	0.44523	66
24	0.44523	0.44872	0.45222	0.45573	0.45924	0.46277	0.46631	65
25	0.46631	0.46985	0.47341	0.47698	0.48055	0.48414	0.48773	64
26	0.48773	0.49134	0.49495	0.49858	0.50222	0.50587	0.50953	63
27	0.50953	0.51320	0.51688	0.52057	0.52427	0.52798	0.53171	62
28	0.53171	0.53545	0.53920	0.54296	0.54674	0.55051	0.55431	61
29	0.55431	0.55812	0.56194	0.56577	0.56962	0.57348	0.57735	60
30	0.57735	0.58124	0.58513	0.58905	0.59297	0.59691	0.60086	59
31	0.60086	0.60483	0.60881	0.61280	0.61681	0.62083	0.62487	58
32	0.62487	0.62892	0.63299	0.63707	0.64117	0.64528	0.64941	57
33	0.64941	0.65355	0.65771	0.66189	0.66608	0.67028	0.67451	56
34	0.67451	0.67875	0.68301	0.68728	0.69157	0.69588	0.70021	55
35	0.70021	0.70455	0.70891	0.71329	0.71769	0.72211	0.72654	54
36	0.72654	0.73100	0.73547	0.73996	0.74447	0.74900	0.75355	53
37	0.75355	0.75812	0.76272	0.76733	0.77196	0.77661	0.78129	52
38	0.78129	0.78598	0.79070	0.79544	0.80020	0.80498	0.80978	51
39	0.80978	0.81461	0.81946	0.82434	0.82923	0.83415	0.83910	50
40	0.83910	0.84407	0.84906	0.85408	0.85912	0.86419	0.86929	49
41	0.86929	0.87441	0.87955	0.88473	0.88992	0.89515	0.90040	48
42	0.90040	0.90569	0.91099	0.91633	0.92170	0.92709	0.93252	47
43	0.93252	0.93797	0.94345	0.94896	0.95451	0.96008	0.96569	46
44	0.96569	0.97133	0.97700	0.98270	0.98843	0.99420	1.00000	45
Tangents	60′	50′	40′	30′	20′	10′	0′ COTANGENTS	Degrees

Degrees	COTANGENTS 0′	10′	20′	30′	40′	50′	60′	Tangents
0	∞	343.77371	171.88540	114.58865	85.93979	68.75009	57.28996	89
1	57.28996	49.10388	42.96408	38.18846	34.36777	31.24158	28.63625	88
2	28.63625	26.43160	24.54176	22.90377	21.47040	20.20555	19.08114	87
3	19.08114	18.07498	17.16934	16.34986	15.60478	14.92442	14.30067	86
4	14.30067	13.72674	13.19688	12.70621	12.25051	11.82617	11.43005	85
5	11.43005	11.05943	10.71191	10.38540	10.07803	9.78817	9.51436	84
6	9.51436	9.25530	9.00983	8.77689	8.55555	8.34496	8.14435	83
7	8.14435	7.95302	7.77035	7.59575	7.42871	7.26873	7.11537	82
8	7.11537	6.96823	6.82694	6.69116	6.56055	6.43484	6.31375	81
9	6.31375	6.19703	6.08444	5.97576	5.87080	5.76937	5.67128	80
10	5.67128	5.57638	5.48451	5.39552	5.30928	5.22566	5.14455	79
11	5.14455	5.06584	4.98940	4.91516	4.84300	4.77286	4.70463	78
12	4.70463	4.63825	4.57363	4.51071	4.44942	4.38969	4.33148	77
13	4.33148	4.27471	4.21933	4.16530	4.11256	4.06107	4.01078	76
14	4.01078	3.96165	3.91364	3.86671	3.82083	3.77595	3.73205	75
15	3.73205	3.68909	3.64705	3.60588	3.56557	3.52609	3.48741	74
16	3.48741	3.44951	3.41236	3.37594	3.34023	3.30521	3.27085	73
17	3.27085	3.23714	3.20406	3.17159	3.13972	3.10842	3.07768	72
18	3.07768	3.04749	3.01783	2.98869	2.96004	2.93189	2.90421	71
19	2.90421	2.87700	2.85023	2.82391	2.79802	2.77254	2.74748	70
20	2.74748	2.72281	2.69853	2.67462	2.65109	2.62791	2.60509	69
21	2.60509	2.58261	2.56046	2.53865	2.51715	2.49597	2.47509	68
22	2.47509	2.45451	2.43422	2.41421	2.39449	2.37504	2.35585	67
23	2.35585	2.33693	2.31826	2.29984	2.28167	2.26374	2.24604	66
24	2.24604	2.22857	2.21132	2.19430	2.17749	2.16090	2.14451	65
25	2.14451	2.12832	2.11233	2.09654	2.08094	2.06553	2.05030	64
26	2.05030	2.03526	2.02039	2.00569	1.99116	1.97680	1.96261	63
27	1.96261	1.94858	1.93470	1.92098	1.90741	1.89400	1.88073	62
28	1.88073	1.86760	1.85462	1.84177	1.82907	1.81649	1.80405	61
29	1.80405	1.79174	1.77955	1.76749	1.75556	1.74375	1.73205	60
30	1.73205	1.72047	1.70901	1.69766	1.68643	1.67530	1.66428	59
31	1.66428	1.65337	1.64256	1.63185	1.62125	1.61074	1.60033	58
32	1.60033	1.59002	1.57981	1.56969	1.55966	1.54972	1.53987	57
33	1.53987	1.53010	1.52043	1.51084	1.50133	1.49190	1.48256	56
34	1.48256	1.47330	1.46411	1.45501	1.44598	1.43703	1.42815	55
35	1.42815	1.41934	1.41061	1.40195	1.39336	1.38484	1.37638	54
36	1.37638	1.36800	1.35968	1.35142	1.34323	1.33511	1.32704	53
37	1.32704	1.31904	1.31110	1.30323	1.29541	1.28764	1.27994	52
38	1.27994	1.27230	1.26471	1.25717	1.24969	1.24227	1.23490	51
39	1.23490	1.22758	1.22031	1.21310	1.20593	1.19882	1.19175	50
40	1.19175	1.18474	1.17777	1.17085	1.16398	1.15715	1.15037	49
41	1.15037	1.14363	1.13694	1.13029	1.12369	1.11713	1.11061	48
42	1.11061	1.10414	1.09770	1.09131	1.08496	1.07864	1.07237	47
43	1.07237	1.06613	1.05994	1.05378	1.04766	1.04158	1.03553	46
44	1.03553	1.02952	1.02355	1.01761	1.01170	1.00583	1.00000	45
Cotangents	60′	50′	40′	30′	20′	10′	0′ TANGENTS	Degrees

Courtesy of New Departure—Hyatt Division, General Motors Corp.

NATURAL TRIGONOMETRIC FUNCTIONS—Continued

Degrees	SECANTS 0′	10′	20′	30′	40′	50′	60′	Cosecants
0	1.00000	1.00000	1.00002	1.00004	1.00007	1.00011	1.00015	89
1	1.00015	1.00021	1.00027	1.00034	1.00042	1.00051	1.00061	88
2	1.00061	1.00072	1.00083	1.00095	1.00108	1.00122	1.00137	87
3	1.00137	1.00153	1.00169	1.00187	1.00205	1.00224	1.00244	86
4	1.00244	1.00265	1.00287	1.00309	1.00333	1.00357	1.00382	85
5	1.00382	1.00408	1.00435	1.00463	1.00491	1.00521	1.00551	84
6	1.00551	1.00582	1.00614	1.00647	1.00681	1.00715	1.00751	83
7	1.00751	1.00787	1.00825	1.00863	1.00902	1.00942	1.00983	82
8	1.00983	1.01024	1.01067	1.01111	1.01155	1.01200	1.01247	81
9	1.01247	1.01294	1.01342	1.01391	1.01440	1.01491	1.01543	80
10	1.01543	1.01595	1.01649	1.01703	1.01758	1.01815	1.01872	79
11	1.01872	1.01930	1.01989	1.02049	1.02110	1.02171	1.02234	78
12	1.02234	1.02298	1.02362	1.02428	1.02494	1.02562	1.02630	77
13	1.02630	1.02700	1.02770	1.02842	1.02914	1.02987	1.03061	76
14	1.03061	1.03137	1.03213	1.03290	1.03368	1.03447	1.03528	75
15	1.03528	1.03609	1.03691	1.03774	1.03858	1.03944	1.04030	74
16	1.04030	1.04117	1.04206	1.04295	1.04385	1.04477	1.04569	73
17	1.04569	1.04663	1.04757	1.04853	1.04950	1.05047	1.05146	72
18	1.05146	1.05246	1.05347	1.05449	1.05552	1.05657	1.05762	71
19	1.05762	1.05869	1.05976	1.06085	1.06195	1.06306	1.06418	70
20	1.06418	1.06531	1.06645	1.06761	1.06878	1.06995	1.07115	69
21	1.07115	1.07235	1.07356	1.07479	1.07602	1.07727	1.07853	68
22	1.07853	1.07981	1.08109	1.08239	1.08370	1.08503	1.08636	67
23	1.08636	1.08771	1.08907	1.09044	1.09183	1.09323	1.09464	66
24	1.09464	1.09606	1.09750	1.09895	1.10041	1.10189	1.10338	65
25	1.10338	1.10488	1.10640	1.10793	1.10947	1.11103	1.11260	64
26	1.11260	1.11419	1.11579	1.11740	1.11903	1.12067	1.12233	63
27	1.12233	1.12400	1.12568	1.12738	1.12910	1.13083	1.13257	62
28	1.13257	1.13433	1.13610	1.13789	1.13970	1.14152	1.14335	61
29	1.14335	1.14521	1.14707	1.14896	1.15085	1.15277	1.15470	60
30	1.15470	1.15665	1.15861	1.16059	1.16259	1.16460	1.16663	59
31	1.16663	1.16868	1.17075	1.17283	1.17493	1.17704	1.17918	58
32	1.17918	1.18133	1.18350	1.18569	1.18790	1.19012	1.19236	57
33	1.19236	1.19463	1.19691	1.19920	1.20152	1.20386	1.20622	56
34	1.20622	1.20859	1.21099	1.21341	1.21584	1.21830	1.22077	55
35	1.22077	1.22327	1.22579	1.22833	1.23089	1.23347	1.23607	54
36	1.23607	1.23869	1.24134	1.24400	1.24669	1.24940	1.25214	53
37	1.25214	1.25489	1.25767	1.26047	1.26330	1.26615	1.26902	52
38	1.26902	1.27191	1.27483	1.27778	1.28075	1.28374	1.28676	51
39	1.28676	1.28980	1.29287	1.29597	1.29909	1.30223	1.30541	50
40	1.30541	1.30861	1.31183	1.31509	1.31837	1.32168	1.32501	49
41	1.32501	1.32838	1.33177	1.33519	1.33864	1.34212	1.34563	48
42	1.34563	1.34917	1.35274	1.35634	1.35997	1.36363	1.36733	47
43	1.36733	1.37105	1.37481	1.37860	1.38242	1.38628	1.39016	46
44	1.39016	1.39409	1.39804	1.40203	1.40606	1.41012	1.41421	45
Secants	60′	50′	40′	30′	20′	10′	0′ COSECANTS	Degrees

Degrees	COSECANTS 0′	10′	20′	30′	40′	50′	60′	Secants
0	∞	343.77516	171.88831	114.59301	85.94561	68.75736	57.29869	89
1	57.29869	49.11406	42.97571	38.20155	34.38232	31.25758	28.65371	88
2	28.65371	26.45051	24.56212	22.92559	21.49368	20.23028	19.10732	87
3	19.10732	18.10262	17.19843	16.38041	15.63679	14.95788	14.33559	86
4	14.33559	13.76312	13.23472	12.74550	12.29125	11.86837	11.47371	85
5	11.47371	11.10455	10.75849	10.43343	10.12752	9.83912	9.56677	84
6	9.56677	9.30917	9.06515	8.83367	8.61379	8.40466	8.20551	83
7	8.20551	8.01565	7.83443	7.66130	7.49571	7.33719	7.18530	82
8	7.18530	7.03962	6.89979	6.76547	6.63633	6.51208	6.39245	81
9	6.39245	6.27719	6.16607	6.05886	5.95536	5.85539	5.75877	80
10	5.75877	5.66533	5.57493	5.48740	5.40263	5.32049	5.24084	79
11	5.24084	5.16359	5.08863	5.01585	4.94517	4.87649	4.80973	78
12	4.80973	4.74482	4.68167	4.62023	4.56041	4.50216	4.44541	77
13	4.44541	4.39012	4.33622	4.28366	4.23239	4.18238	4.13357	76
14	4.13357	4.08591	4.03938	3.99393	3.94952	3.90613	3.86370	75
15	3.86370	3.82223	3.78166	3.74198	3.70315	3.66515	3.62796	74
16	3.62796	3.59154	3.55587	3.52094	3.48671	3.45317	3.42030	73
17	3.42030	3.38808	3.35649	3.32551	3.29512	3.26531	3.23607	72
18	3.23607	3.20737	3.17920	3.15155	3.12440	3.09774	3.07155	71
19	3.07155	3.04584	3.02057	2.99574	2.97135	2.94737	2.92380	70
20	2.92380	2.90063	2.87785	2.85545	2.83342	2.81175	2.79043	69
21	2.79043	2.76945	2.74881	2.72850	2.70851	2.68884	2.66947	68
22	2.66947	2.65040	2.63162	2.61313	2.59491	2.57698	2.55930	67
23	2.55930	2.54190	2.52474	2.50784	2.49119	2.47477	2.45859	66
24	2.45859	2.44264	2.42692	2.41142	2.39614	2.38107	2.36620	65
25	2.36620	2.35154	2.33708	2.32282	2.30875	2.29487	2.28117	64
26	2.28117	2.26766	2.25432	2.24116	2.22817	2.21535	2.20269	63
27	2.20269	2.19019	2.17786	2.16568	2.15366	2.14178	2.13005	62
28	2.13005	2.11847	2.10704	2.09574	2.08458	2.07356	2.06267	61
29	2.06267	2.05191	2.04128	2.03077	2.02039	2.01014	2.00000	60
30	2.00000	1.98998	1.98008	1.97029	1.96062	1.95106	1.94160	59
31	1.94160	1.93226	1.92302	1.91388	1.90485	1.89591	1.88709	58
32	1.88708	1.87834	1.86970	1.86116	1.85271	1.84435	1.83608	57
33	1.83608	1.82790	1.81981	1.81180	1.80388	1.79604	1.78829	56
34	1.78829	1.78062	1.77303	1.76552	1.75808	1.75073	1.74345	55
35	1.74345	1.73624	1.72911	1.72205	1.71506	1.70815	1.70130	54
36	1.70130	1.69452	1.68782	1.68117	1.67460	1.66809	1.66164	53
37	1.66164	1.65526	1.64894	1.64268	1.63648	1.63035	1.62427	52
38	1.62427	1.61825	1.61229	1.60639	1.60054	1.59475	1.58902	51
39	1.58902	1.58333	1.57771	1.57213	1.56661	1.56114	1.55572	50
40	1.55572	1.55036	1.54504	1.53977	1.53455	1.52938	1.52425	49
41	1.52425	1.51918	1.51415	1.50916	1.50422	1.49933	1.49448	48
42	1.49448	1.48967	1.48491	1.48019	1.47551	1.47087	1.46628	47
43	1.46628	1.46173	1.45721	1.45274	1.44831	1.44391	1.43956	46
44	1.43956	1.43524	1.43096	1.42672	1.42251	1.41835	1.41421	45
Cosecants	60′	50′	40′	30′	20′	10′	0′ SECANTS	Degrees

Courtesy of New Departure—Hyatt Division, General Motors Corp.

Appendix E

cam scale layout

CAM SCALE LAYOUT METHOD. In Figure P-13 is shown a convenient device for laying out a cam profile. It is particularly useful for a preliminary layout, as a very accurate delineation of the profile may be obtained with little or no calculation. This scale is copyrighted by Commercial Cam and Machine Co., Chicago, Ill.

As a substitute, if the scale is not available, an ordinary full or half size drafting scale may be used. Divisions for the commonly used curves are given in Table P-2.

Methods of using the cam scale for profile layout of various types of cam motion are shown in Fig. P-14 through P-18.

Division	Parabolic	Harmonic	Cycloidal	Mod. Trap.	Mod. Sine
0	0	0	0	0	0
1	13/64	1/4	1/16	3/32	7/64
2	51/64	61/64	31/64	39/64	11/16
3	1-51/64	2-1/16	1-31/64	1-39/64	1-25/32
4	3-13/64	3-29/64	3-1/16	3-3/32	3-9/32
5	5	5	5	5	5
6	6-5/64	6-35/64	6-15/16	6-29/32	6-23/32
7	8-13/64	7-15/16	8-33/64	8-25/64	8-7/32
8	9-13/64	9-3/64	9-33/64	9-25/64	9-5/16
9	9-51/64	9-3/4	9-15/16	9-29/32	9-57/64
10	10	10	10	10	10

TABLE P-2

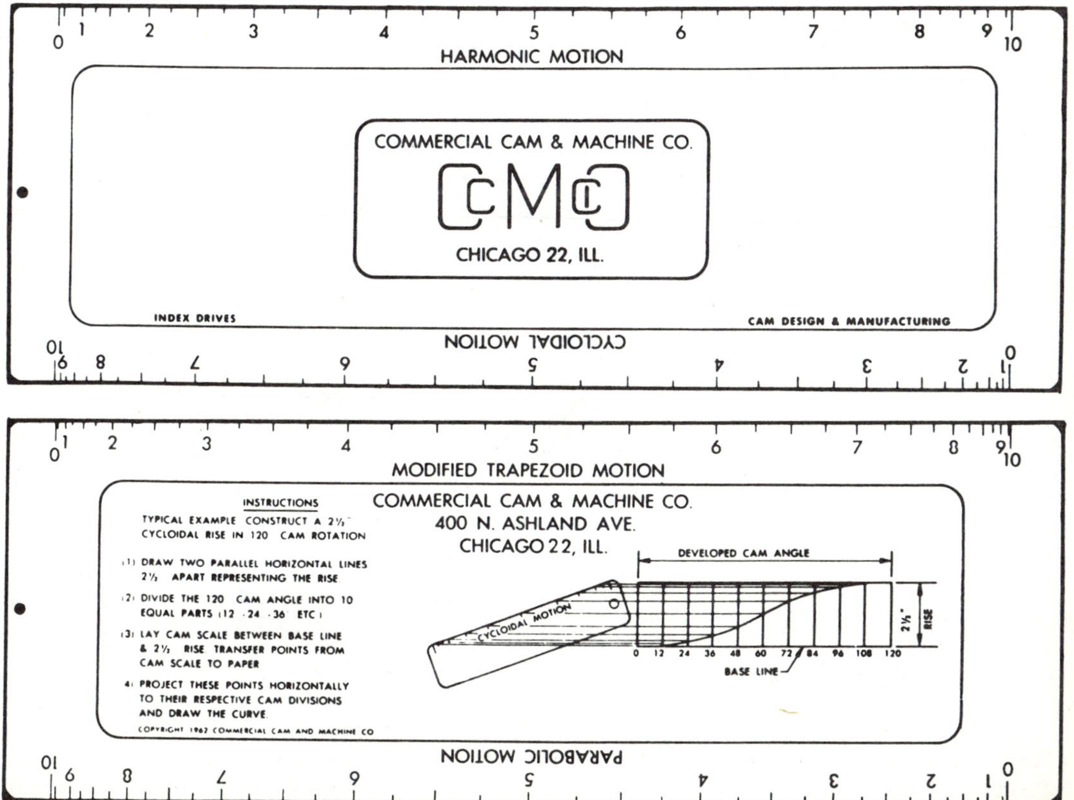

Fig. P-13

Courtesy of CAMCO Division, Emerson Electric Co.

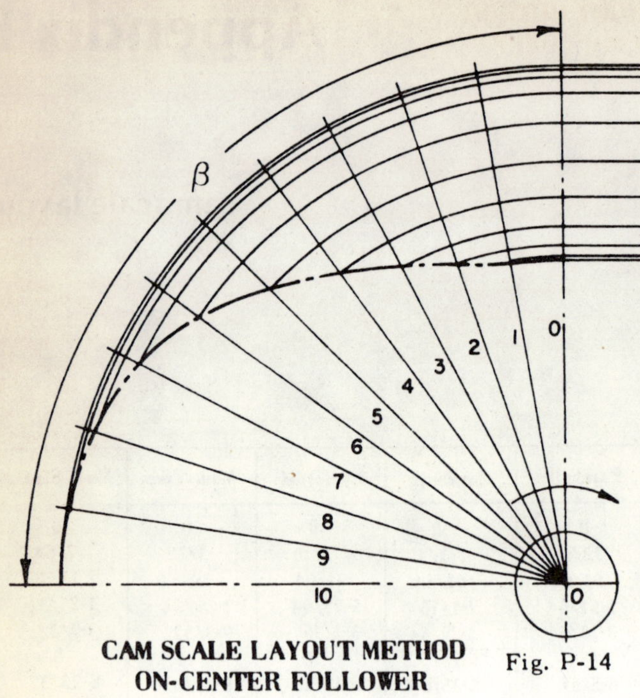

CAM SCALE LAYOUT METHOD ON-CENTER FOLLOWER

Fig. P-14

INSTRUCTIONS

1. Divide rotation angle β into 10 equal parts by radial lines.
2. Draw two parallel horizontal lines through extremities of displacement h
3. Lay cam scale between these lines. Transfer points from cam scale to layout.
4. Project these points horizontally to zero radial line.
5. With O as center, inscribe radius through each point, intersecting corresponding radial line.
6. Draw smooth curve through these intersections to establish cam pitch curve.

Courtesy of CAMCO Division, Emerson Electric Co.

CAM SCALE LAYOUT METHOD OFFSET FOLLOWER

Fig. P-15

INSTRUCTIONS

1. Divide cam rotation angle β into 10 equal parts by radial lines.
2. Draw offset lines parallel to each radial line and tangent to offset circle.
3. Draw two parallel horizontal lines through the extremities of displacement h.
4. Lay cam scale between these lines. Transfer points from cam scale to layout.
5. Project each point horizontally to offset line zero.
6. With O as center inscribe radius through each point, intersecting corresponding offset line.
7. Draw smooth curve through these intersections to establish the cam pitch curve.

CAM SCALE LAYOUT METHOD
SWINGING ARM FOLLOWER

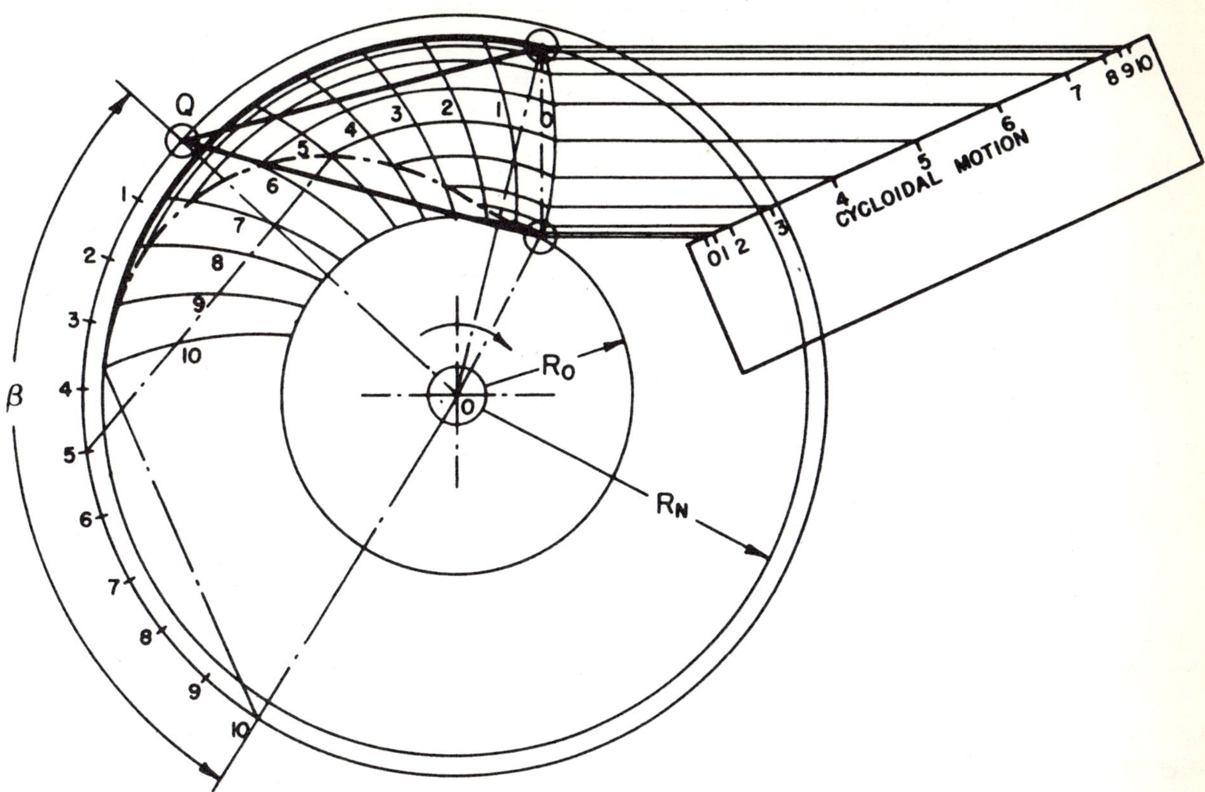

Fig. P-16

Courtesy of CAMCO Division, Emerson Electric Co.

INSTRUCTIONS

1. Draw concentric circles with radii R_O and R_N through extremities of swinging arm angular displacement.
2. Draw circle through fulcrum point Q with radius OQ.
3. Lay out cam rotation angle β from fulcrum point Q in opposite direction to cam rotation.
4. Divide arc of angle β into 10 equal parts.
5. With each point as a center, inscribe radius equal to length of swinging arm.
6. Draw two parallel lines perpendicular to line through extremities of swinging arm angular displacement.
7. Lay cam scale between these lines and transfer points from cam scale to layout.
8. Project these points to arc of swinging arm.
9. With O as center inscribe arc through each point, intersecting corresponding swinging arm arcs.
10. Draw smooth curve through these intersections to establish cam pitch curve.

CAM SCALE LAYOUT METHOD
CYCLOIDAL AND HARMONIC COMBINATION

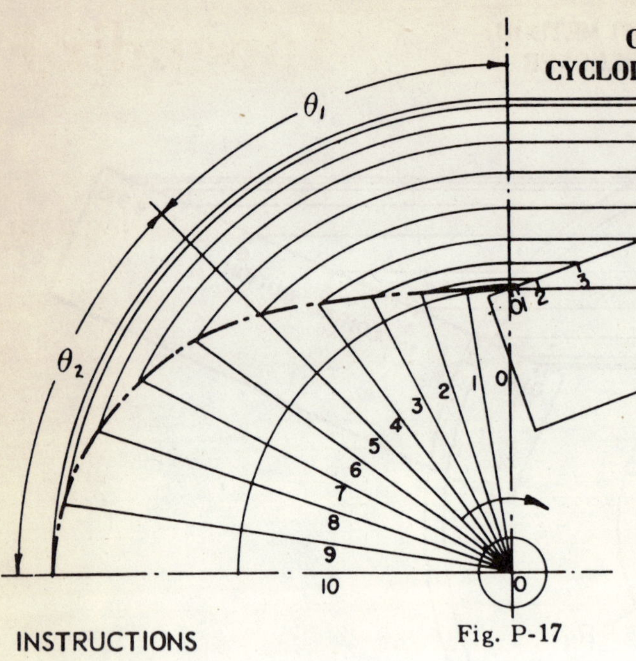

Fig. P-17

INSTRUCTIONS

1. Lay out rotation angles θ_1 and θ_2 and divide each into 5 equal parts by radial lines.

2. Draw three horizontal parallel lines A, B and C representing calculated values h_1 and h_2

3. Lay cycloidal cam scale with zero on base line A and point 5 on line B. Transfer points from cam scale to layout.

4. Lay harmonic cam scale with point 5 on line B and point 10 on line C. Transfer points from cam scale to layout.

5. Project these points to the zero radial line.

6. With O as center, inscribe radius through each point intersecting corresponding radial line.

7. Draw smooth curve through these intersections to establish the cam pitch curve.

CAM SCALE LAYOUT METHOD
VELOCITY-ADJUSTED CURVE

Courtesy of CAMCO Division,
Emerson Electric Co.

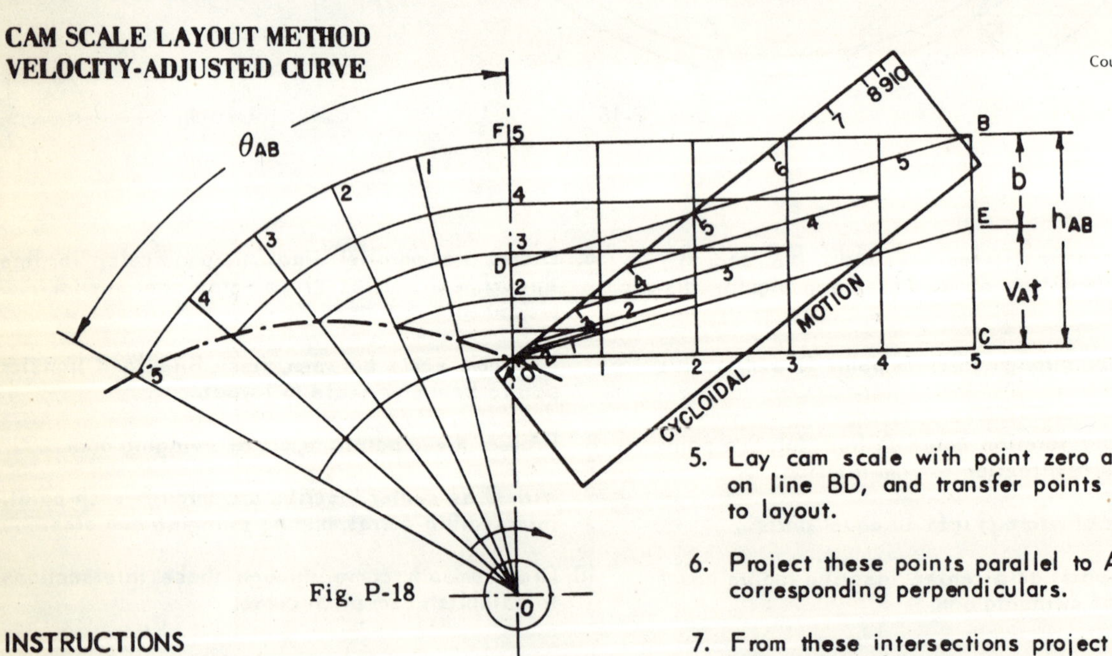

Fig. P-18

INSTRUCTIONS

1. Draw line AC any convenient length and divide into 5 equal parts.

2. Erect perpendiculars at each division point and complete rectangle ACBF.

3. Draw line AE, making CE equal to calculated value $v_A t$.

4. Draw line BD parallel to AE.

5. Lay cam scale with point zero at A and point 5 on line BD, and transfer points from cam scale to layout.

6. Project these points parallel to AE, intersecting corresponding perpendiculars.

7. From these intersections project points horizontally to line AF.

8. Lay out rotation angle θ_{AB} and divide into 5 equal parts by radial lines.

9. Inscribe radii with O as center through projected points on line AF, intersecting corresponding radial lines.

10. Draw smooth curve through these intersections to establish cam pitch curve.

Appendix F

gearing data

RULES AND FORMULAS
For Calculating Dimensions of Diametral Pitch Gears

DIMENSION WANTED	RULE	FORMULA
Circular Pitch (CP)	Divide 3.1416 by Diametral Pitch	$CP = \frac{3.1416}{DP}$
Circular Pitch (CP)	Multiply Pitch Diameter by 3.1416 and divide Product by Number of Teeth	$CP = \frac{PD \times 3.1416}{N}$
Diametral Pitch (DP)	Divide 3.1416 by Circular Pitch	$DP = \frac{3.1416}{CP}$
Diametral Pitch (DP)	Divide Number of Teeth by Pitch Diameter	$DP = \frac{N}{PD}$
Pitch Diameter (PD)	Divide Number of Teeth by Diametral Pitch	$PD = \frac{N}{DP}$
Pitch Diameter (PD)	Multiply Number of Teeth by Circular Pitch and divide Product by 3.1416	$PD = \frac{N \times CP}{3.1416}$
Center Distance (CDi)	Add Number of Teeth in Gear and Pinion, and divide Sum by Twice Diametral Pitch	$CDi = \frac{Ng + Np}{2DP}$
Center Distance (CDi)	Multiply Sum of Number of Teeth in Gear and Pinion by Circular Pitch, and divide Product by 6.2832	$CDi = \frac{(Ng + Np)\,CP}{6.2832}$
Addendum (A)	Divide 1 by Diametral Pitch (For Full-length Teeth)	$A = \frac{1}{DP}$
Addendum (A)	Divide Circular Pitch by 3.1416 (For Full-length Teeth)	$A = \frac{CP}{3.1416}$
Clearance (C)	Subtract Addendum from Dedendum	See Tables of Gear Tooth Parts
Dedendum (D)	Add Clearance to Addendum	$D = A + C$
Whole Depth (WD)	Add Dedendum to Addendum	$WD = A + D$
Circular Thickness (CTh)	Divide 1.5708 by Diametral Pitch	$CTh = \frac{1.5708}{DP}$
Circular Thickness (CTh)	Divide Circular Pitch by 2	$CTh = \frac{CP}{2}$
Outside Diameter (OD)	Add 2 to Number of Teeth, and divide Sum by Diametral Pitch (For Full-length Teeth)	$OD = \frac{N + 2}{DP}$
Outside Diameter (OD)	Add Two Times the Addendum to the Pitch Diameter	$OD = PD + 2A$
Number of Teeth (N)	Multiply Pitch Diameter by Diametral Pitch	$N = PD \times DP$
Number of Teeth (N)	Multiply Pitch Diameter by 3.1416, and divide Product by Circular Pitch	$N = \frac{PD \times 3.1416}{CP}$
Root Diameter (RD)	Subtract Two Times the Whole Depth from Outside Diameter	$RD = OD—2WD$
Base Circle Diameter $(BCDi)$	Multiply Pitch Diameter by Cosine of Pressure Angle	$BCDi = PD \times CosVP$

Courtesy of The Fellows Gear Shaper Co.

1

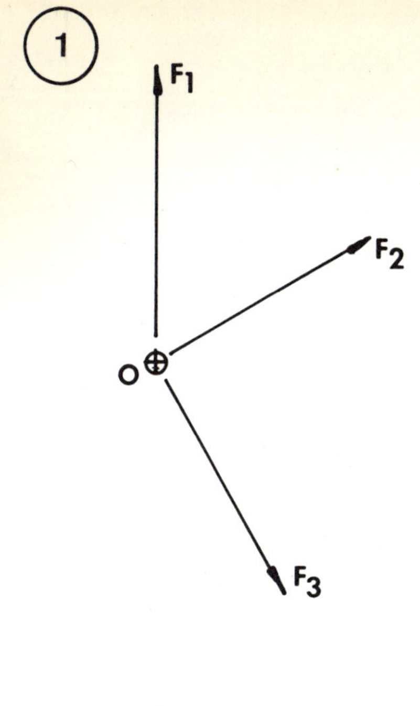

2

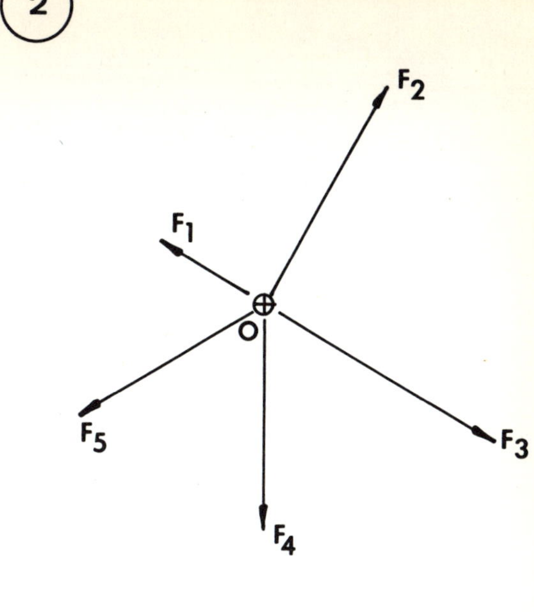

3

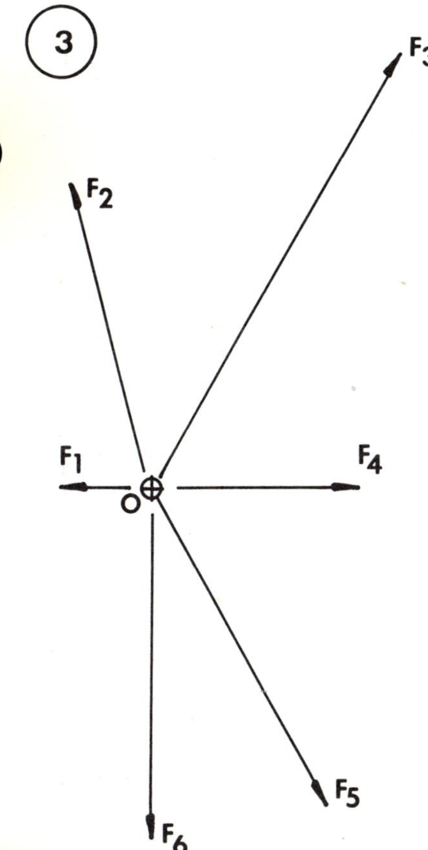

O

MECHANISM DRAFTING AND DESIGN	NAME: SECT. NO.: FILE NO.: DATE:	GRADE	PROBLEM 1A

1

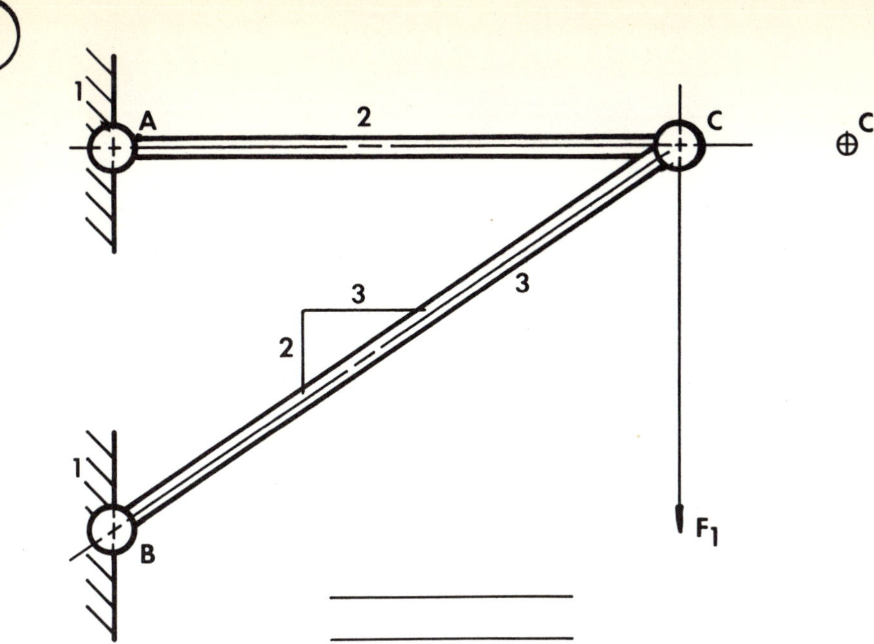

2

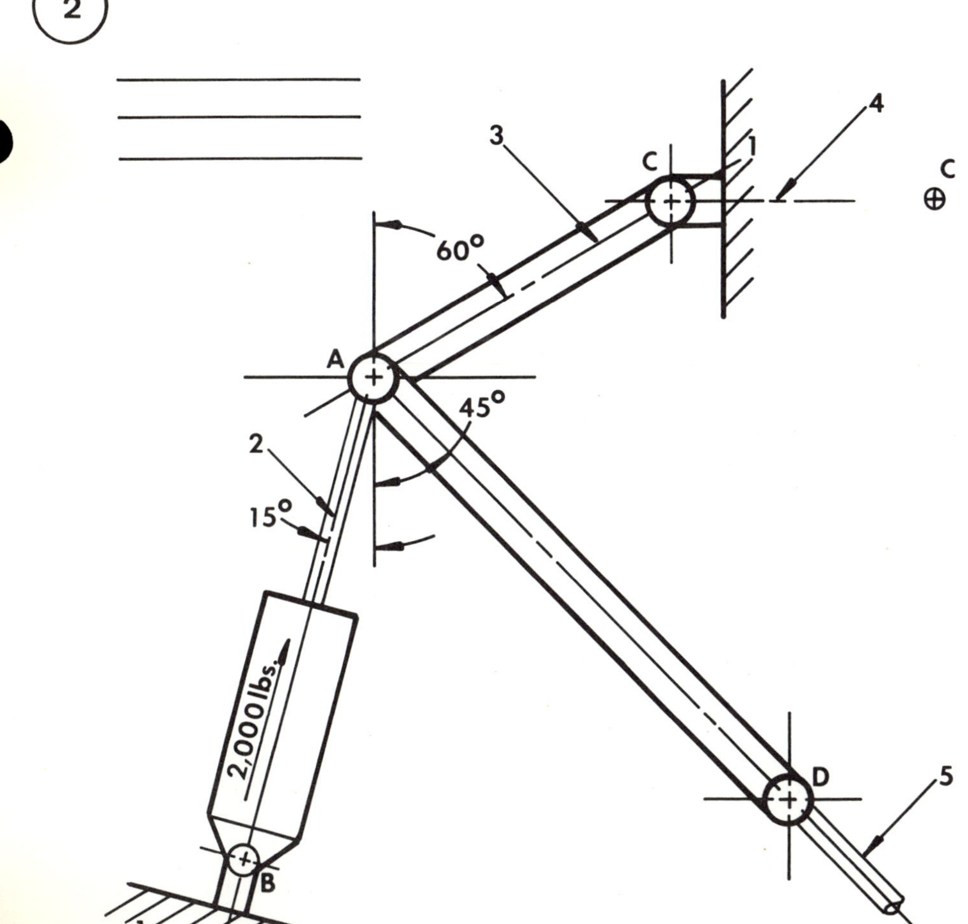

MECHANISM DRAFTING AND DESIGN	NAME: SECT. NO.: FILE NO.: DATE:	GRADE	PROBLEM 1B

C

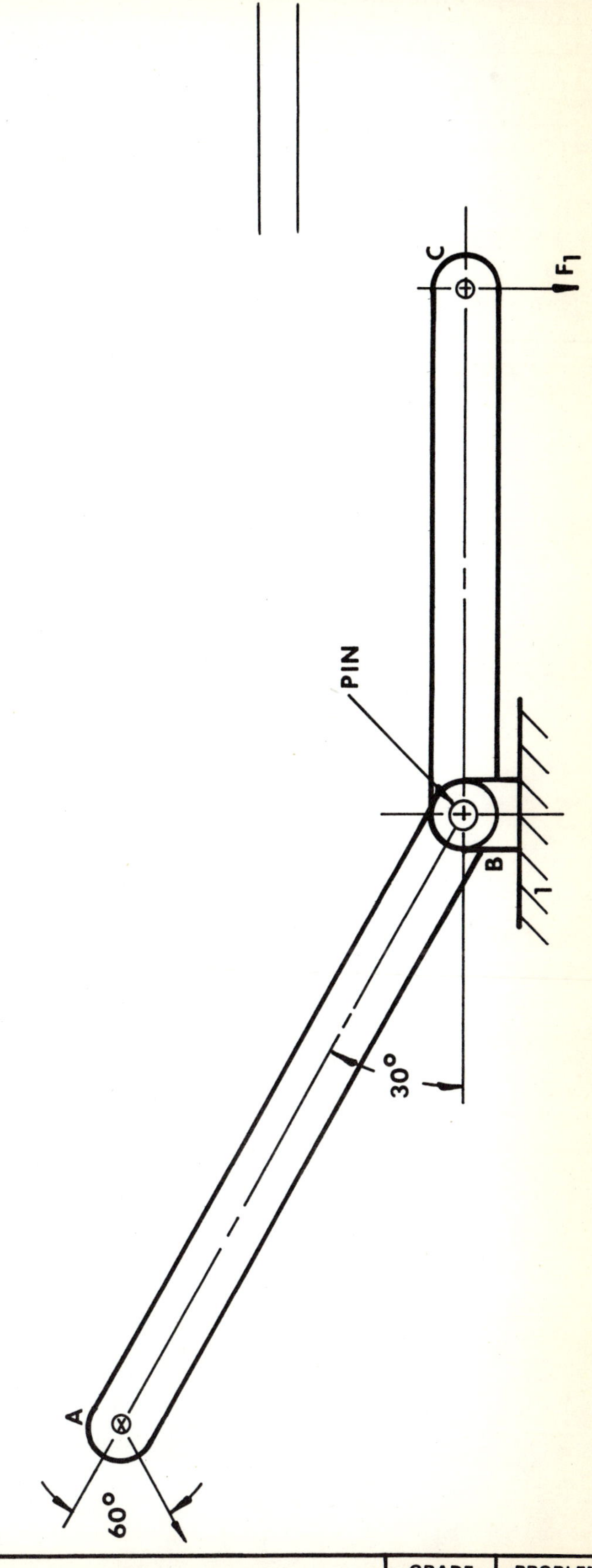

MECHANISM DRAFTING AND DESIGN	NAME: SECT. NO.: FILE NO.: DATE:	GRADE	PROBLEM 1C

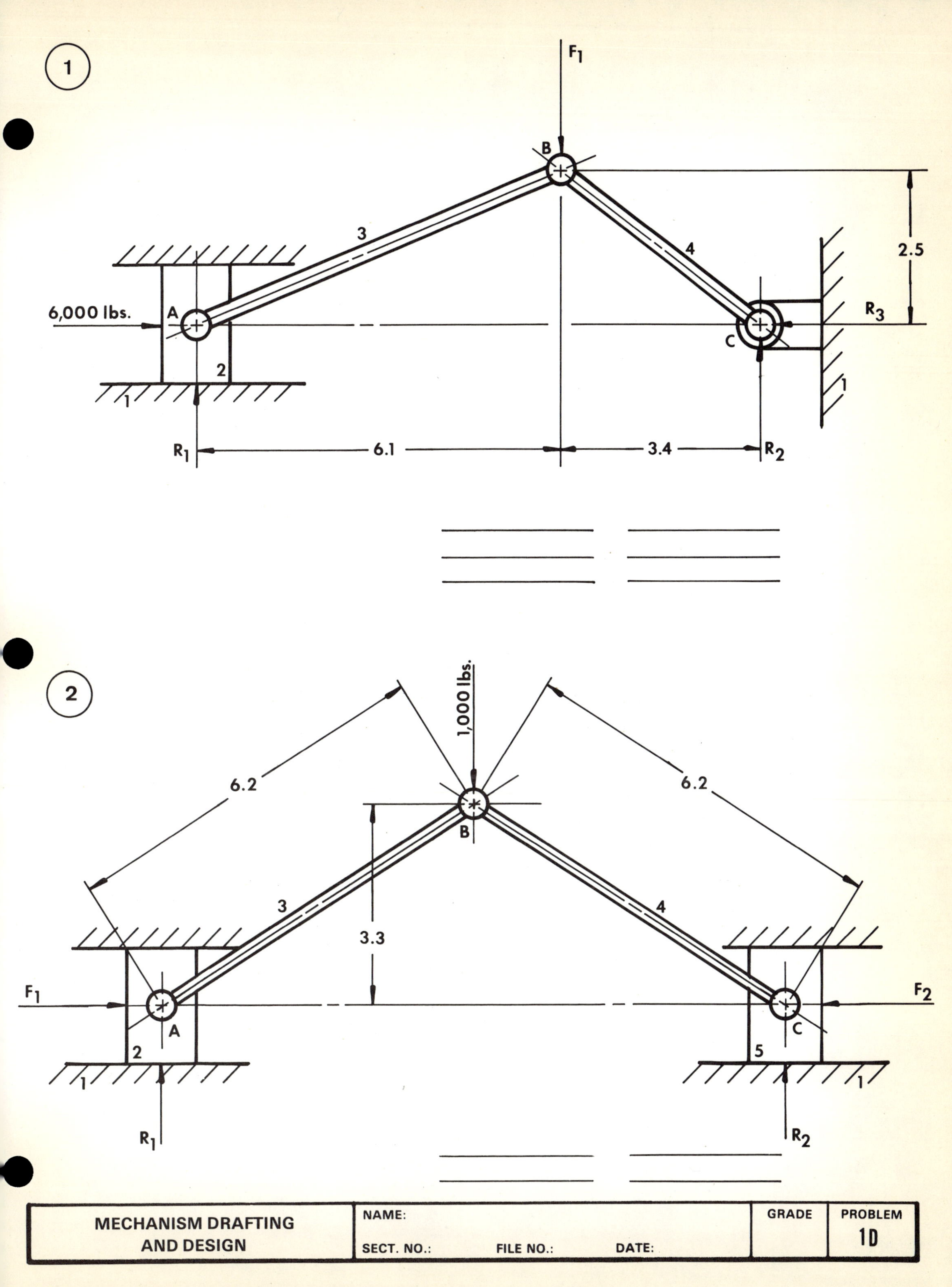

MECHANISM DRAFTING AND DESIGN	NAME: SECT. NO.: FILE NO.: DATE:	GRADE	PROBLEM 1D

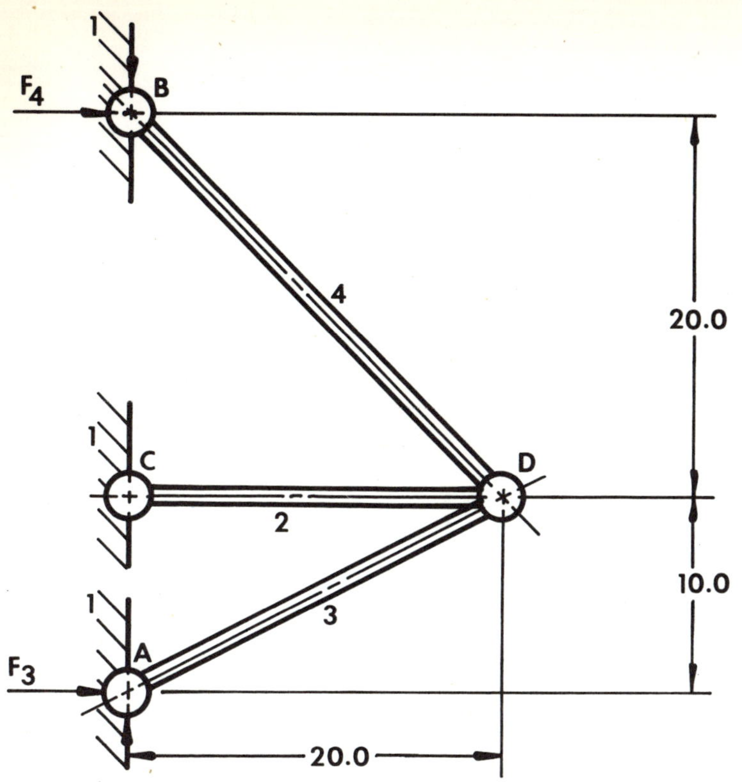

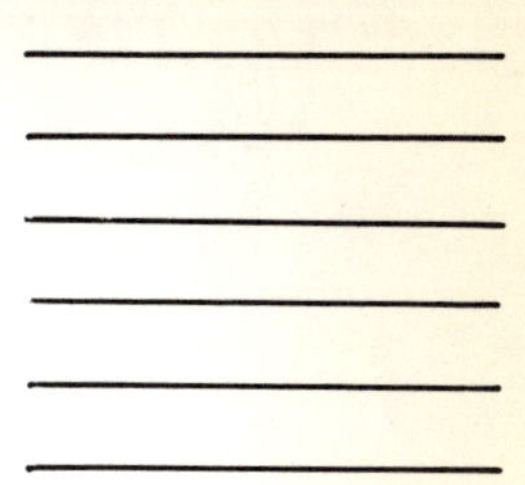

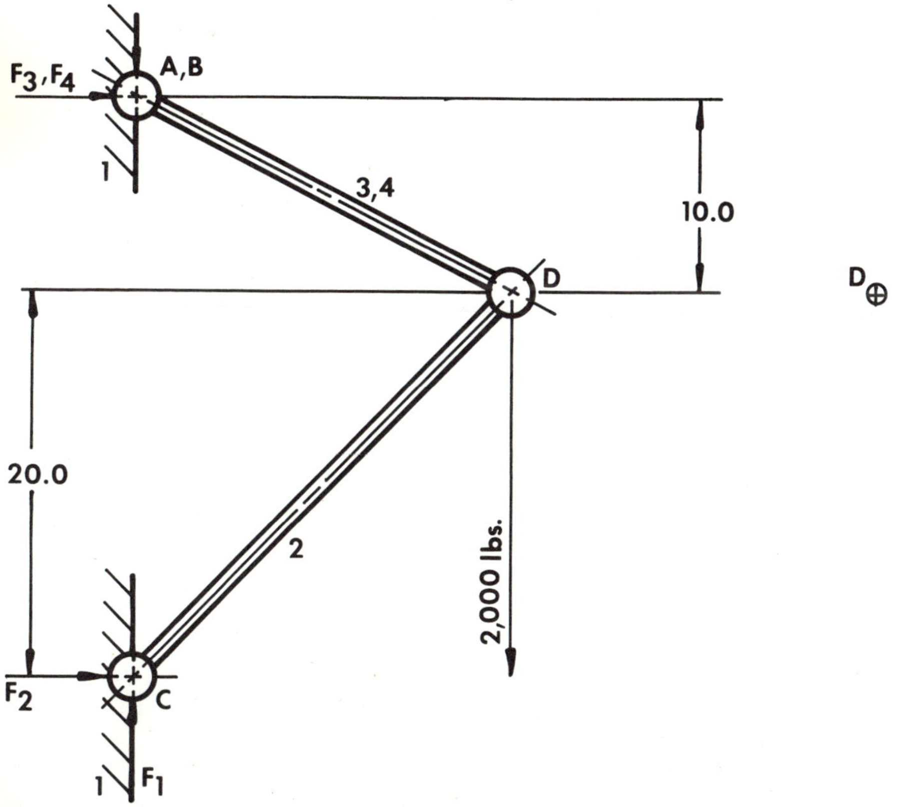

1
F4
B
4
20.0
1
C
D
2
⊕D
10.0
1
3
A
F3
20.0

F3,F4
A,B
1
3,4
10.0
D
D⊕
20.0
2
2,000 lbs.
F2
C
1
F1

MECHANISM DRAFTING AND DESIGN	NAME: SECT. NO.: FILE NO.: DATE:	GRADE	PROBLEM 1E

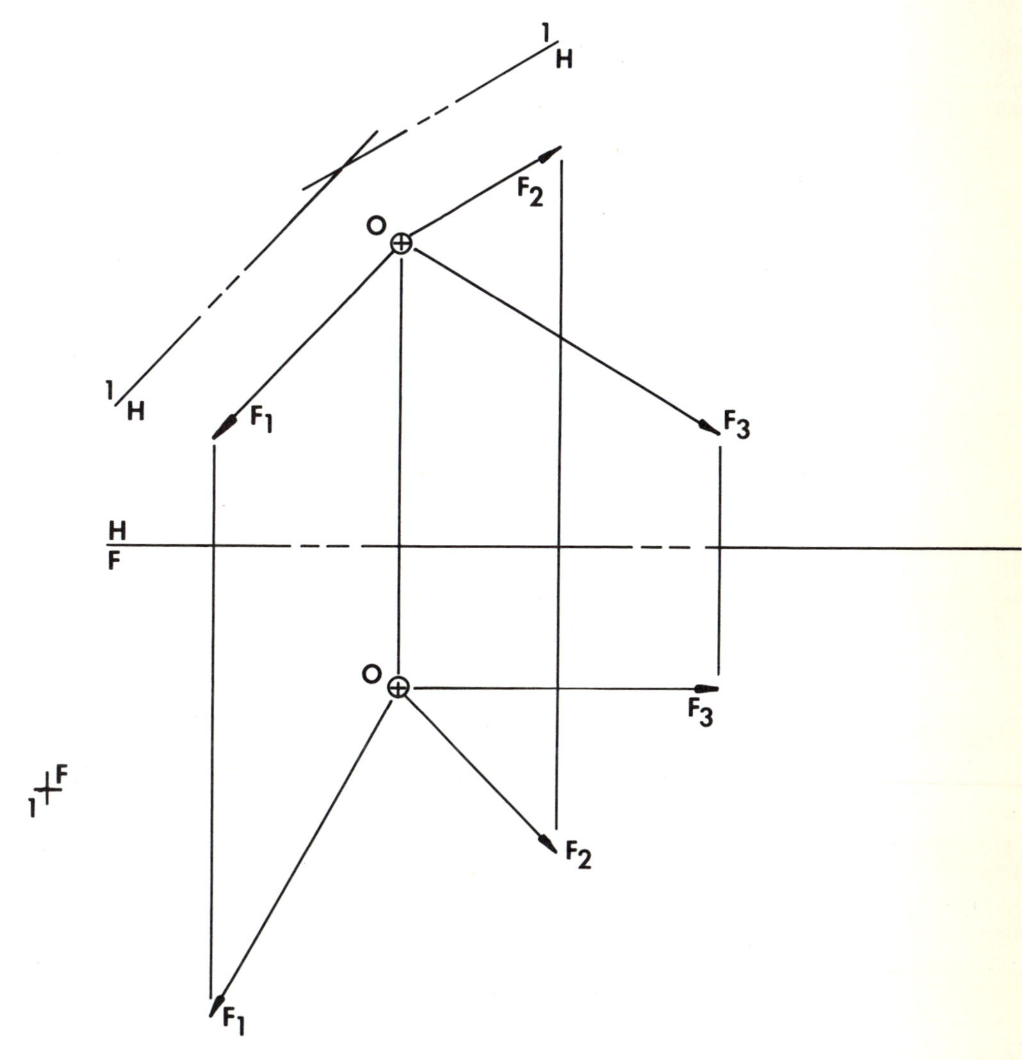

MECHANISM DRAFTING AND DESIGN	NAME: SECT. NO.: FILE NO.: DATE:	GRADE	PROBLEM 1F

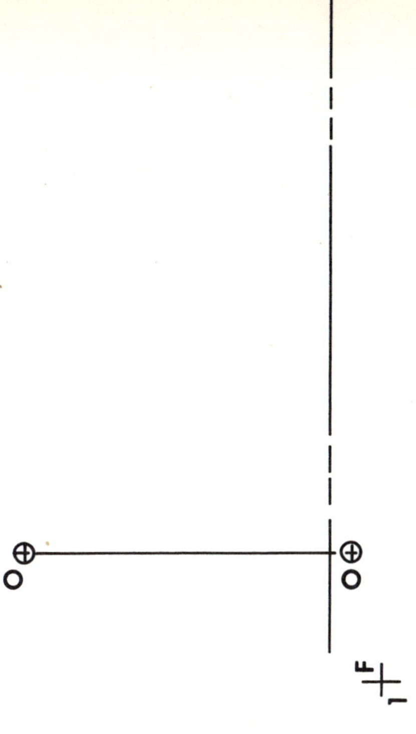

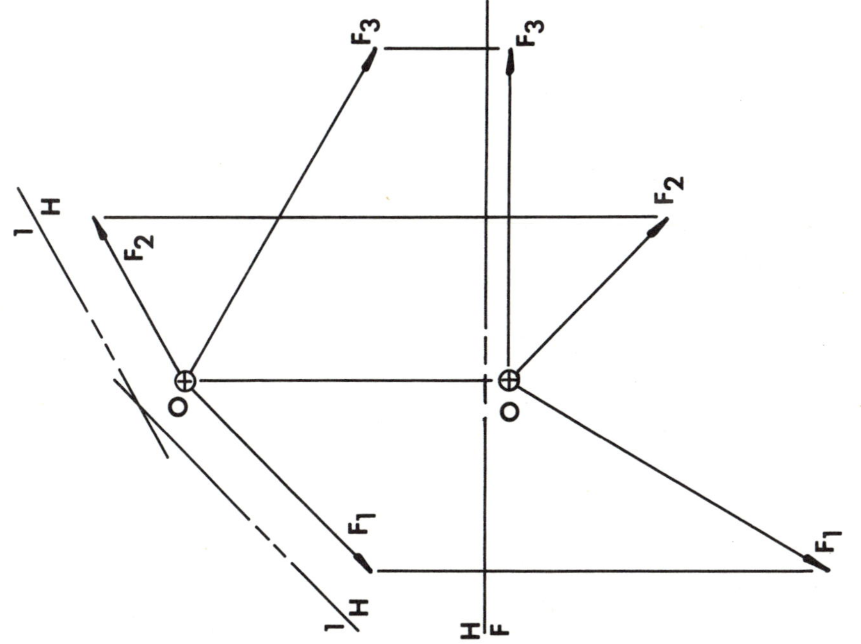

MECHANISM DRAFTING AND DESIGN	NAME:	GRADE	PROBLEM
	SECT. NO.: FILE NO.: DATE:		1G

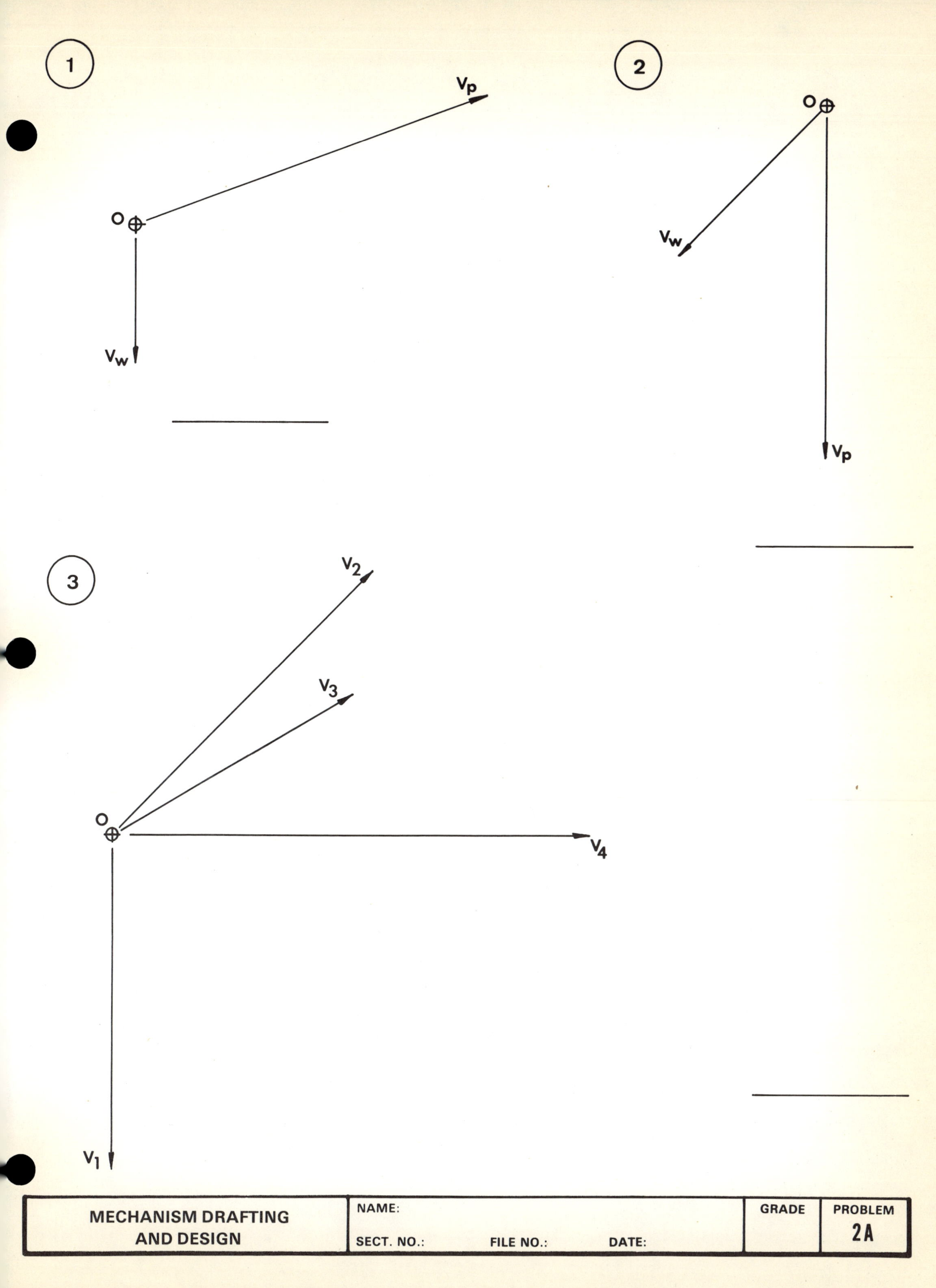
1
V_p
O
V_w
2
O
V_w
V_p
3
V_2
V_3
O
V_4
V_1
MECHANISM DRAFTING AND DESIGN
NAME:
SECT. NO.:
FILE NO.:
DATE:
GRADE
PROBLEM
2A

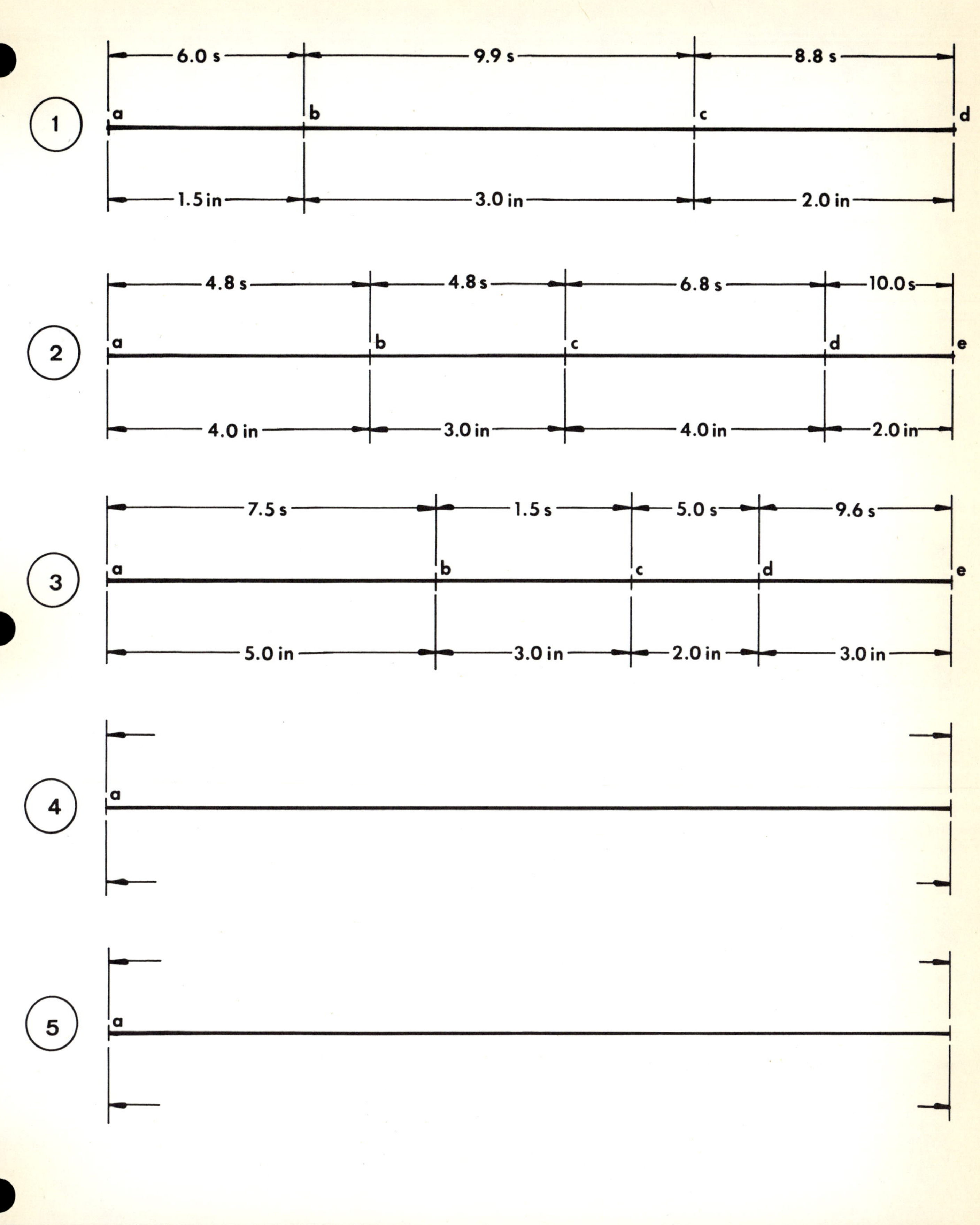

MECHANISM DRAFTING AND DESIGN	NAME: SECT. NO.: FILE NO.: DATE:	GRADE	PROBLEM 2B

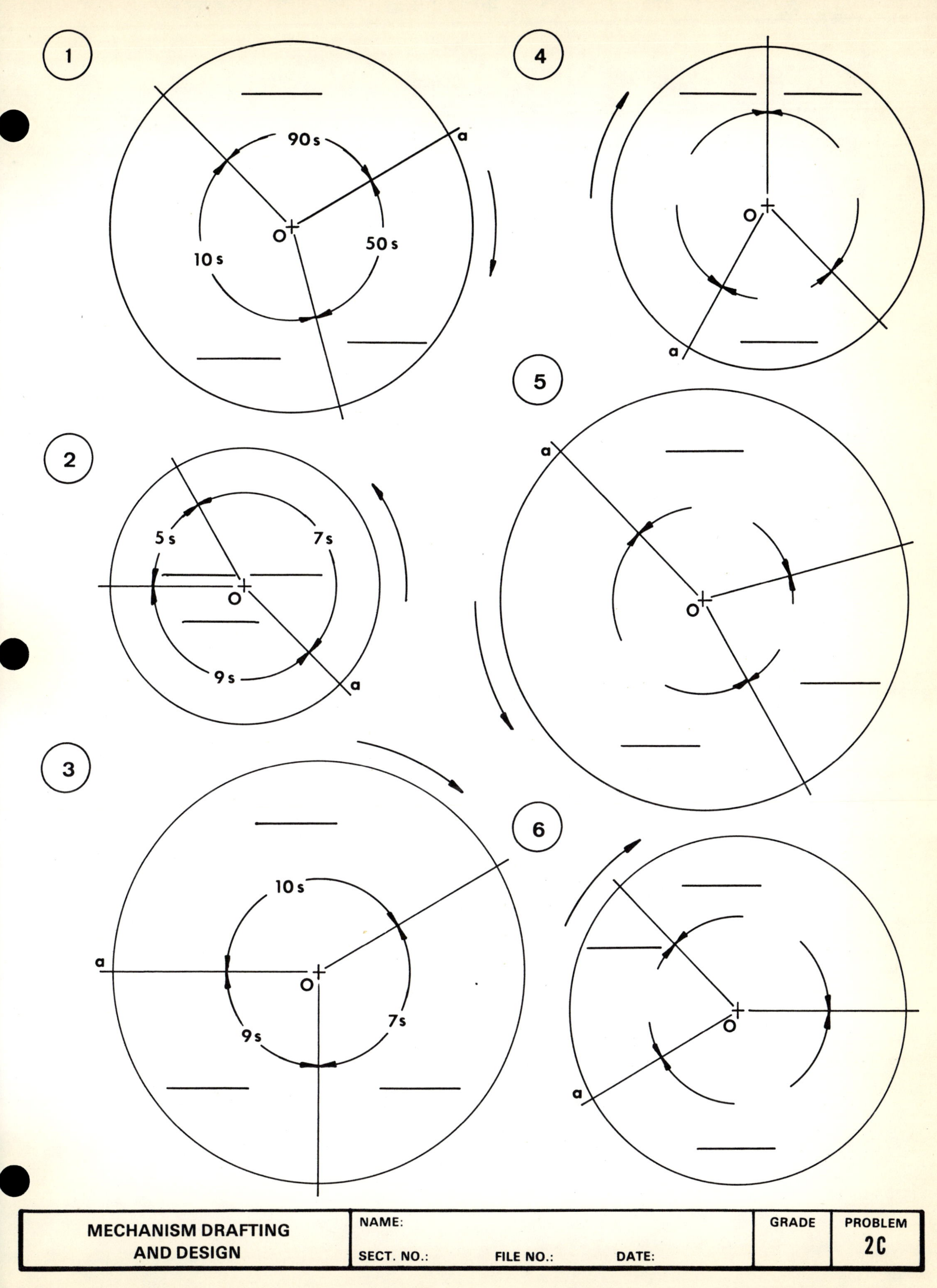

MECHANISM DRAFTING AND DESIGN	NAME: SECT. NO.: FILE NO.: DATE:	GRADE	PROBLEM 2C

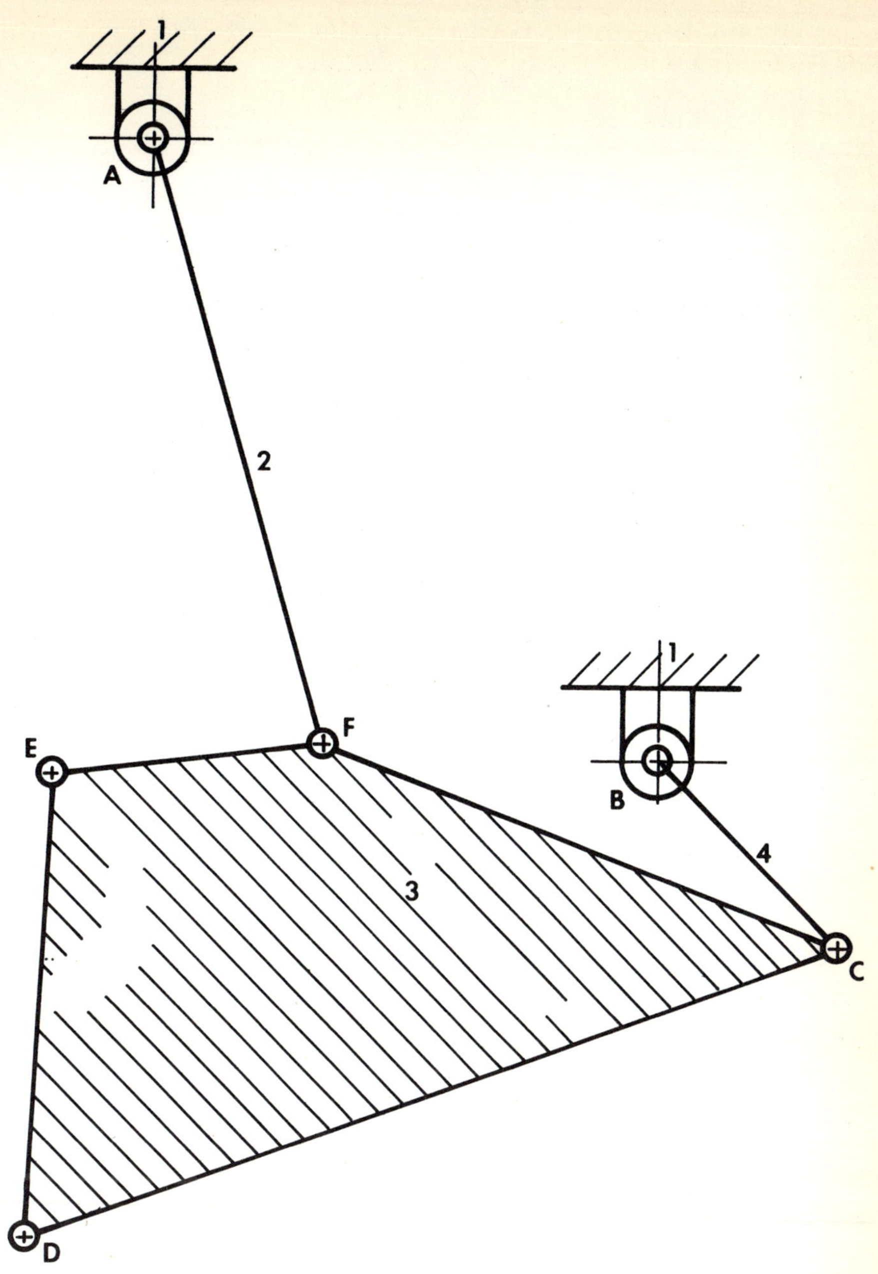

MECHANISM DRAFTING AND DESIGN	NAME: SECT. NO.: FILE NO.: DATE:	GRADE	PROBLEM 3A

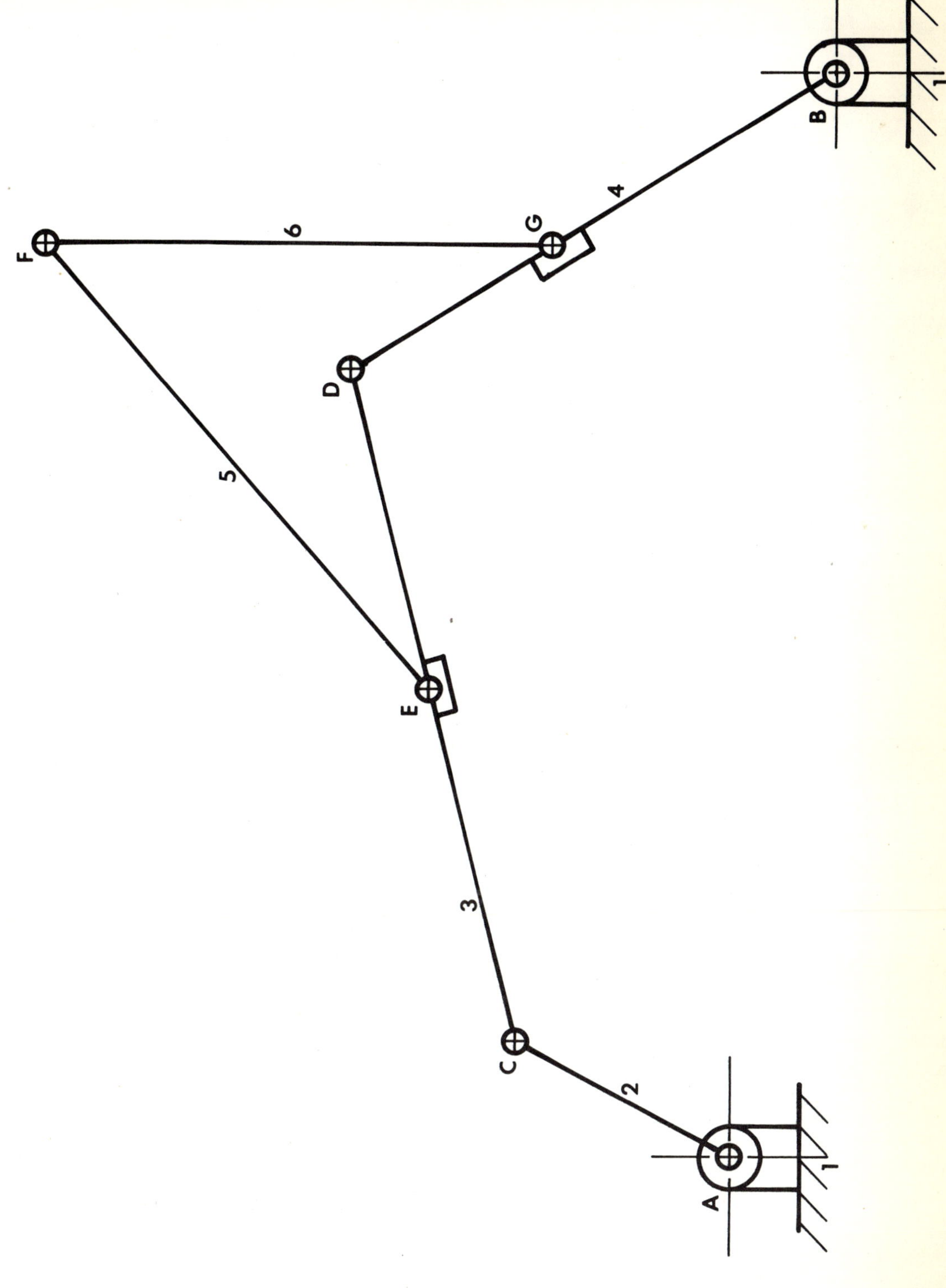

MECHANISM DRAFTING AND DESIGN	NAME: SECT. NO.: FILE NO.: DATE:	GRADE	PROBLEM 3B

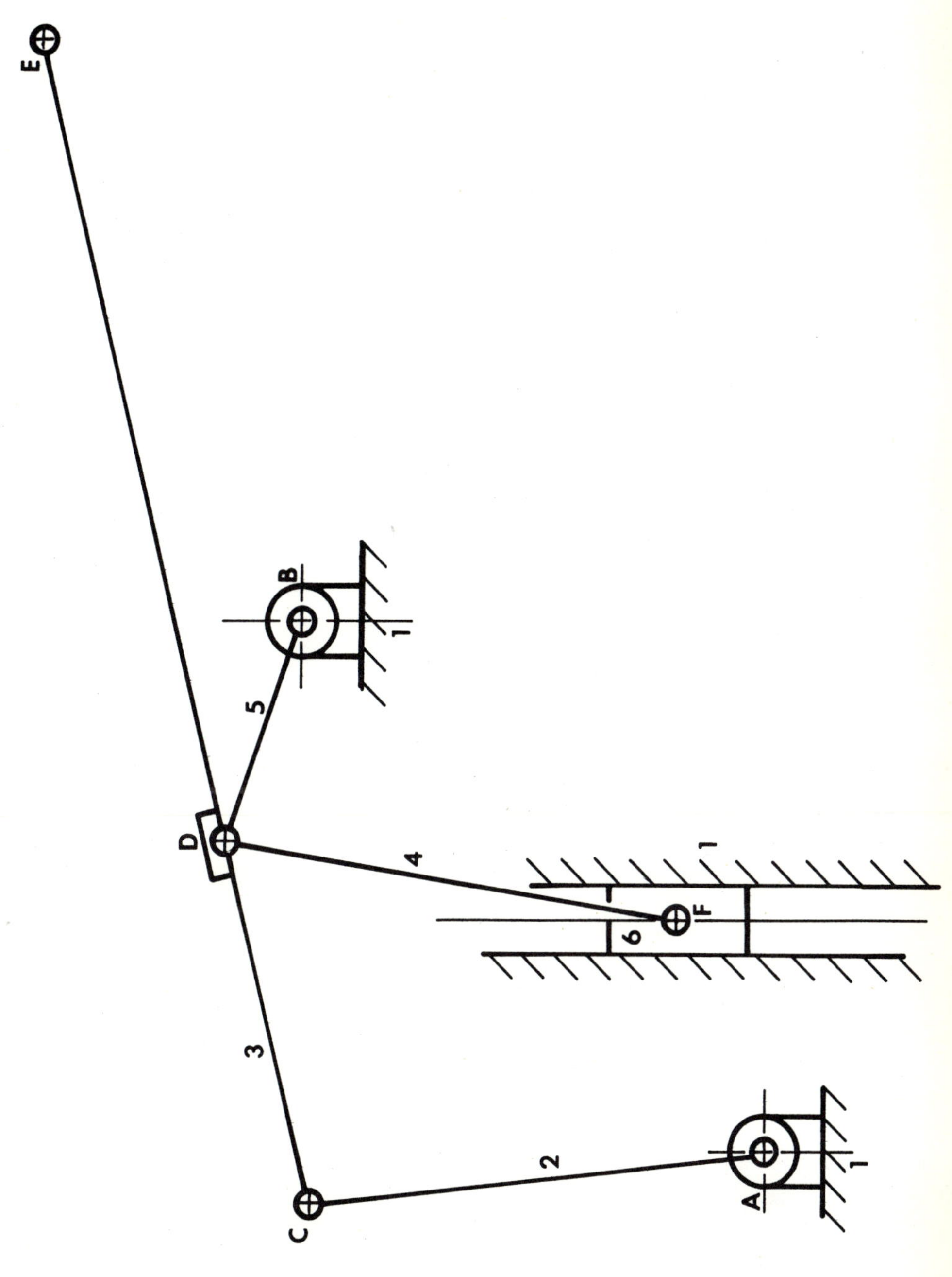

MECHANISM DRAFTING AND DESIGN	NAME: SECT. NO.: FILE NO.: DATE:	GRADE	PROBLEM 3C

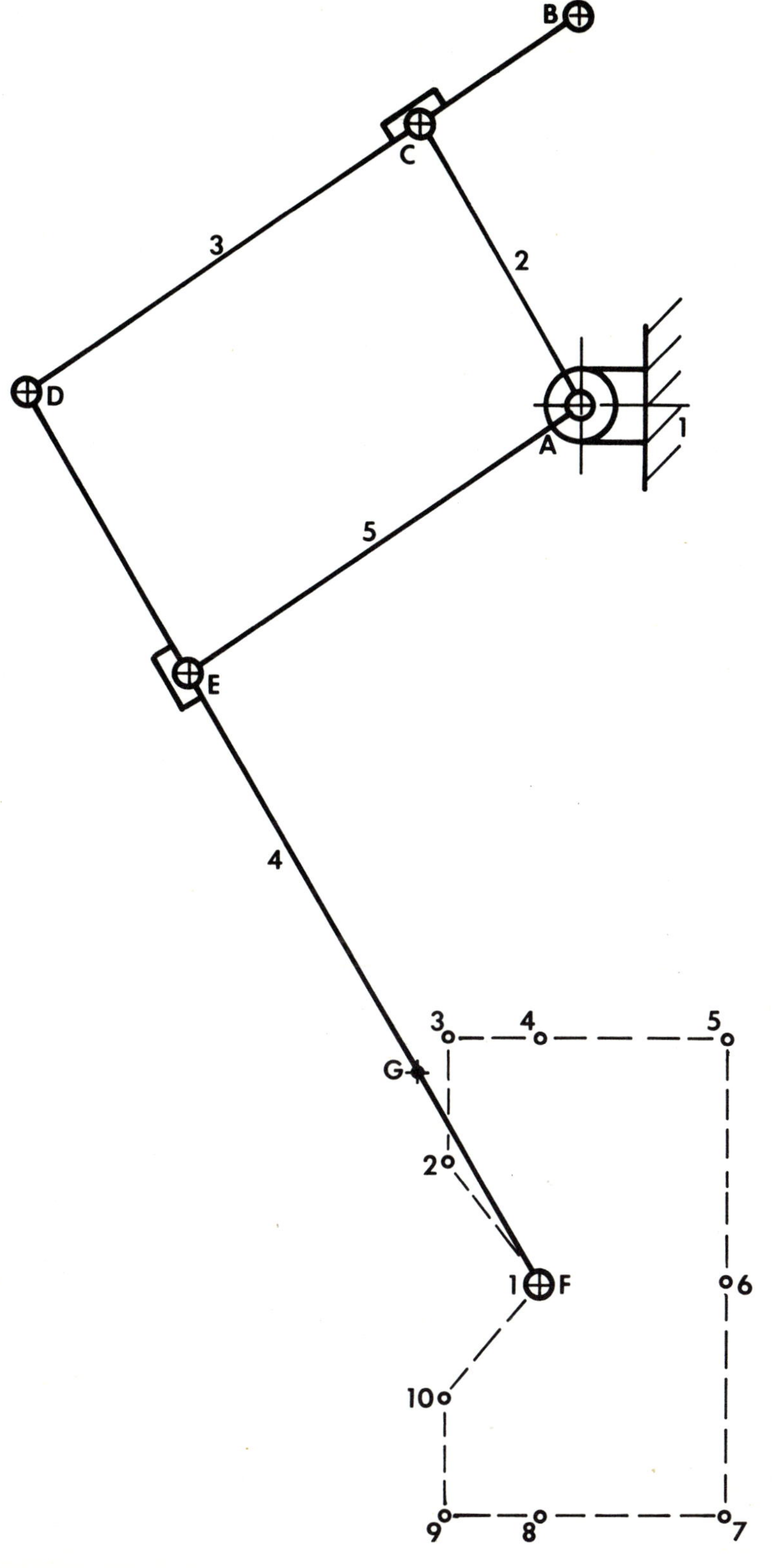

MECHANISM DRAFTING AND DESIGN	NAME: SECT. NO.: FILE NO.: DATE:	GRADE	PROBLEM 3D

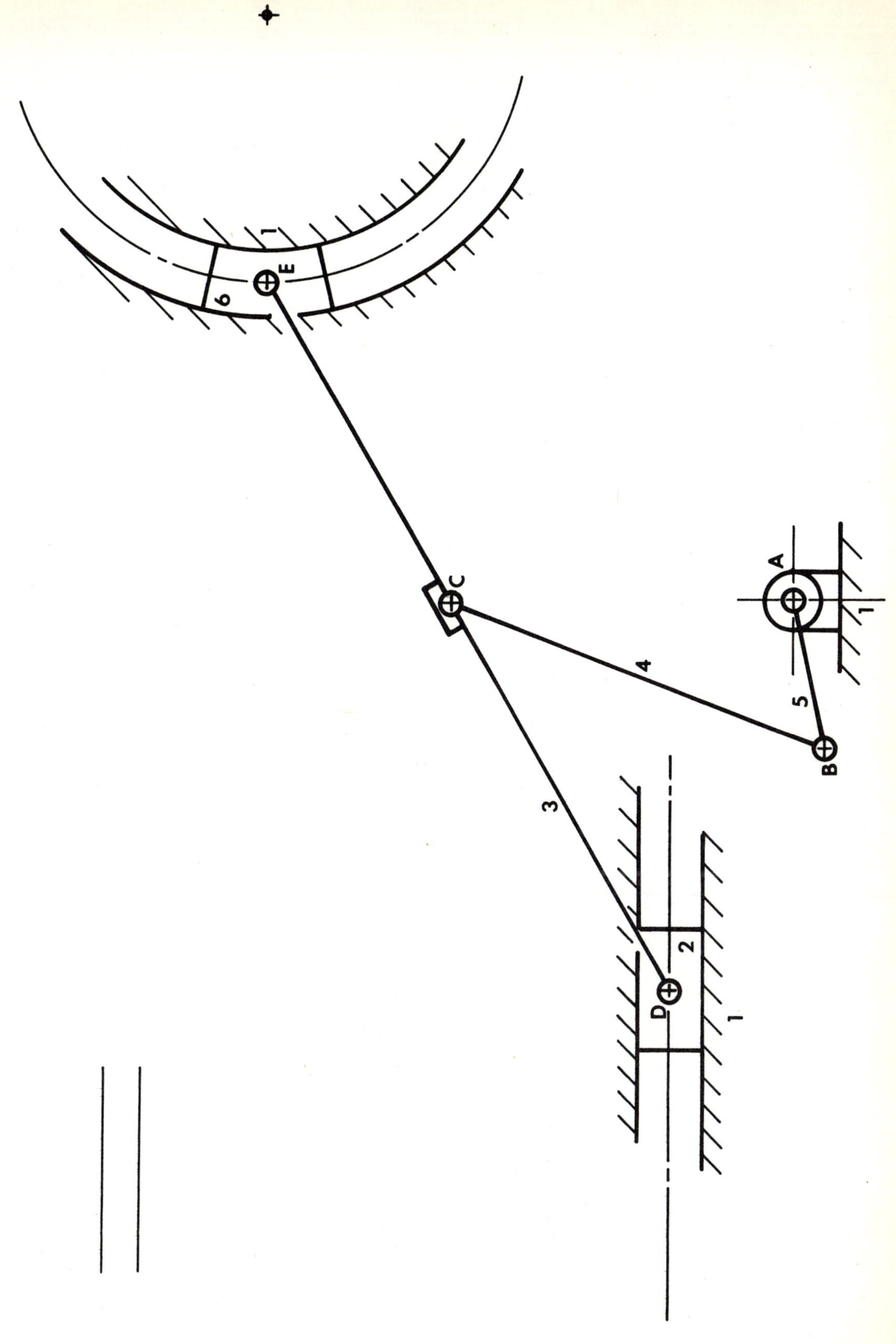

MECHANISM DRAFTING AND DESIGN	NAME: SECT. NO.: FILE NO.: DATE:	GRADE	PROBLEM 3E

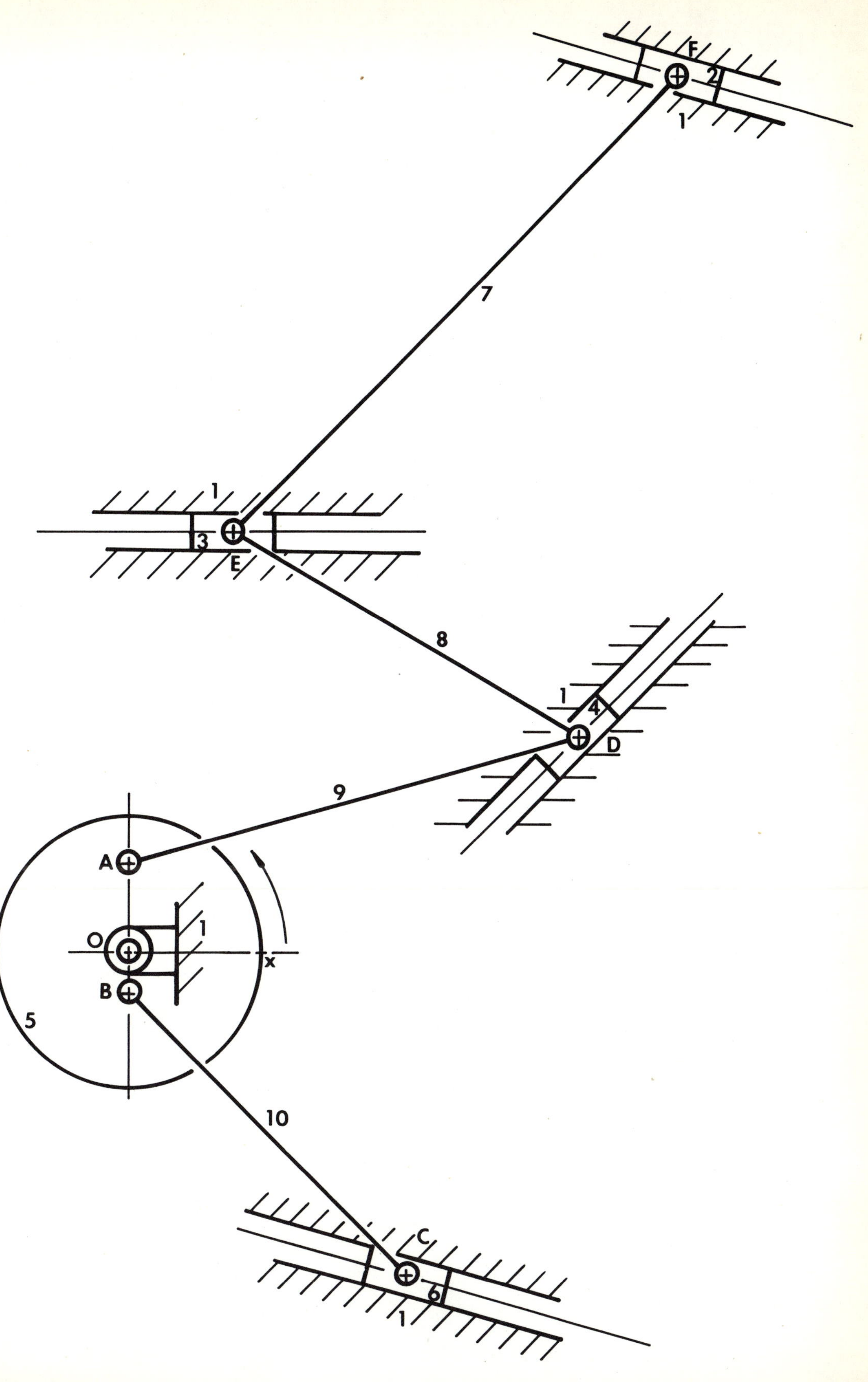

MECHANISM DRAFTING AND DESIGN	NAME: SECT. NO.: FILE NO.: DATE:	GRADE	PROBLEM 3F

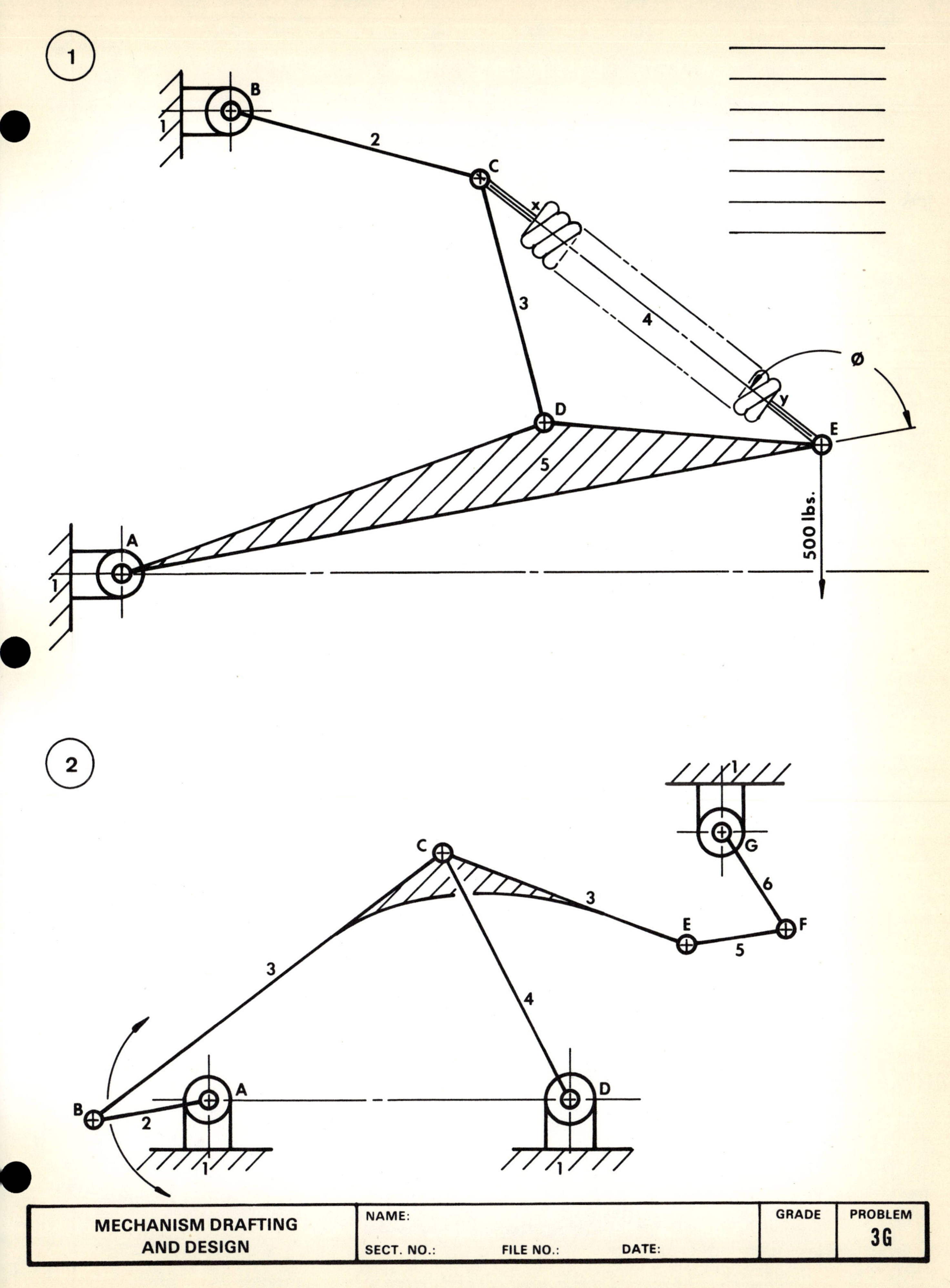

MECHANISM DRAFTING AND DESIGN	NAME: SECT. NO.: FILE NO.: DATE:	GRADE	PROBLEM 3G

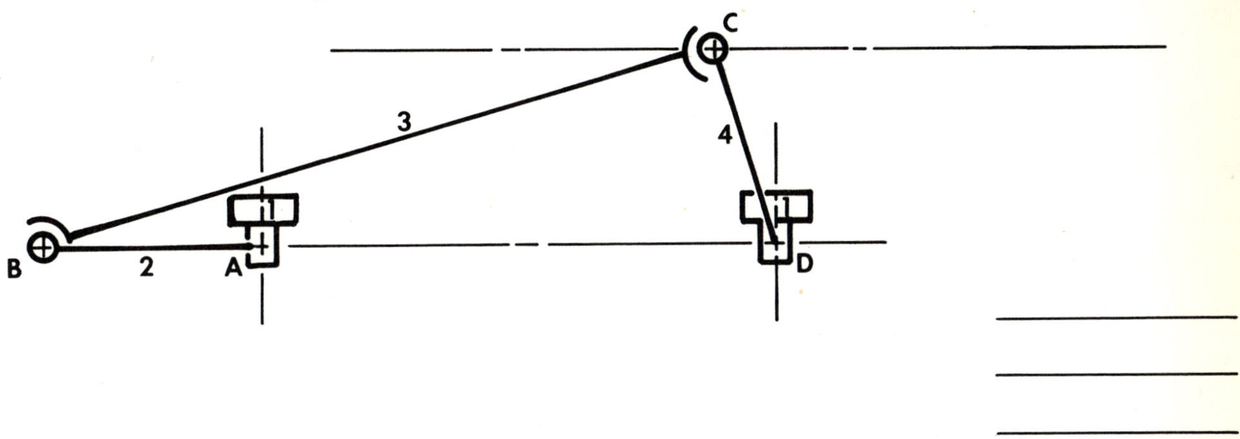
C
3
4
1
1
B
2
A
D

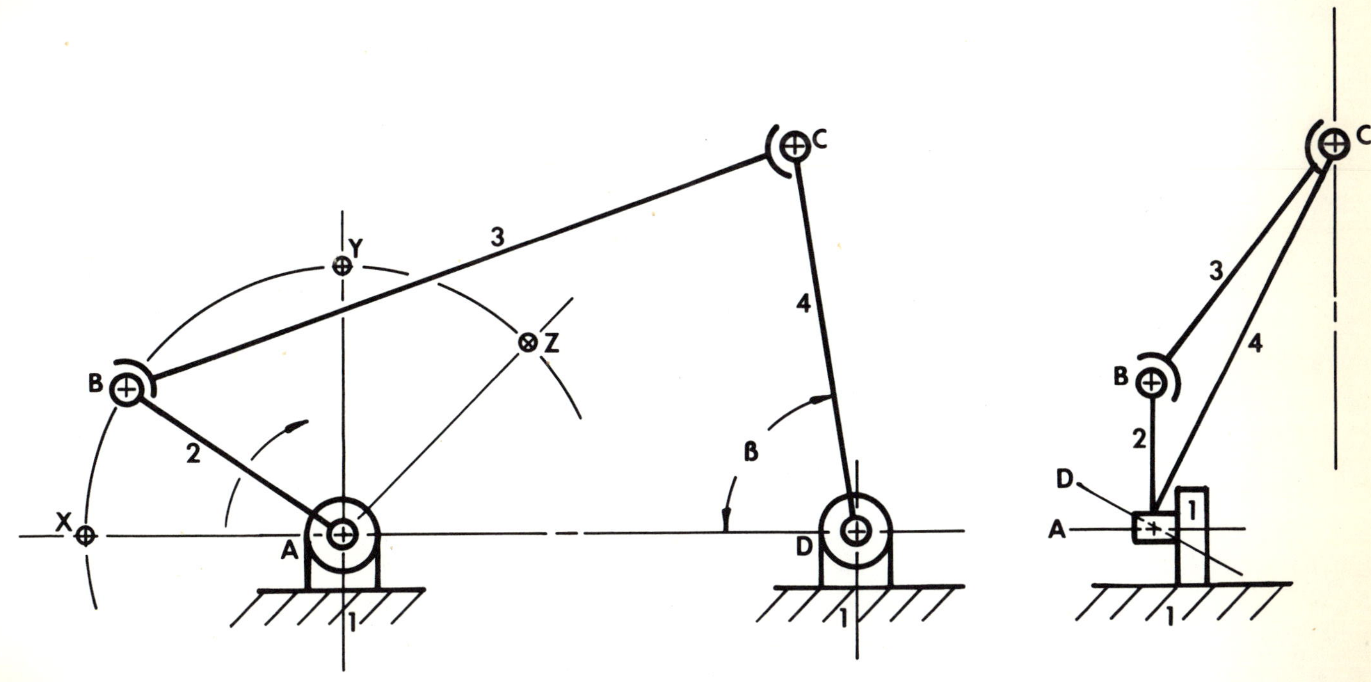
C
Y
3
4
Z
B
2
B
X
A
D
1
1
C
3
4
B
2
D
A
1
1

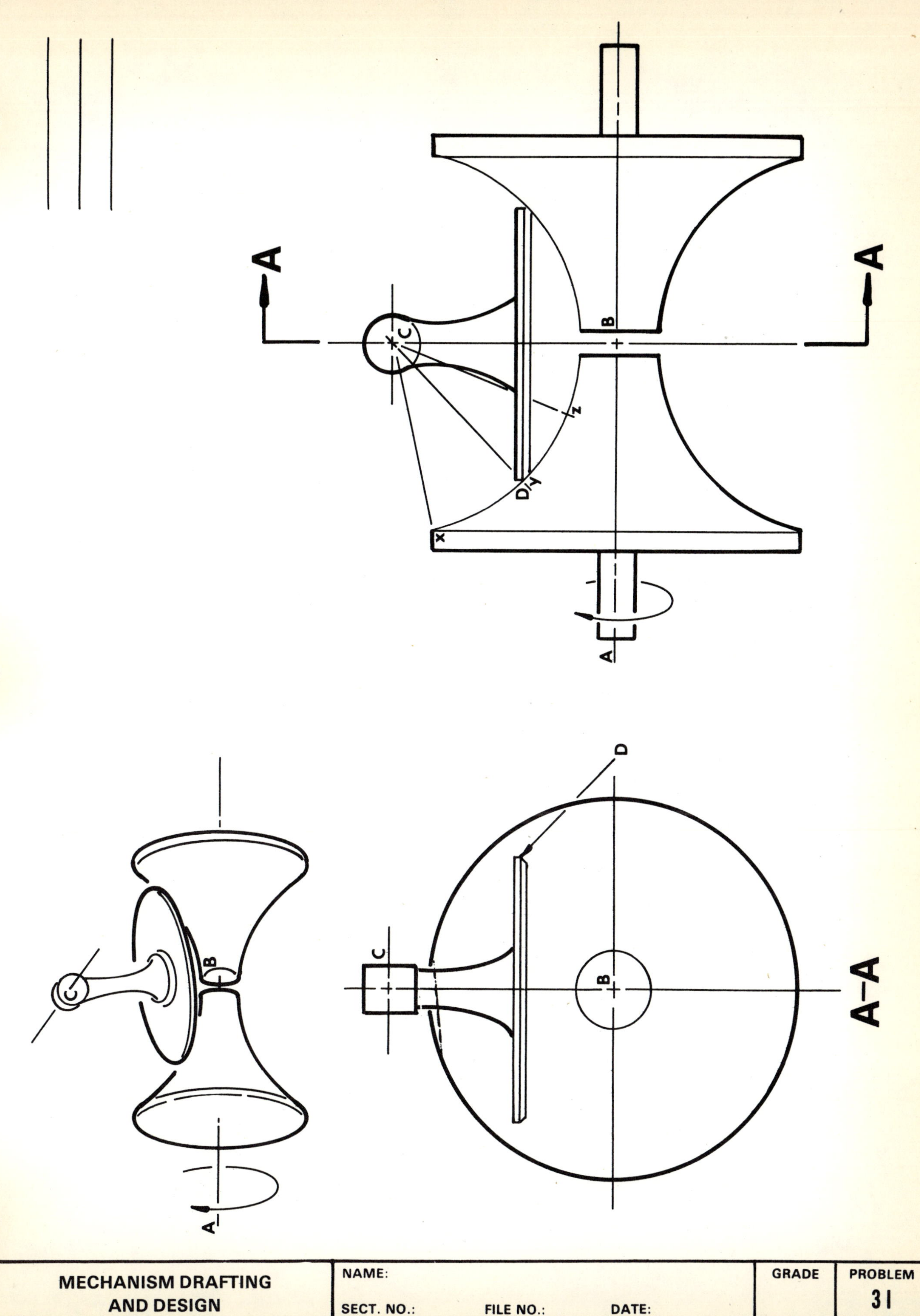

MECHANISM DRAFTING AND DESIGN	NAME: SECT. NO.: FILE NO.: DATE:	GRADE	PROBLEM 31

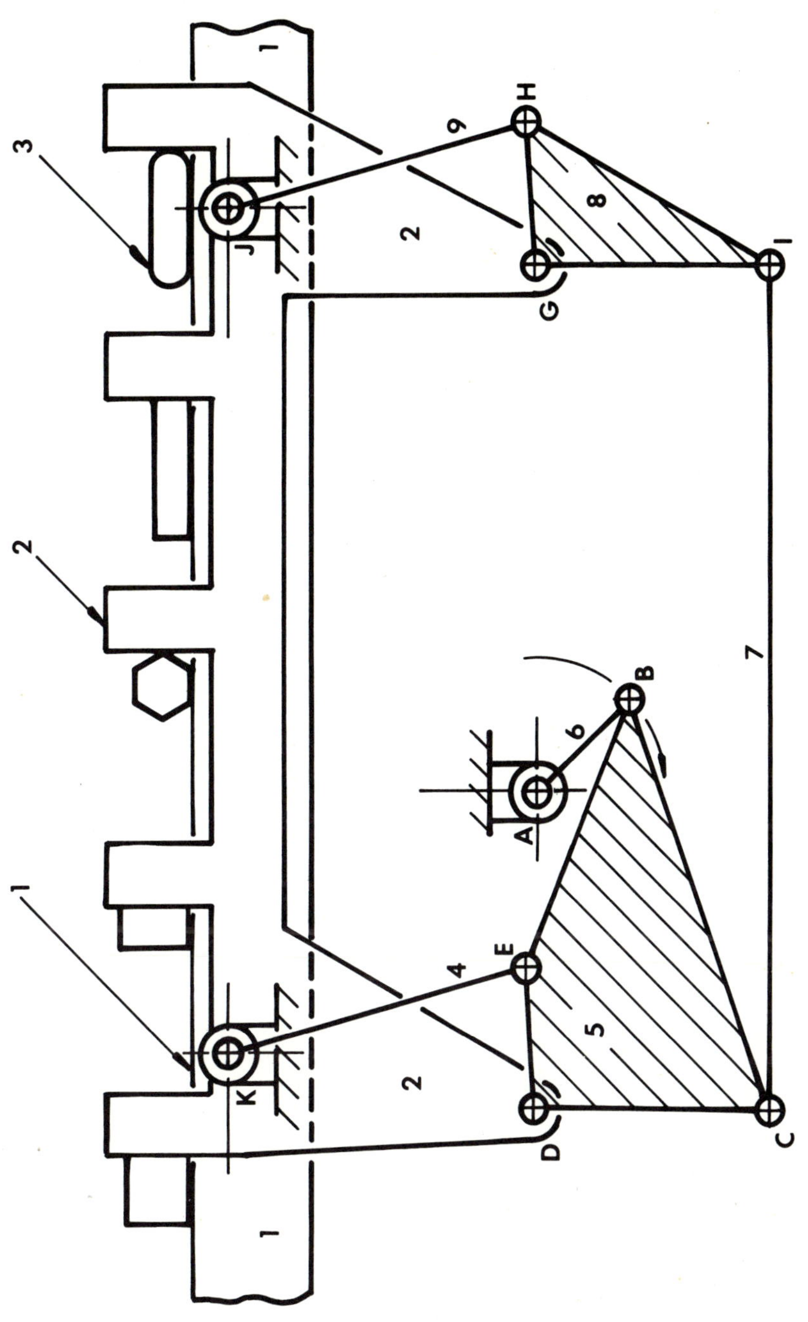

MECHANISM DRAFTING AND DESIGN	NAME:	GRADE	PROBLEM 3J
	SECT. NO.: FILE NO.: DATE:		

1

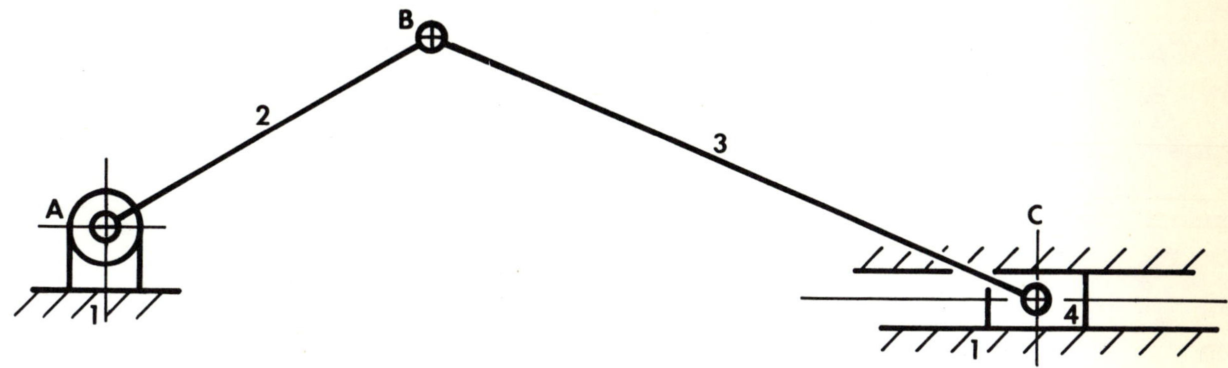

2

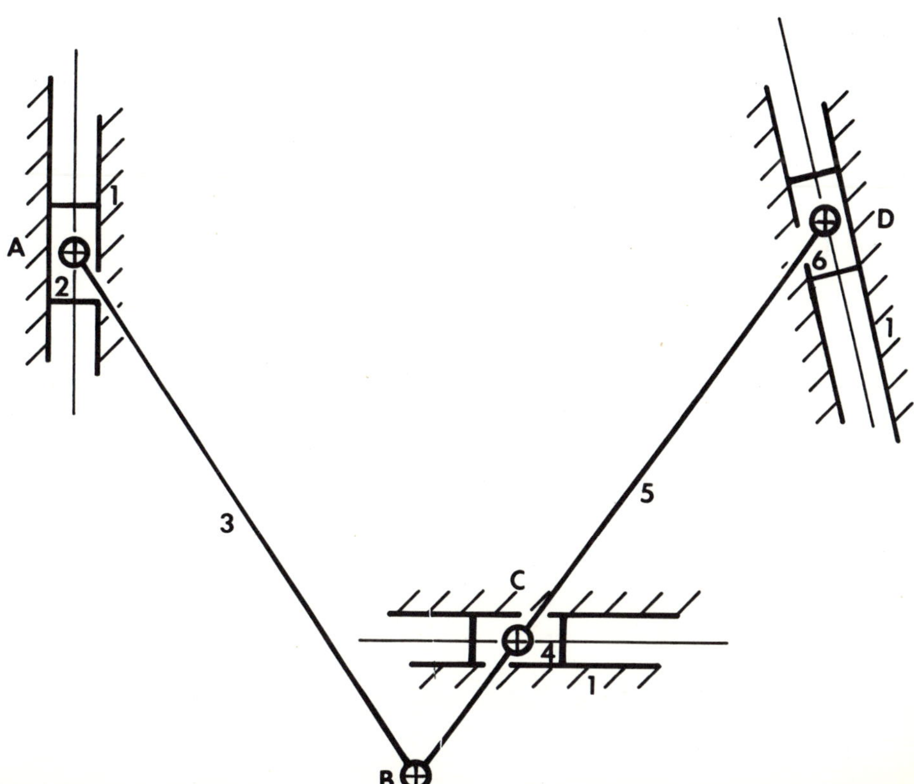

MECHANISM DRAFTING AND DESIGN	NAME: SECT. NO.: FILE NO.: DATE:	GRADE	PROBLEM 4A

1

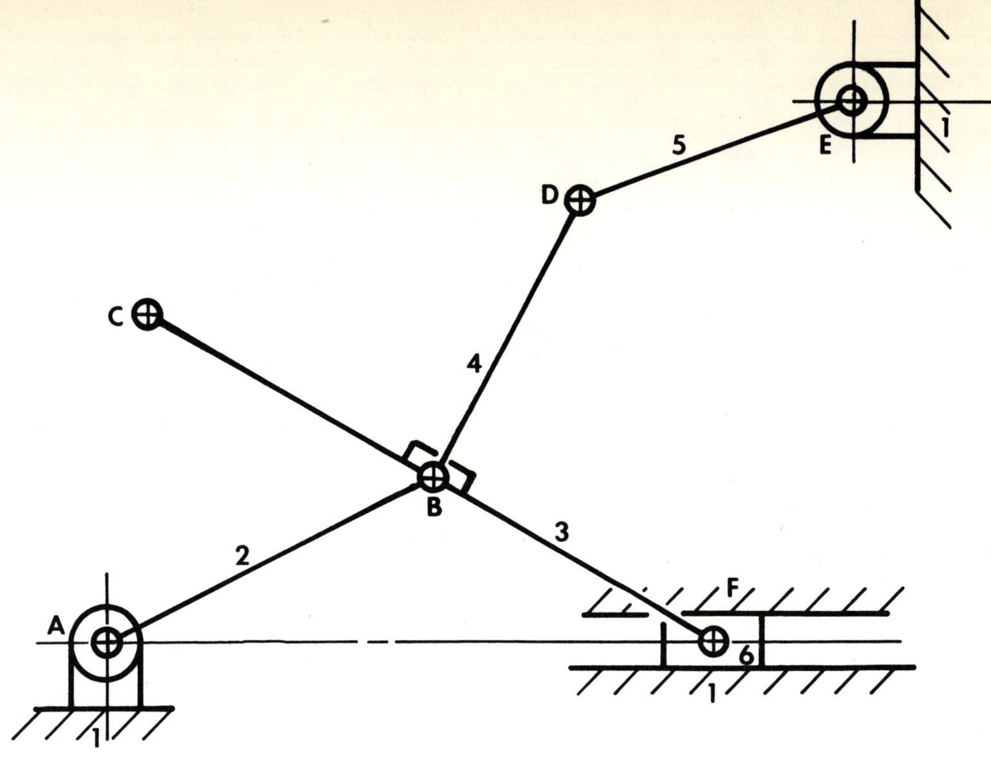

2

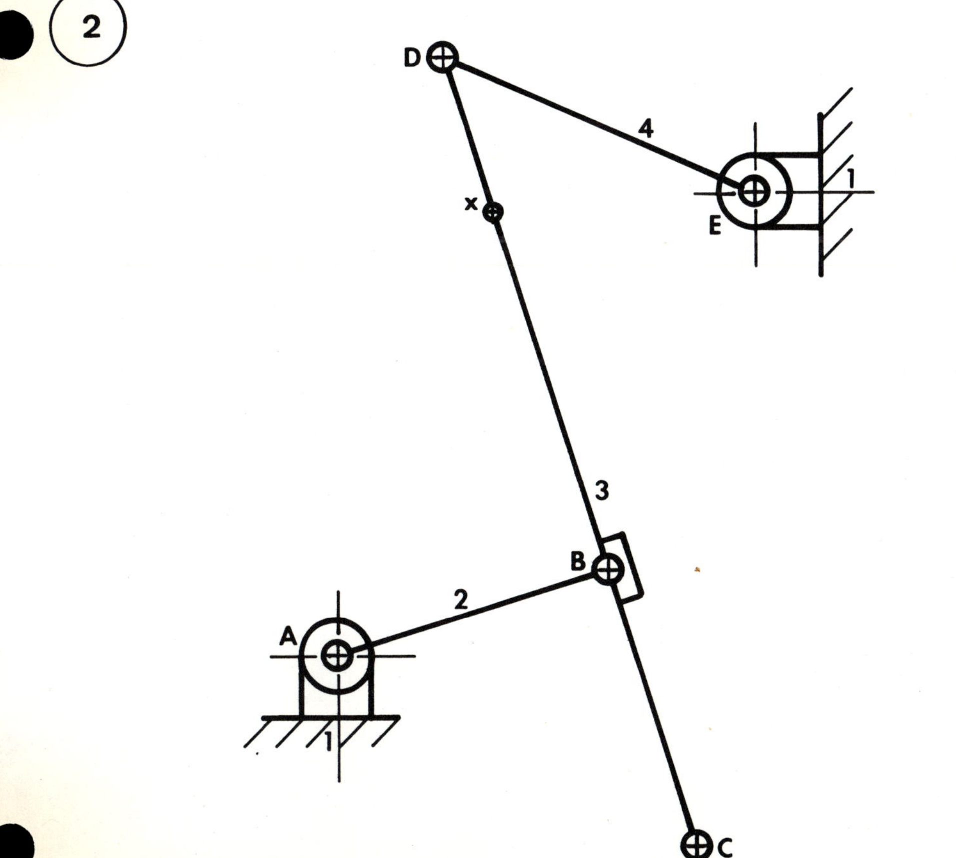

MECHANISM DRAFTING AND DESIGN	NAME: SECT. NO.: FILE NO.: DATE:	GRADE	PROBLEM 4B

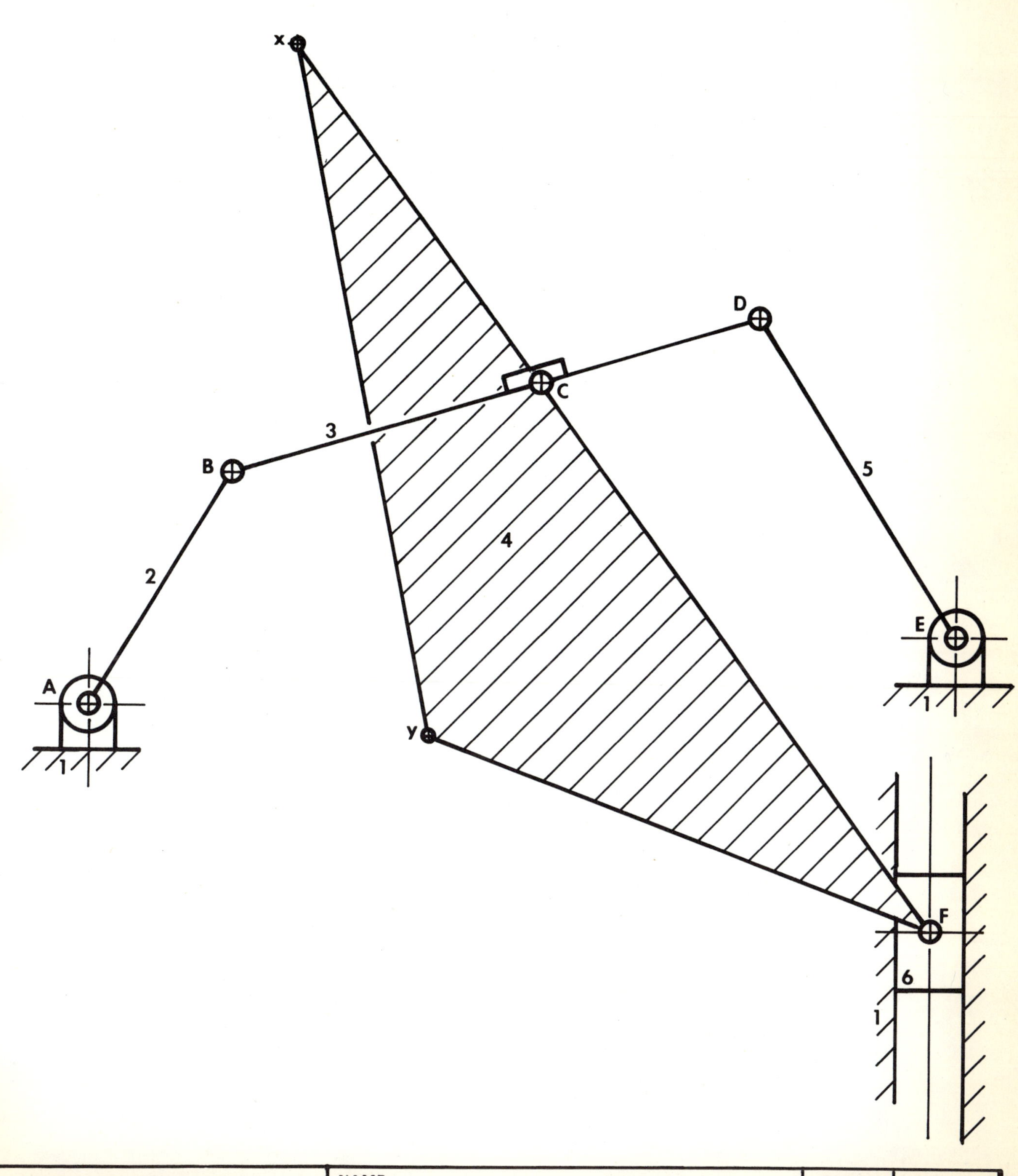

MECHANISM DRAFTING AND DESIGN	NAME: SECT. NO.: FILE NO.: DATE:	GRADE	PROBLEM 4C

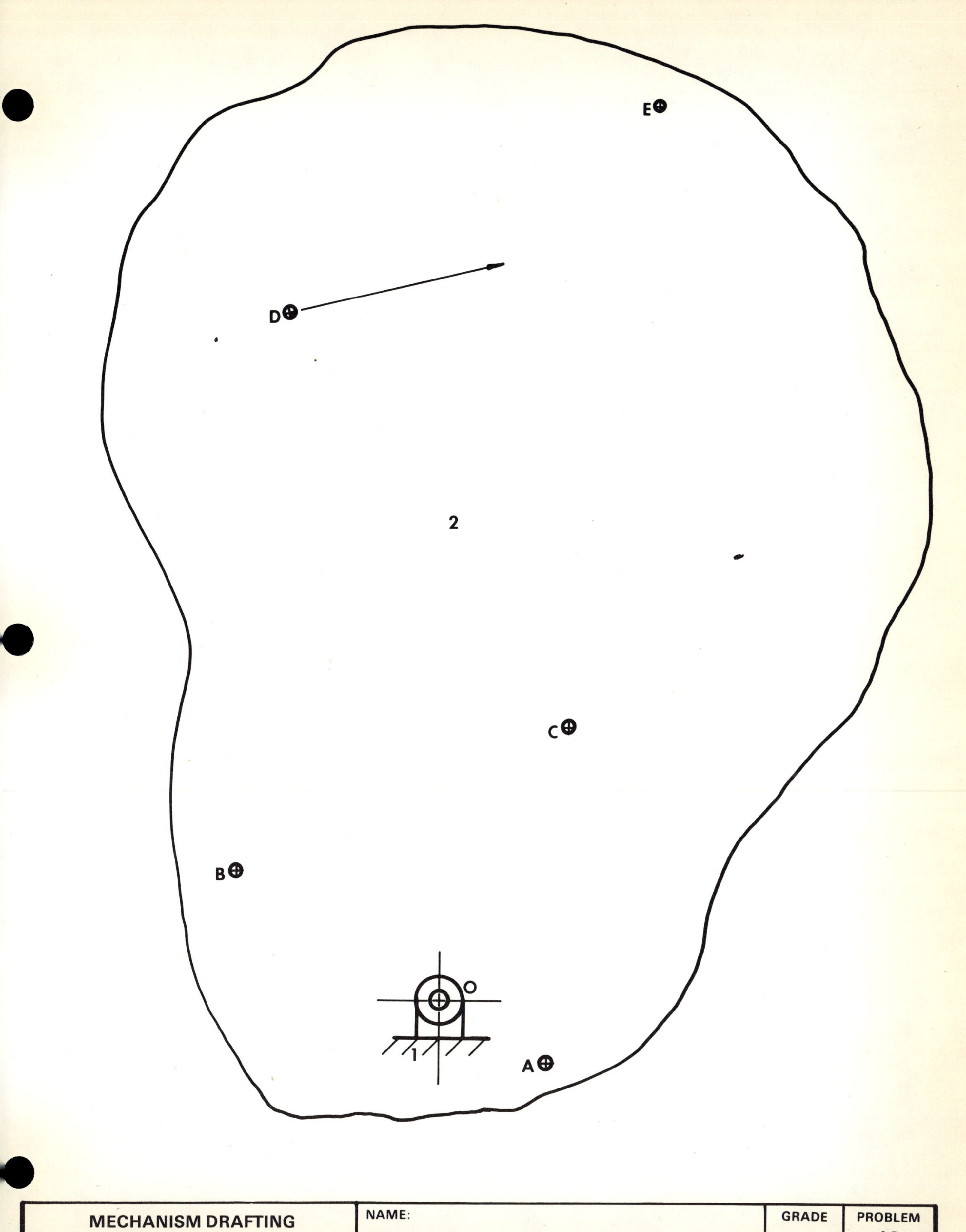

MECHANISM DRAFTING AND DESIGN	NAME: SECT. NO.: FILE NO.: DATE:	GRADE	PROBLEM 4D

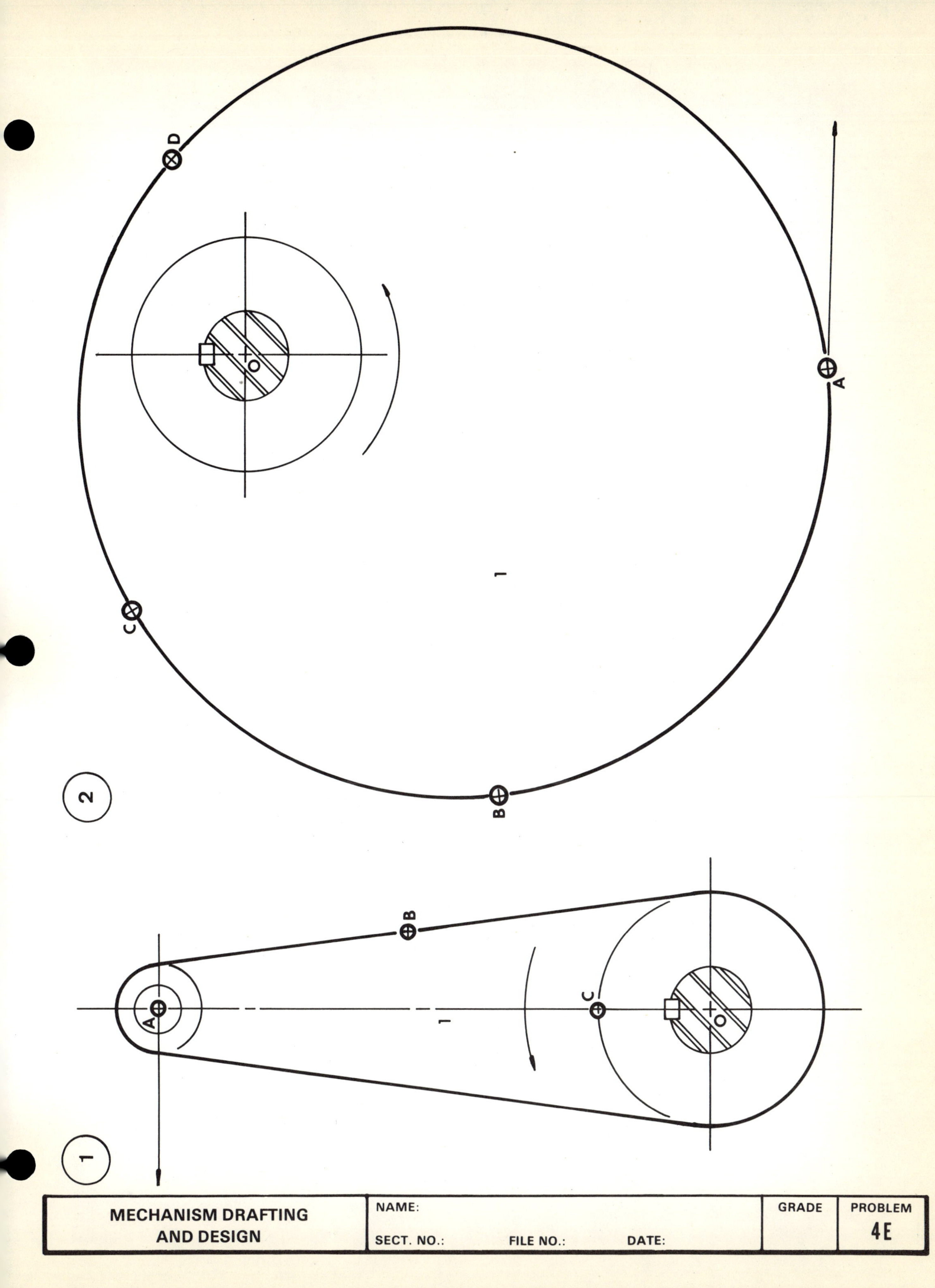

D
A
C
B
O
1
2
A
B
C
O
1
1
MECHANISM DRAFTING
AND DESIGN
NAME:
SECT. NO.:
FILE NO.:
DATE:
GRADE
PROBLEM
4E

1

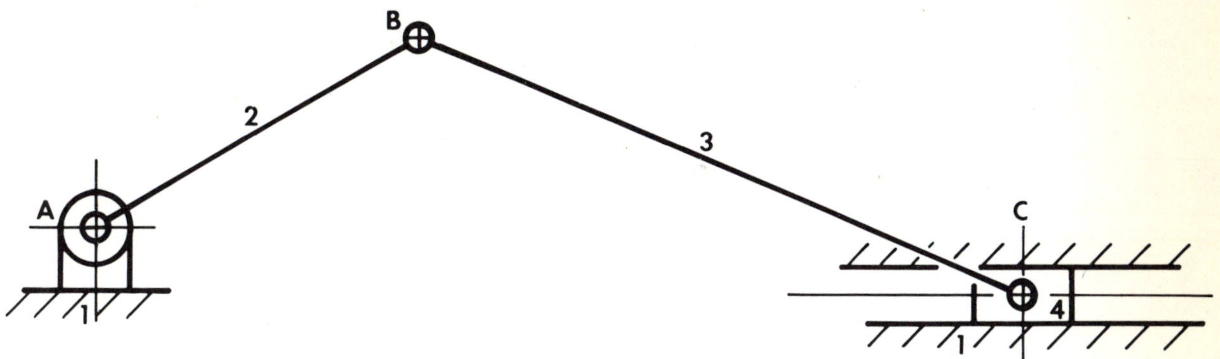

2

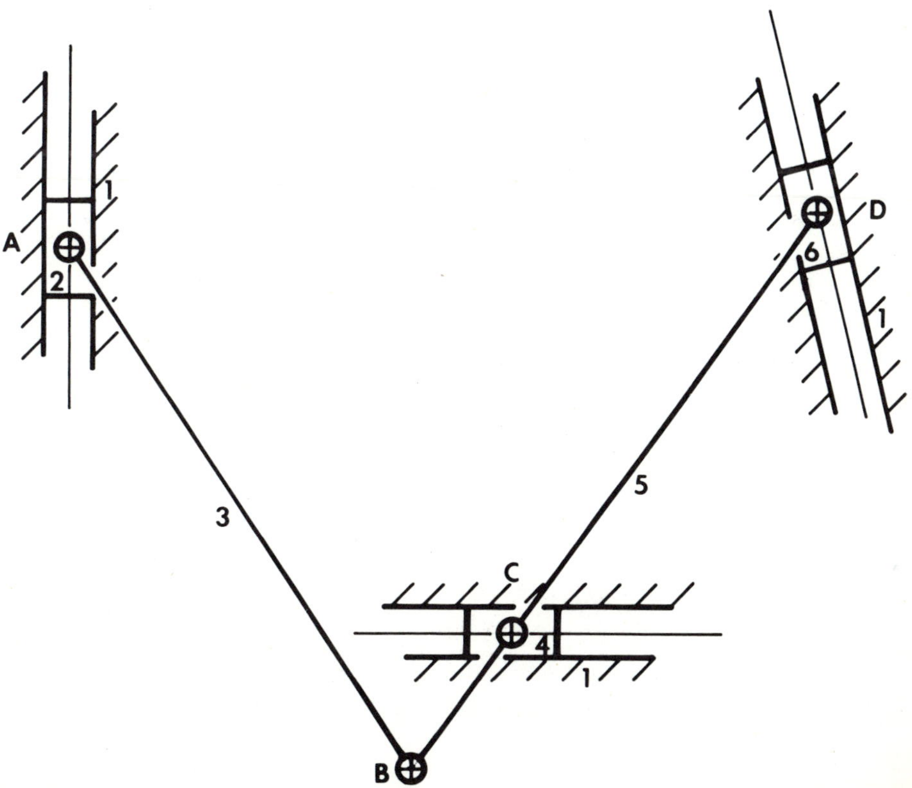

MECHANISM DRAFTING AND DESIGN	NAME: SECT. NO.: FILE NO.: DATE:	GRADE	PROBLEM 5A

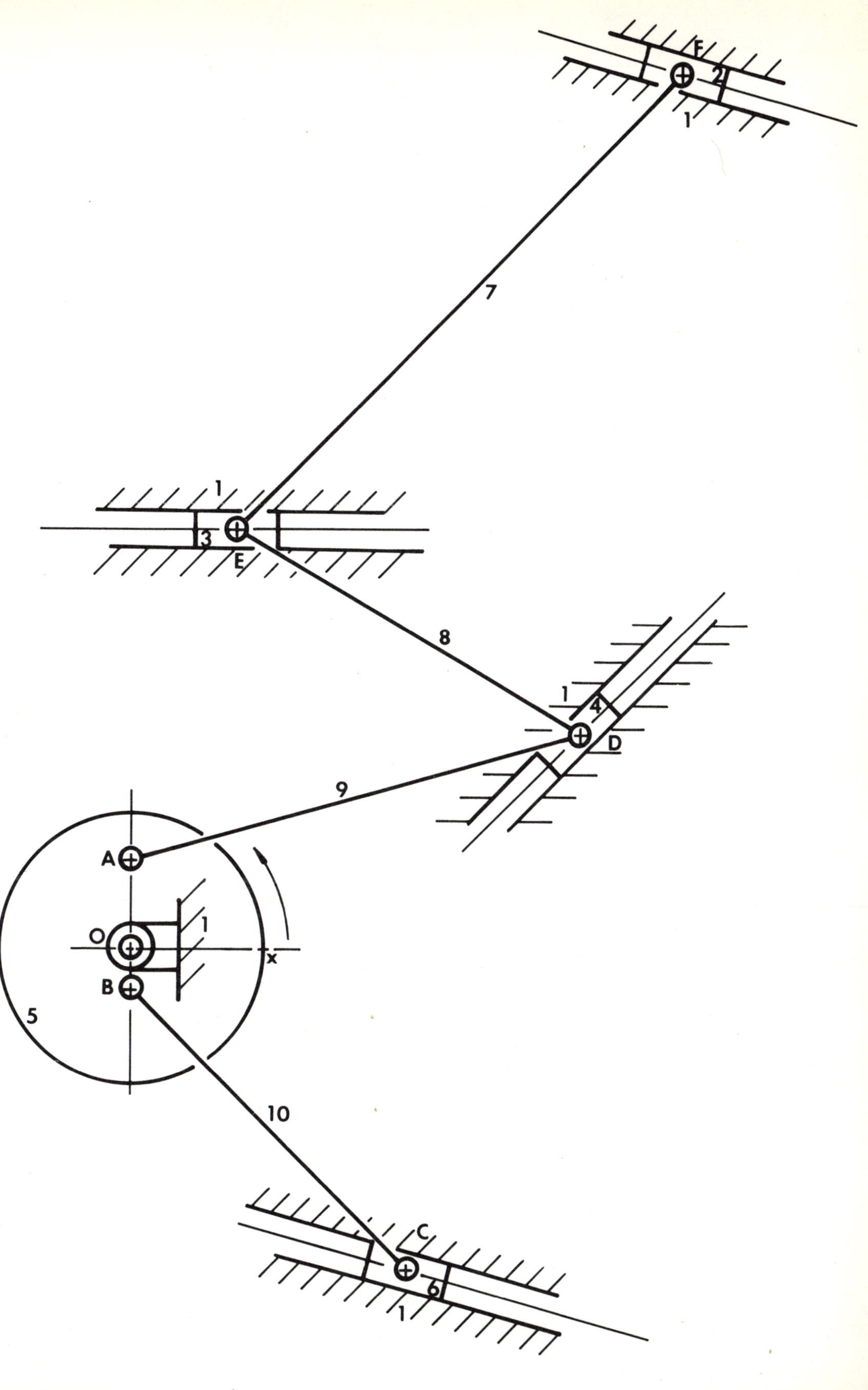

MECHANISM DRAFTING AND DESIGN	NAME: SECT. NO.: FILE NO.: DATE:	GRADE	PROBLEM 5B

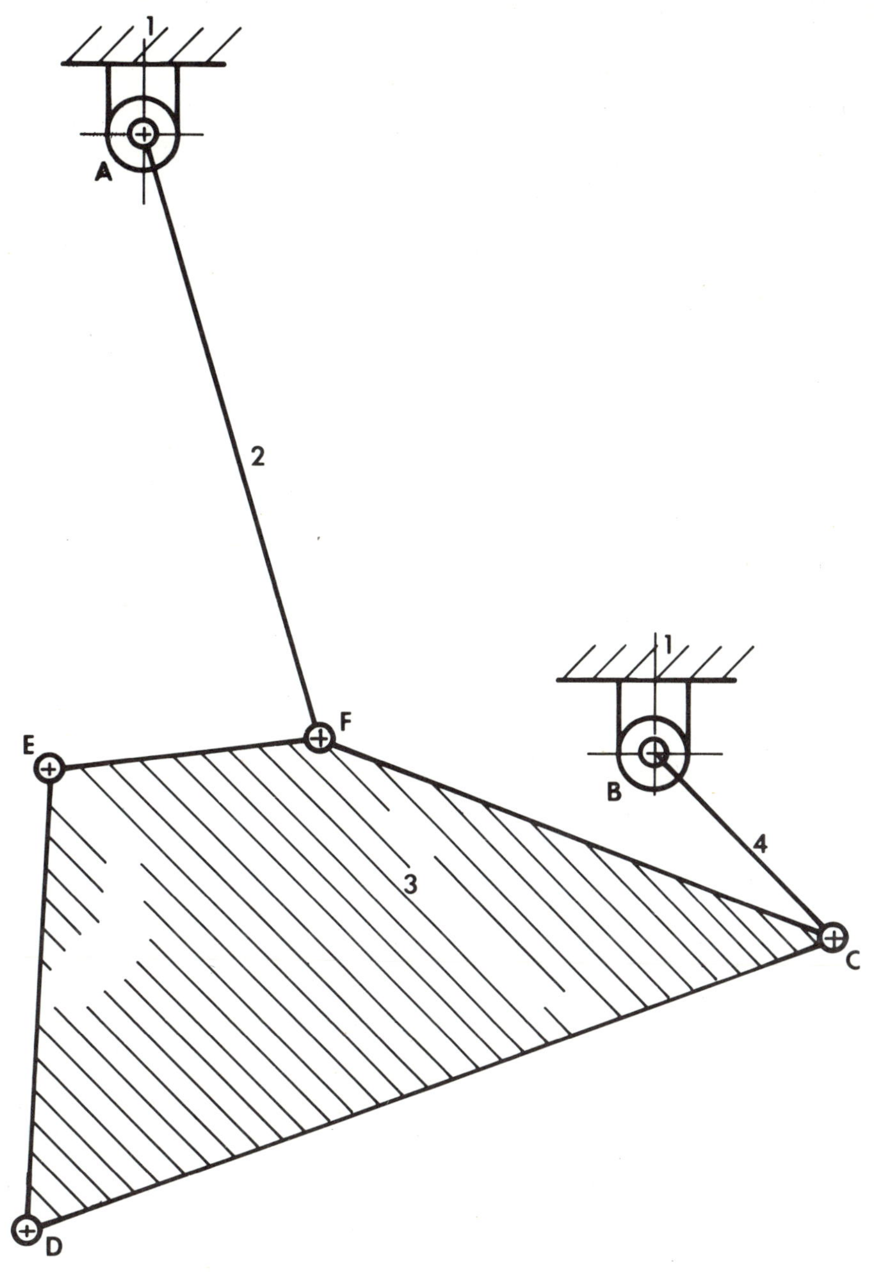

MECHANISM DRAFTING AND DESIGN	NAME: SECT. NO.: FILE NO.: DATE:	GRADE	PROBLEM 5C

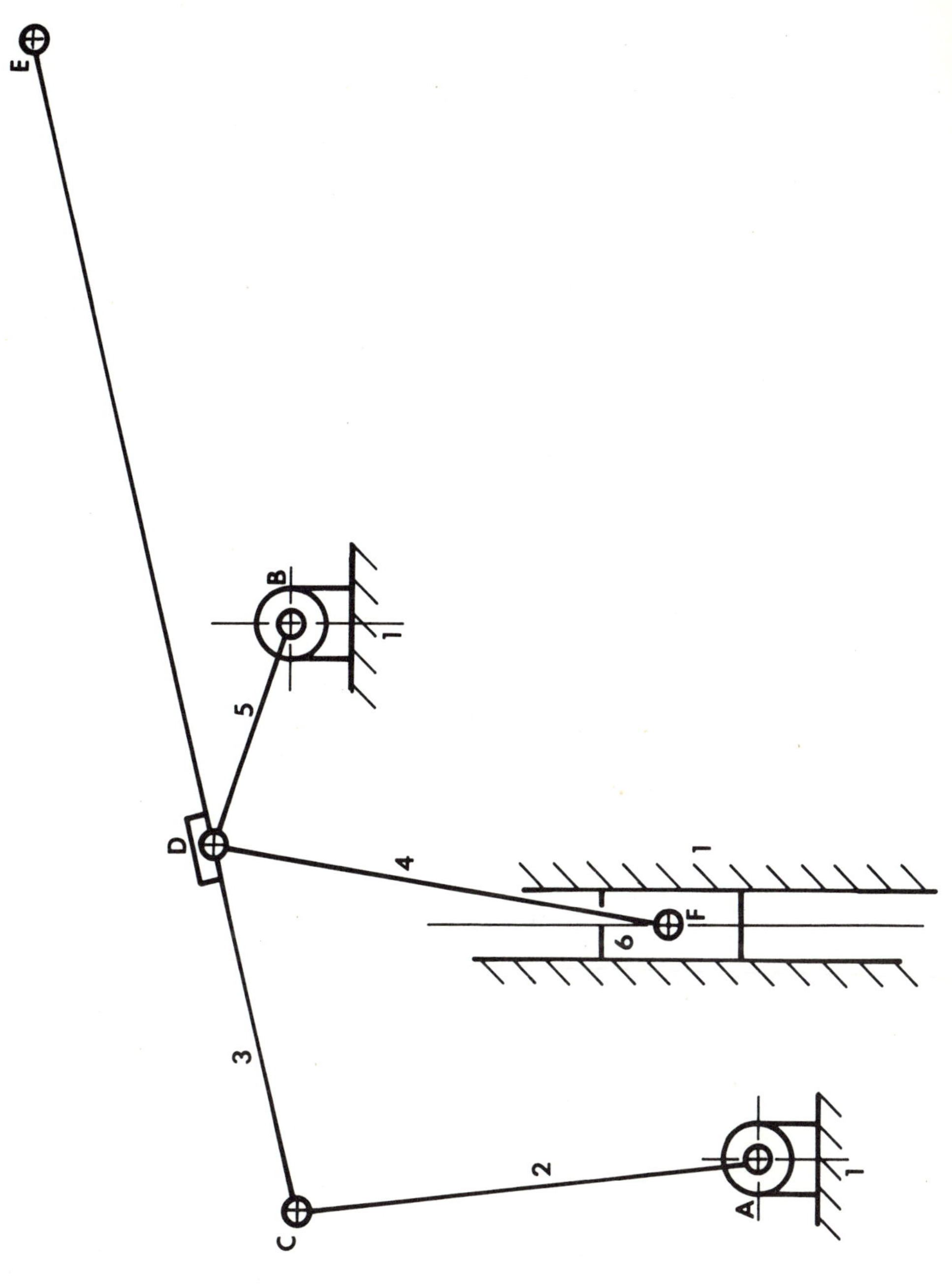

MECHANISM DRAFTING AND DESIGN	NAME: SECT. NO.: FILE NO.: DATE:	GRADE	PROBLEM 5D

1

E 1 5 D C 4 B 2 3 A F 6 1 1

2

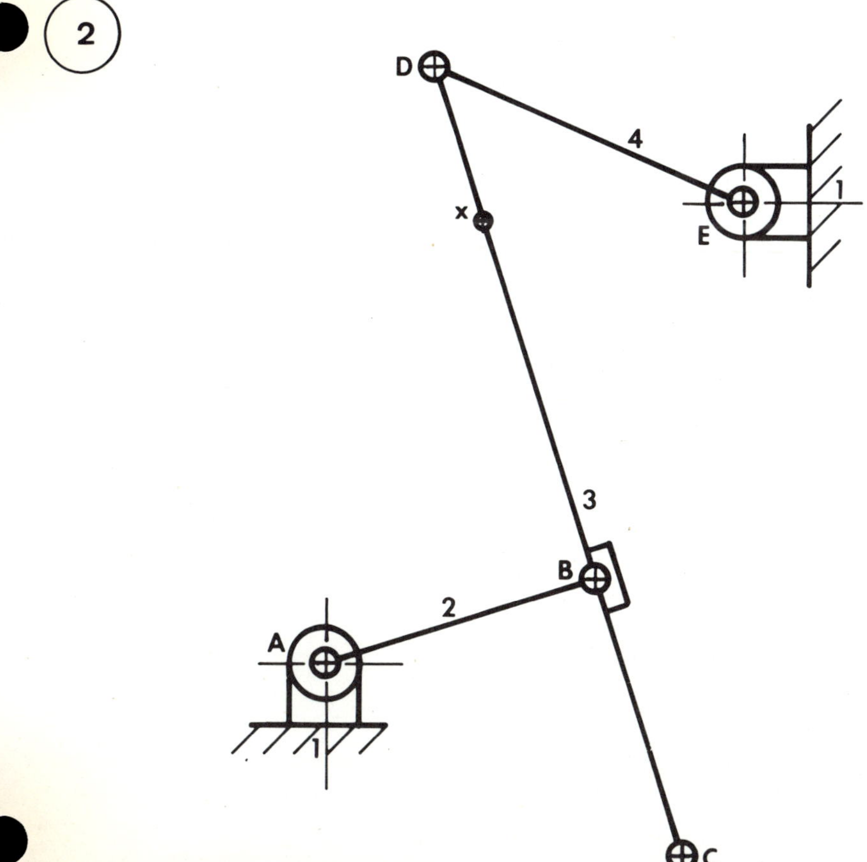

MECHANISM DRAFTING AND DESIGN	NAME: SECT. NO.: FILE NO.: DATE:	GRADE	PROBLEM 5E

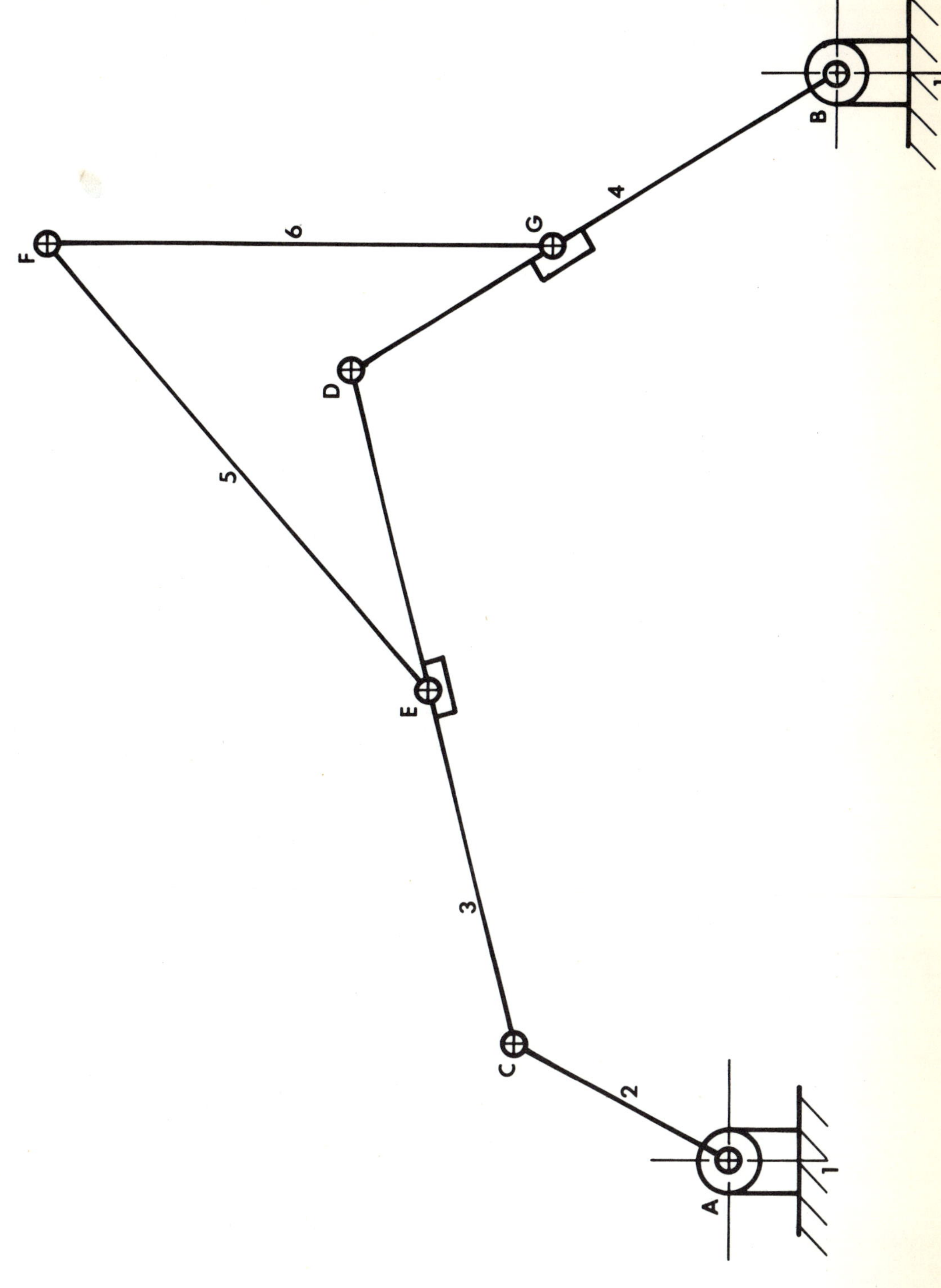

MECHANISM DRAFTING AND DESIGN	NAME: SECT. NO.: FILE NO.: DATE:	GRADE	PROBLEM 5F

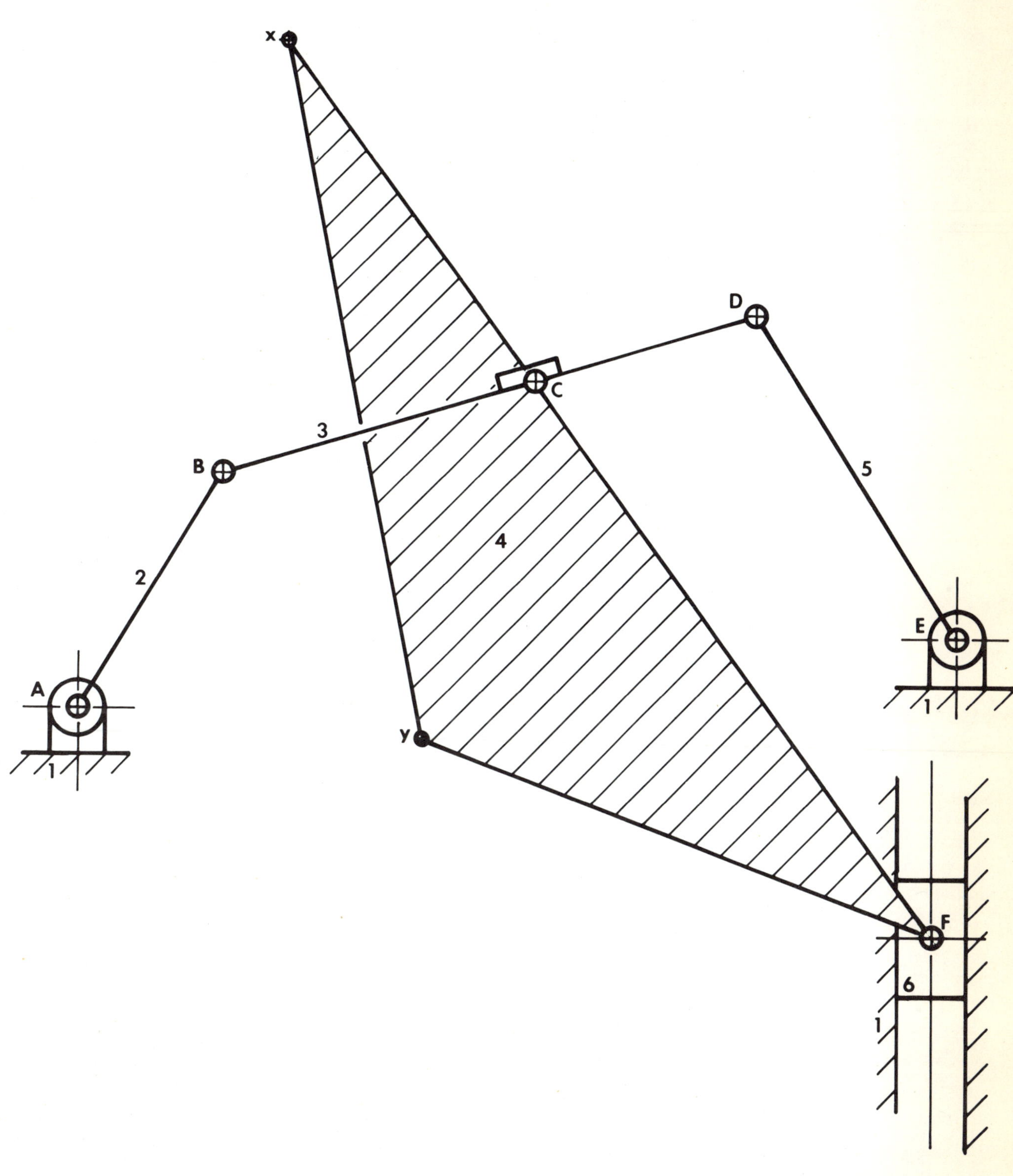

MECHANISM DRAFTING AND DESIGN	NAME: SECT. NO.: FILE NO.: DATE:	GRADE	PROBLEM 5G

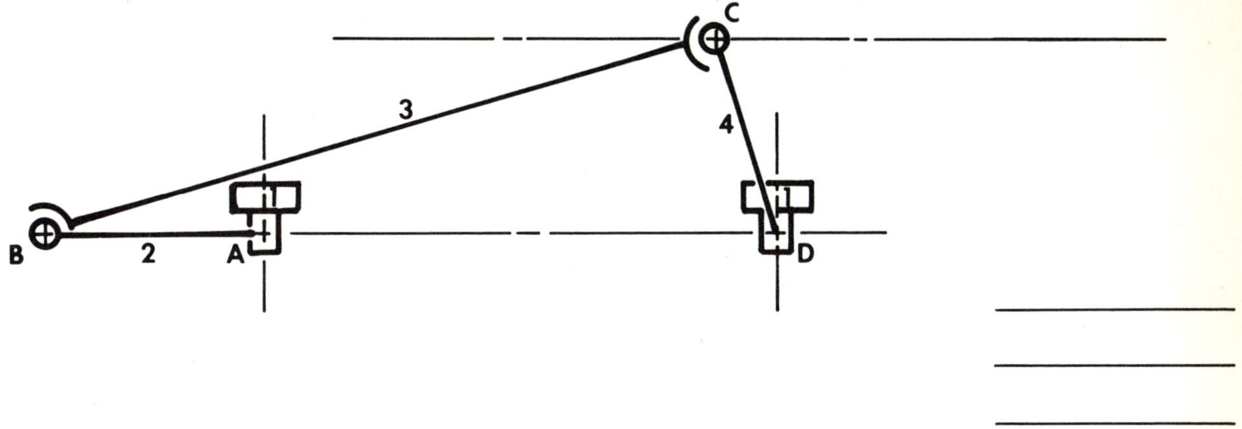

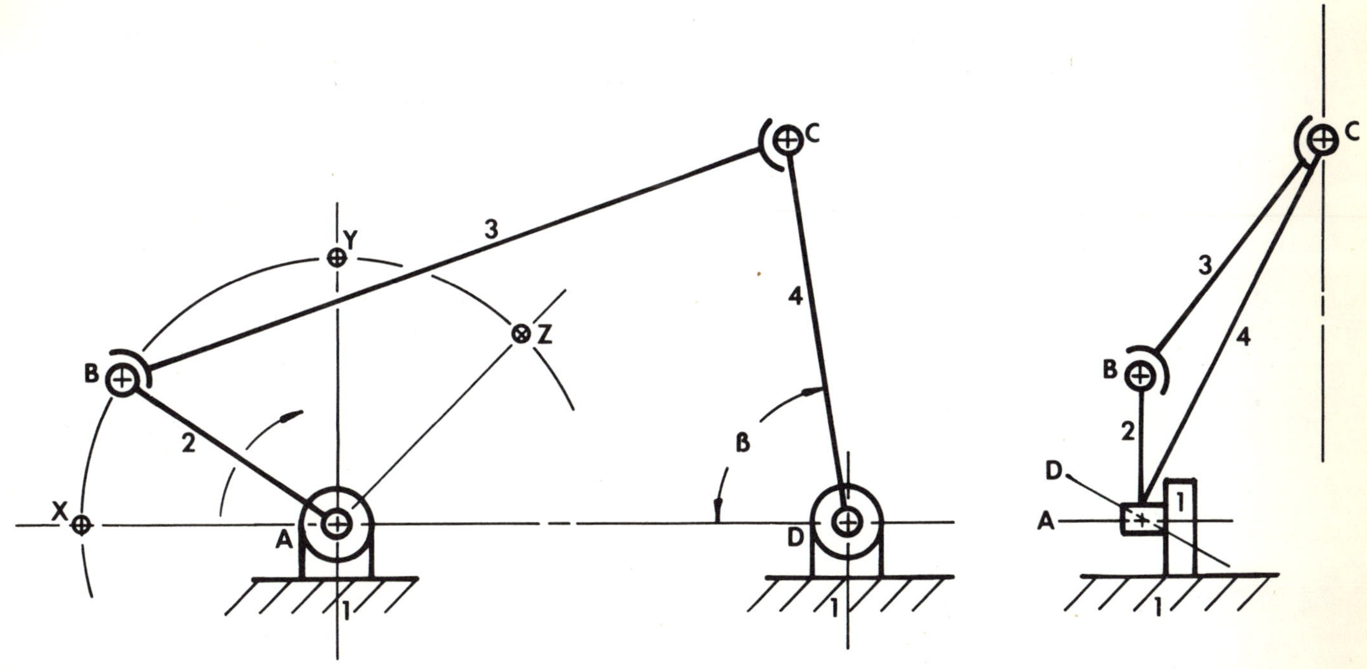

MECHANISM DRAFTING AND DESIGN	NAME: SECT. NO.: FILE NO.: DATE:	GRADE	PROBLEM 5H

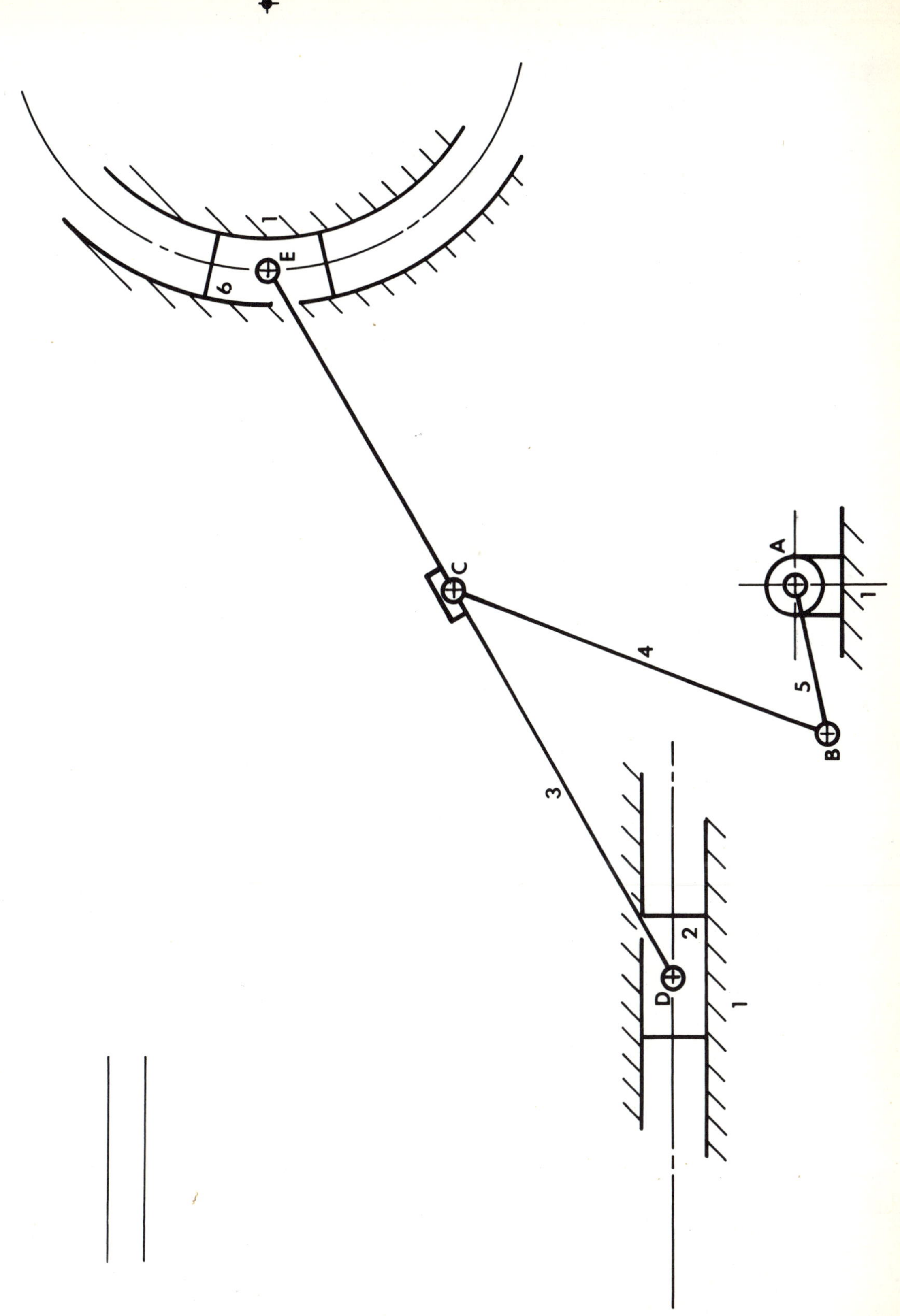

MECHANISM DRAFTING AND DESIGN	NAME: SECT. NO.: FILE NO.: DATE:	GRADE	PROBLEM 51

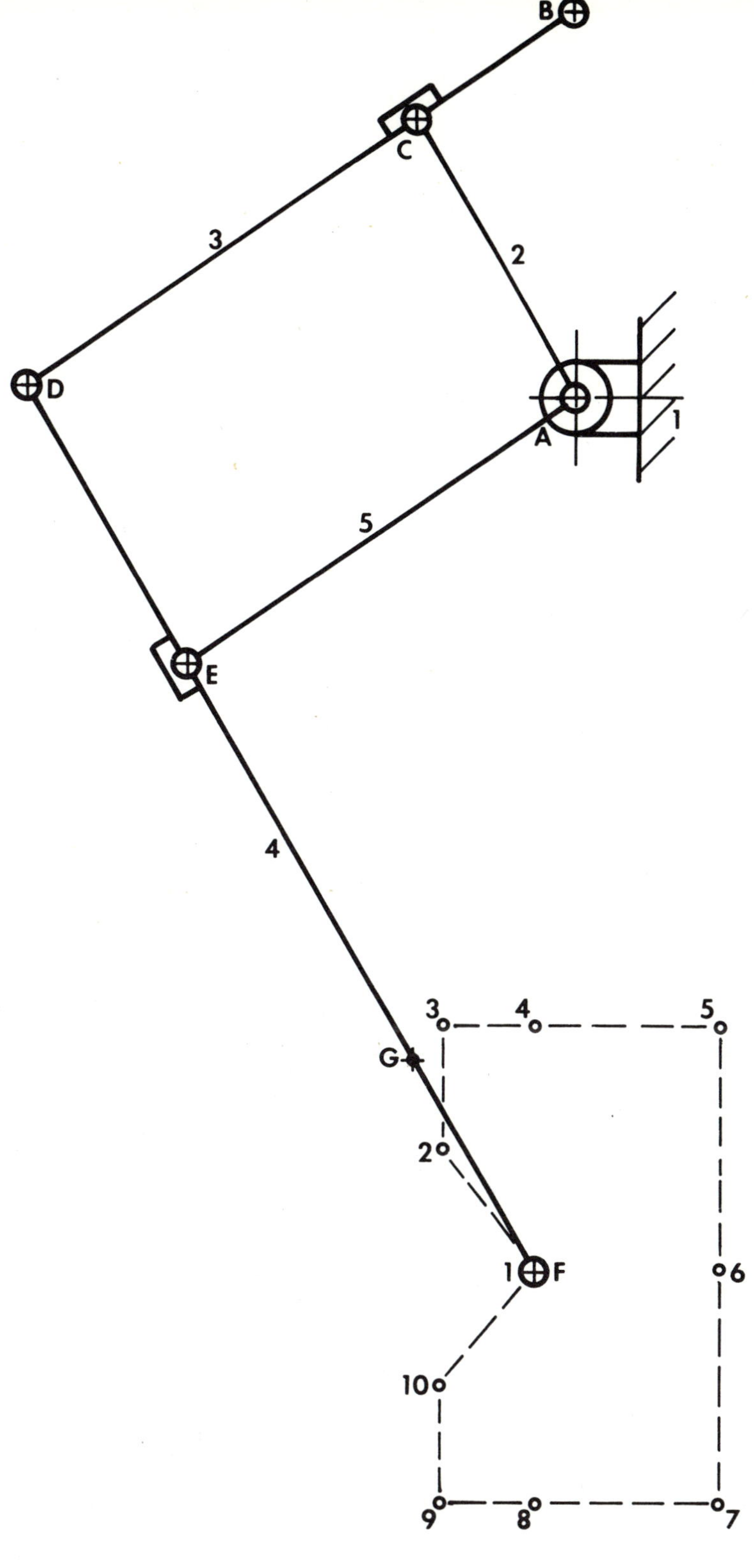

MECHANISM DRAFTING AND DESIGN	NAME: SECT. NO.: FILE NO.: DATE:	GRADE	PROBLEM 5 J

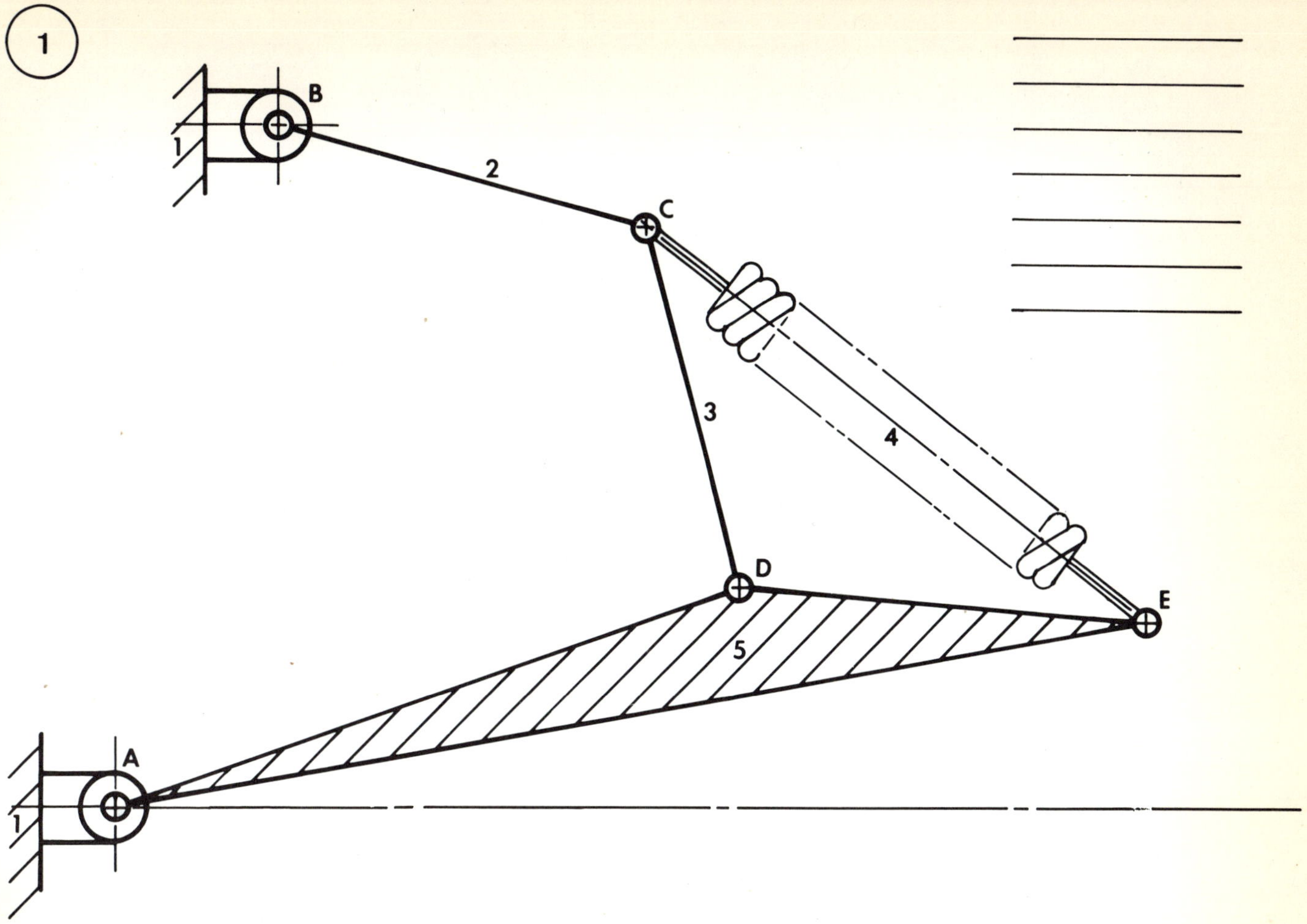

1

B
1
2
C
3
4
D
E
5
A
1

2

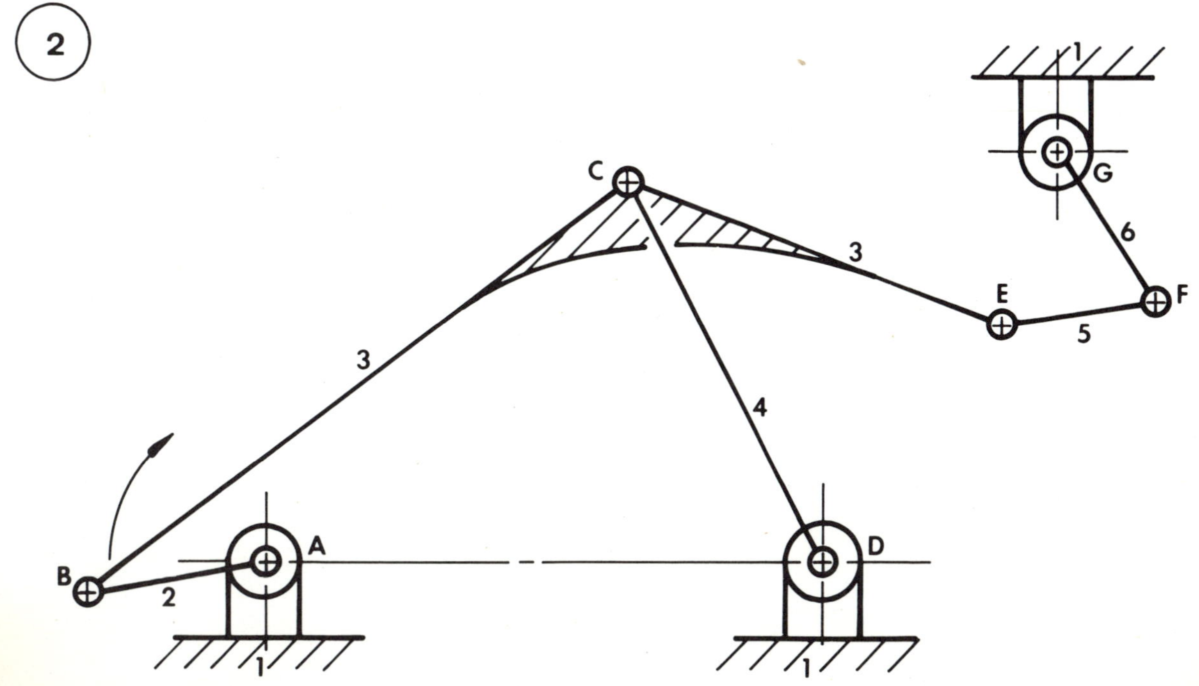

MECHANISM DRAFTING AND DESIGN	NAME: SECT. NO.: FILE NO.: DATE:	GRADE	PROBLEM 5K

1

2

3

4

5

MECHANISM DRAFTING AND DESIGN	NAME: SECT. NO.: FILE NO.: DATE:	GRADE	PROBLEM 6 A

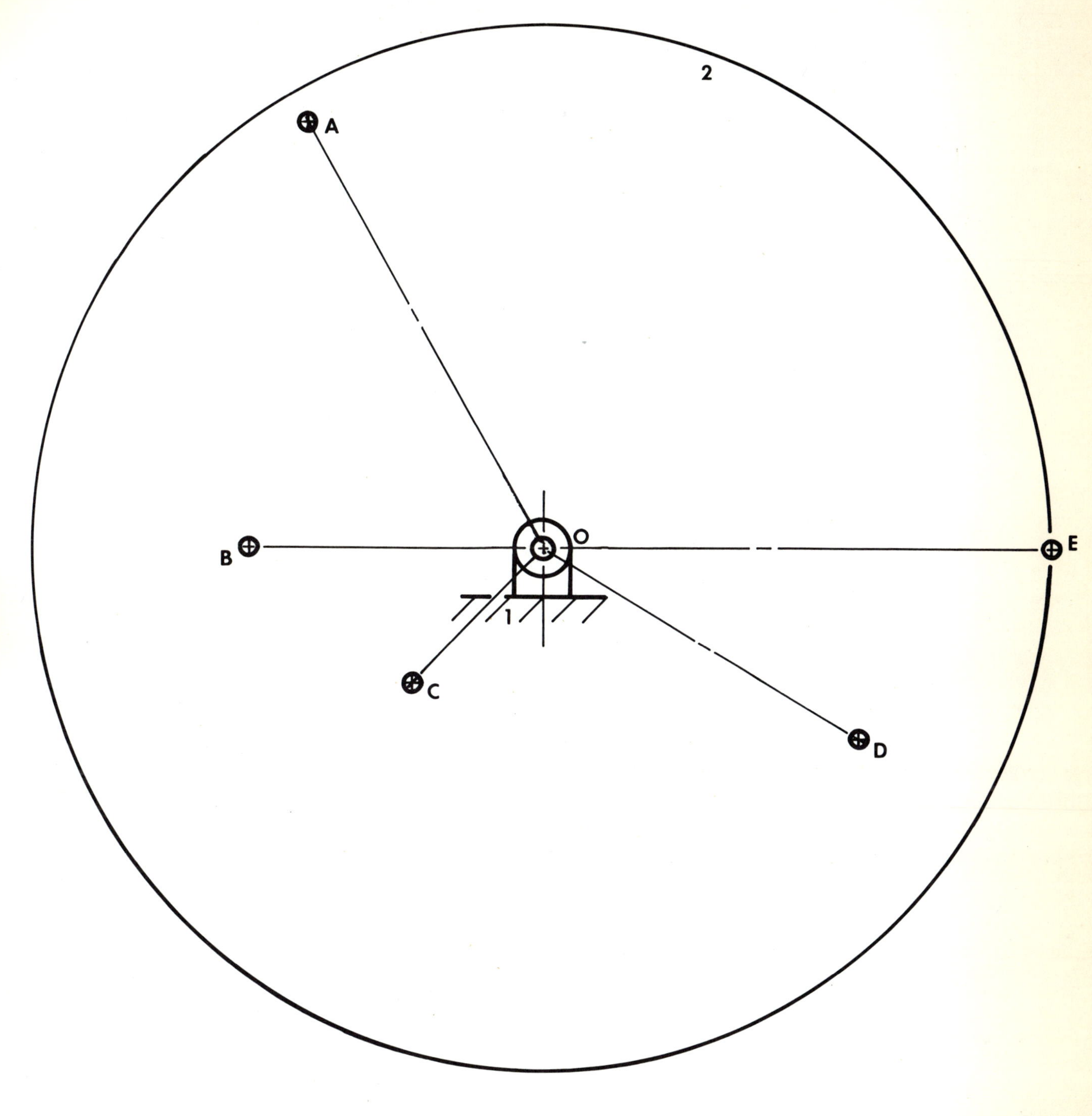

MECHANISM DRAFTING AND DESIGN	NAME: SECT. NO.: FILE NO.: DATE:	GRADE	PROBLEM 6 B

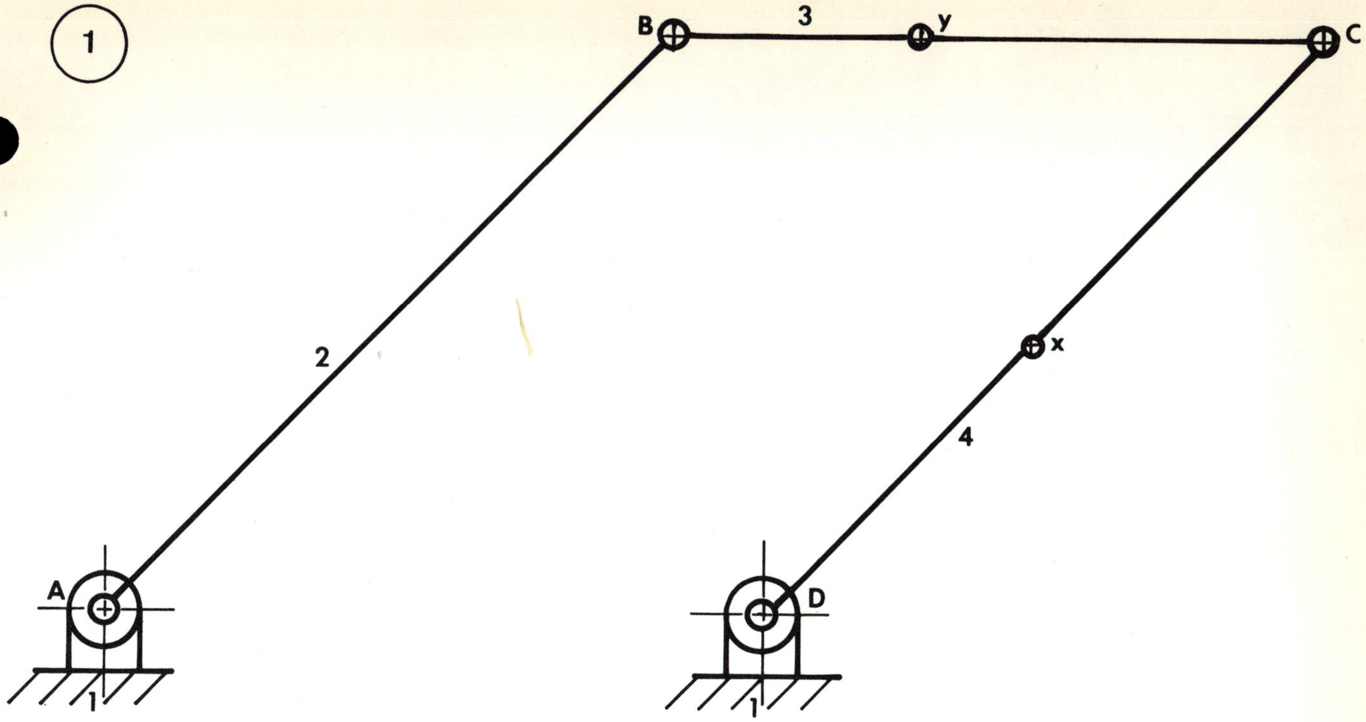

1
B
3
y
C
2
x
4
A
D
1
1

2
B
3
A
2
1

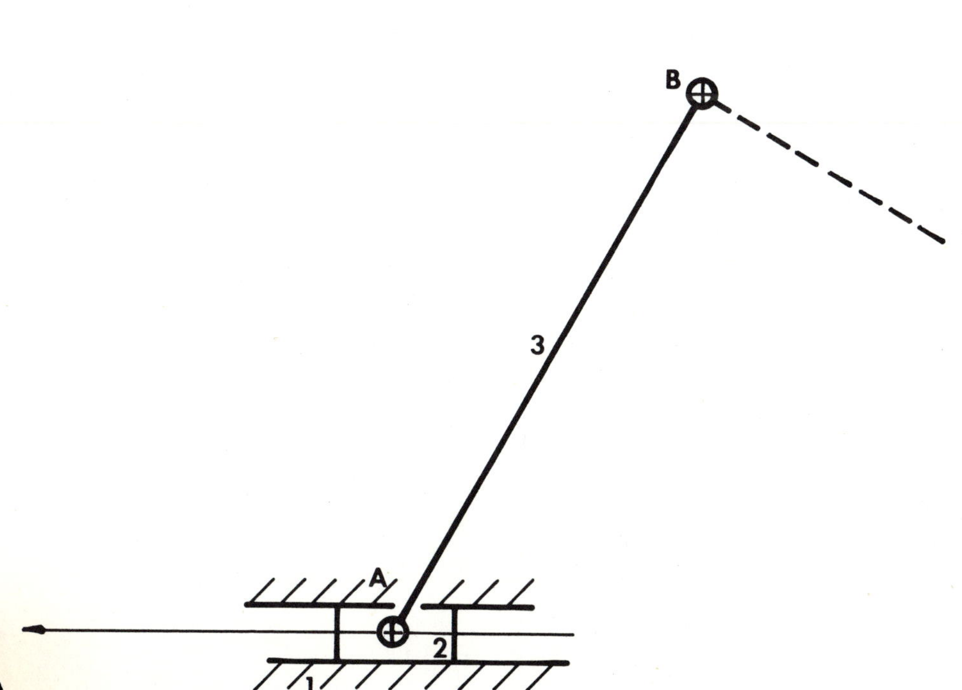

MECHANISM DRAFTING AND DESIGN	NAME: SECT. NO.: FILE NO.: DATE:	GRADE	PROBLEM 6C

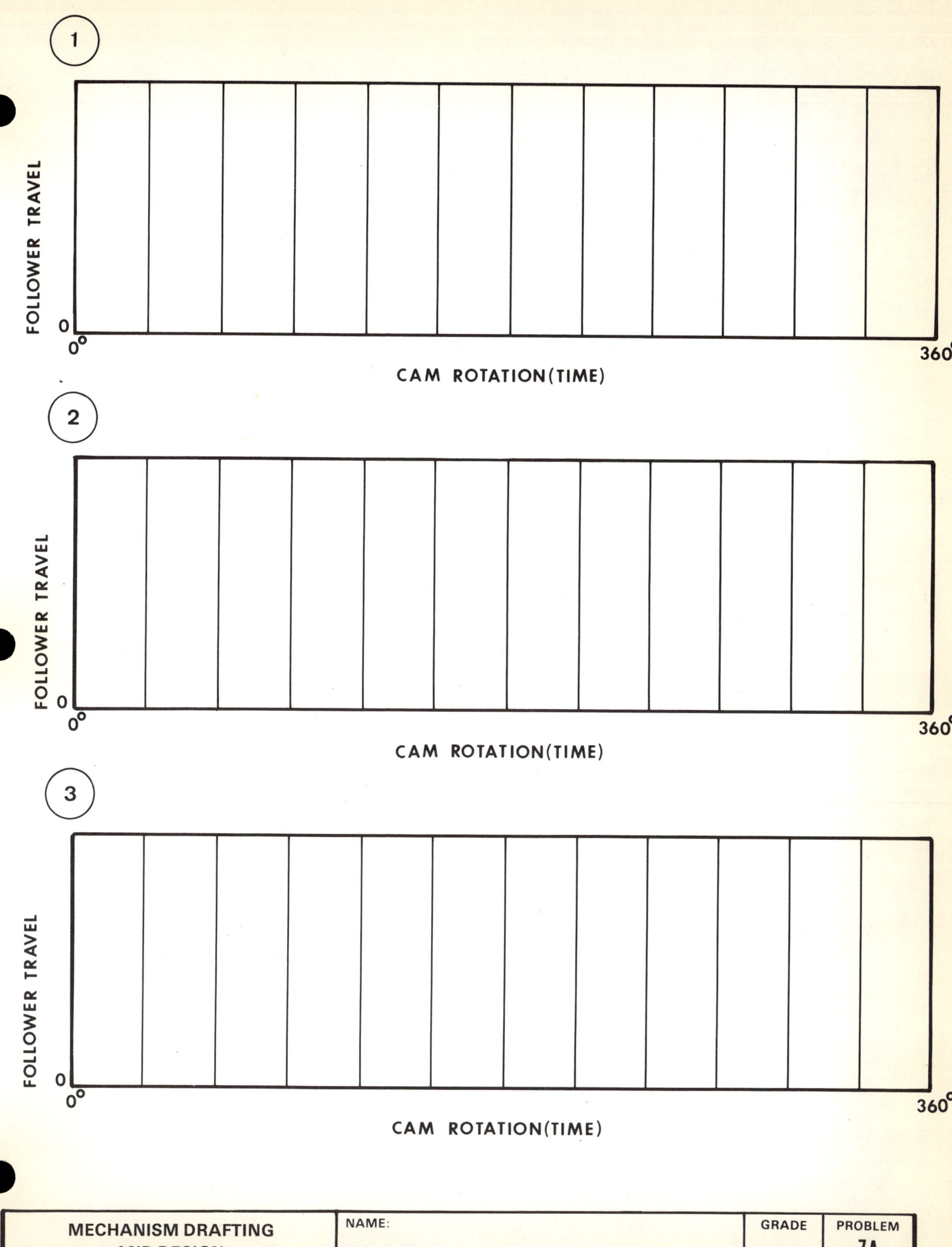
1
FOLLOWER TRAVEL
0
0°
360°
CAM ROTATION(TIME)
2
FOLLOWER TRAVEL
0
0°
360°
CAM ROTATION(TIME)
3
FOLLOWER TRAVEL
0
0°
360°
CAM ROTATION(TIME)
MECHANISM DRAFTING AND DESIGN
NAME:
SECT. NO.:
FILE NO.:
DATE:
GRADE
PROBLEM
7A

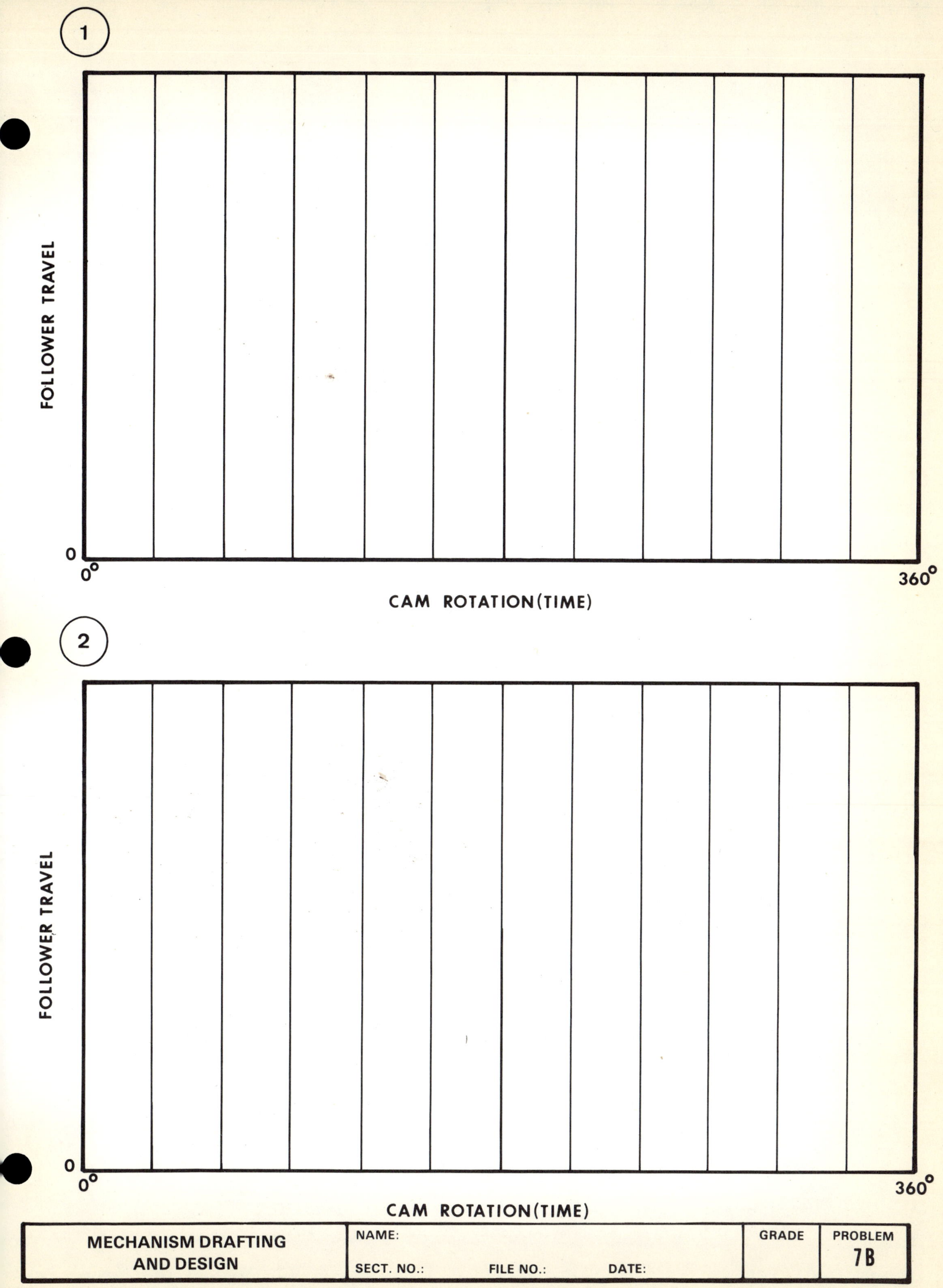

MECHANISM DRAFTING AND DESIGN	NAME: SECT. NO.: FILE NO.: DATE:	GRADE	PROBLEM 7B

360°

CAM ROTATION (TIME)

0°

0°

FOLLOWER TRAVEL

MECHANISM DRAFTING AND DESIGN	NAME: SECT. NO.: FILE NO.: DATE:	GRADE	PROBLEM 7C

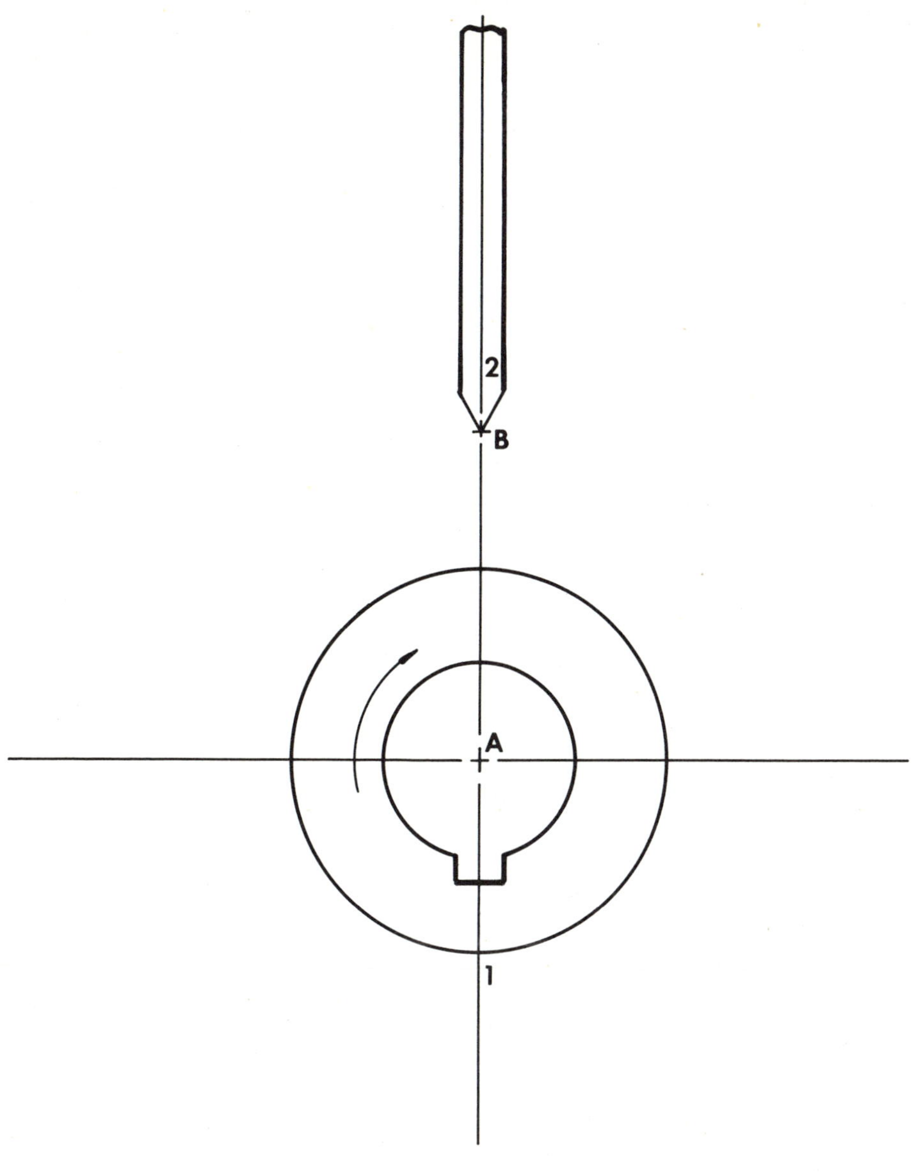

MECHANISM DRAFTING AND DESIGN	NAME: SECT. NO.: FILE NO.: DATE:	GRADE	PROBLEM 7D

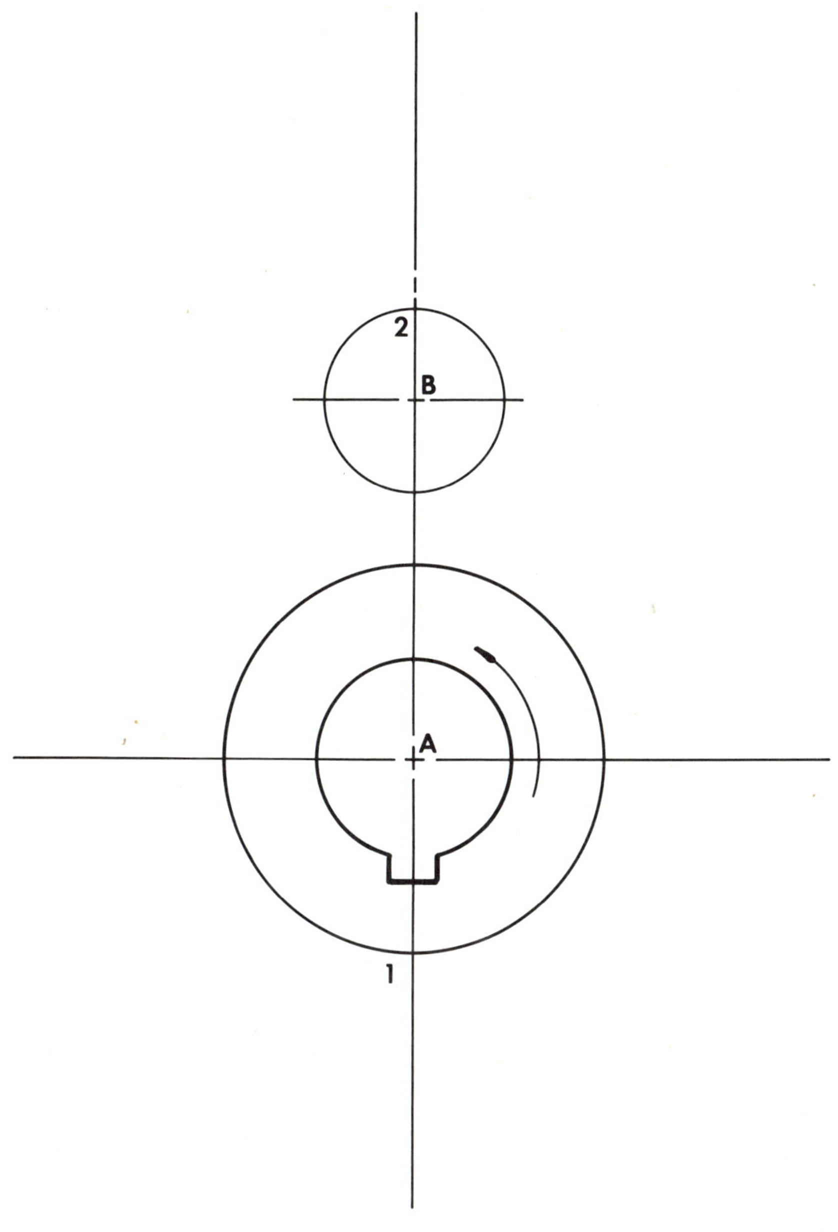

MECHANISM DRAFTING AND DESIGN	NAME: SECT. NO.: FILE NO.: DATE:	GRADE	PROBLEM 7E

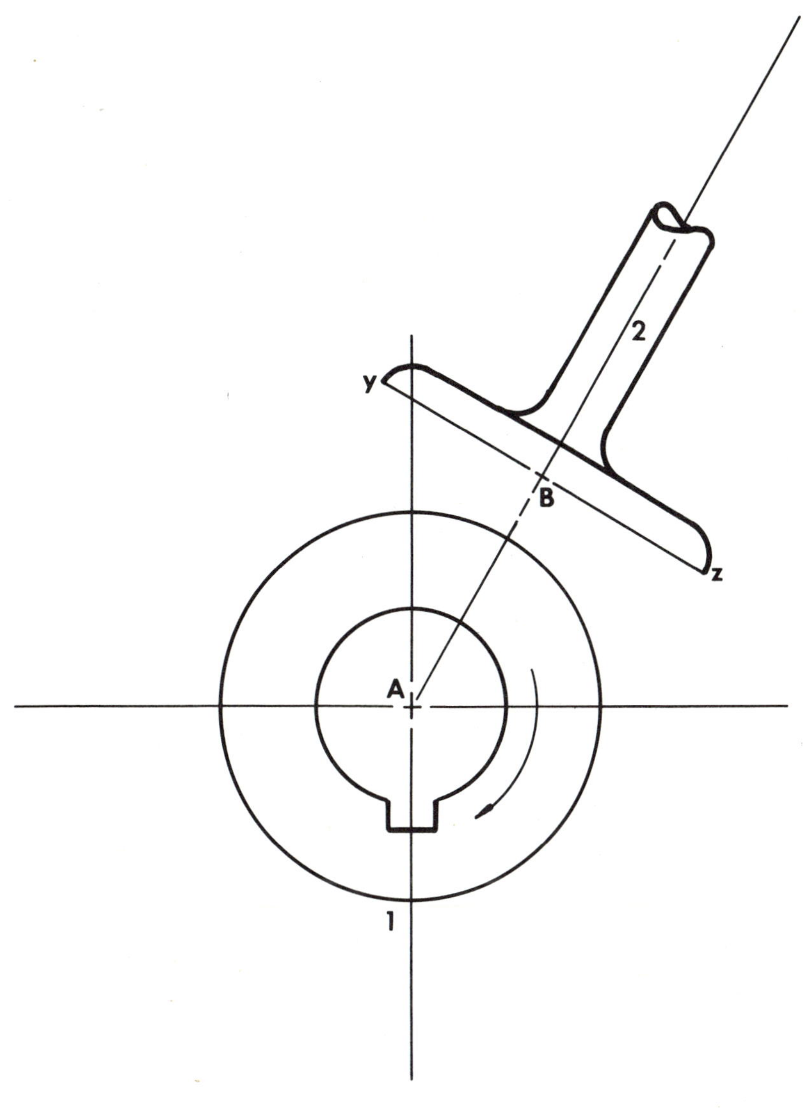

MECHANISM DRAFTING AND DESIGN	NAME: SECT. NO.: FILE NO.: DATE:	GRADE	PROBLEM 7F

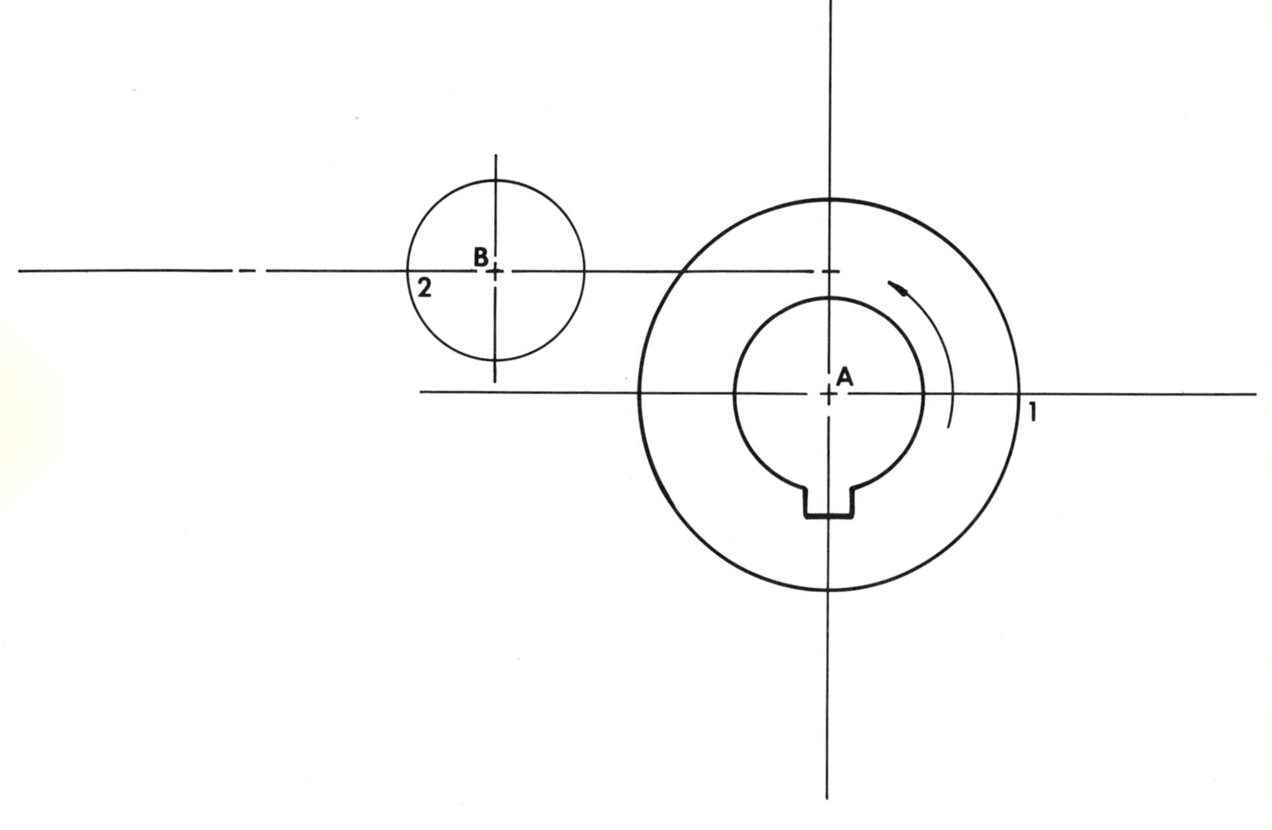

MECHANISM DRAFTING AND DESIGN	NAME: SECT. NO.: FILE NO.: DATE:	GRADE	PROBLEM 7G

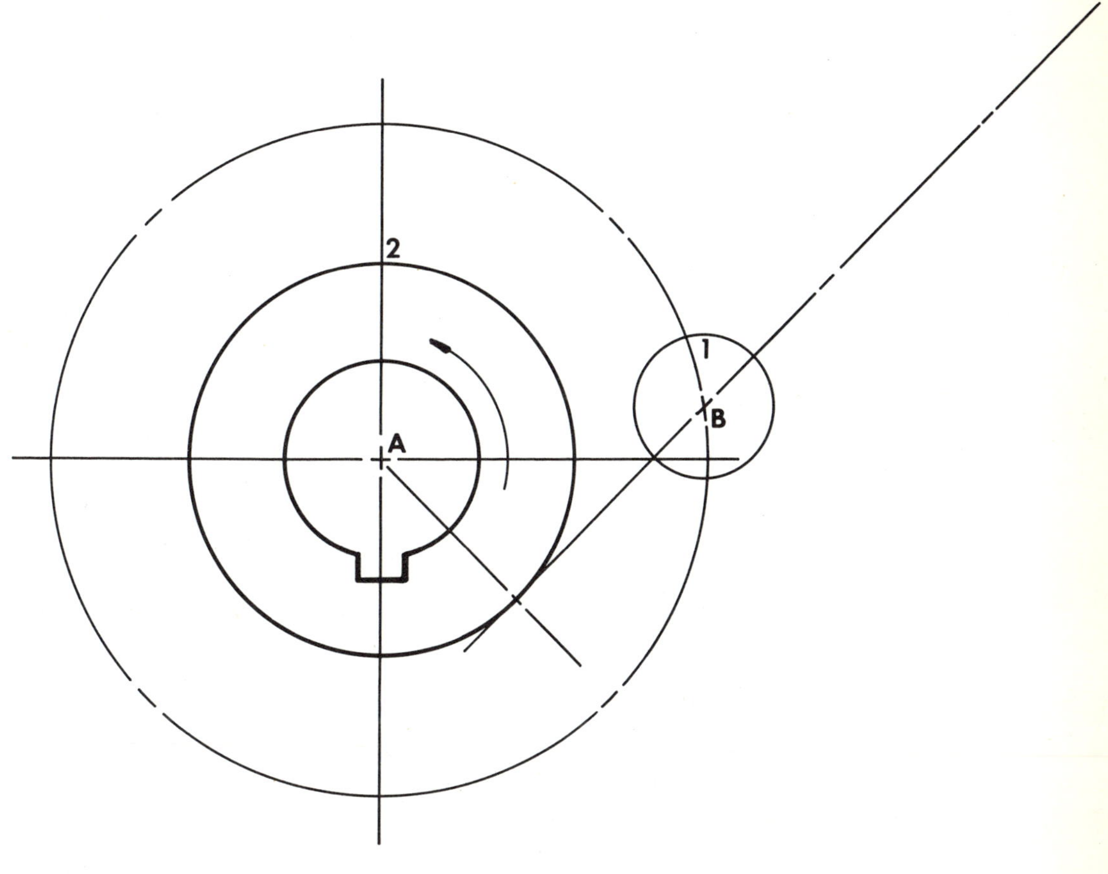

MECHANISM DRAFTING AND DESIGN	NAME: SECT. NO.: FILE NO.: DATE:	GRADE	PROBLEM 7H

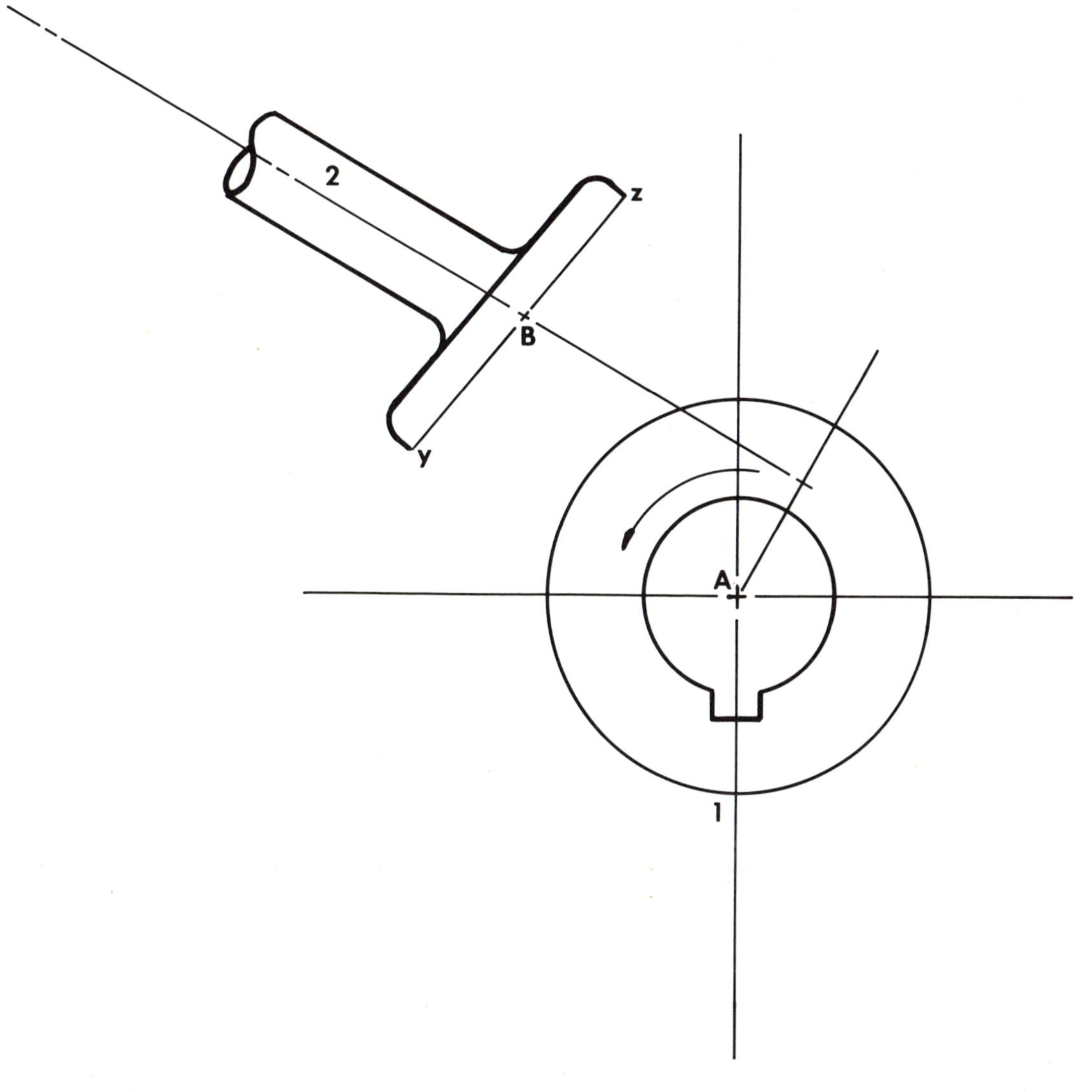

MECHANISM DRAFTING AND DESIGN	NAME: SECT. NO.: FILE NO.: DATE:	GRADE	PROBLEM 71

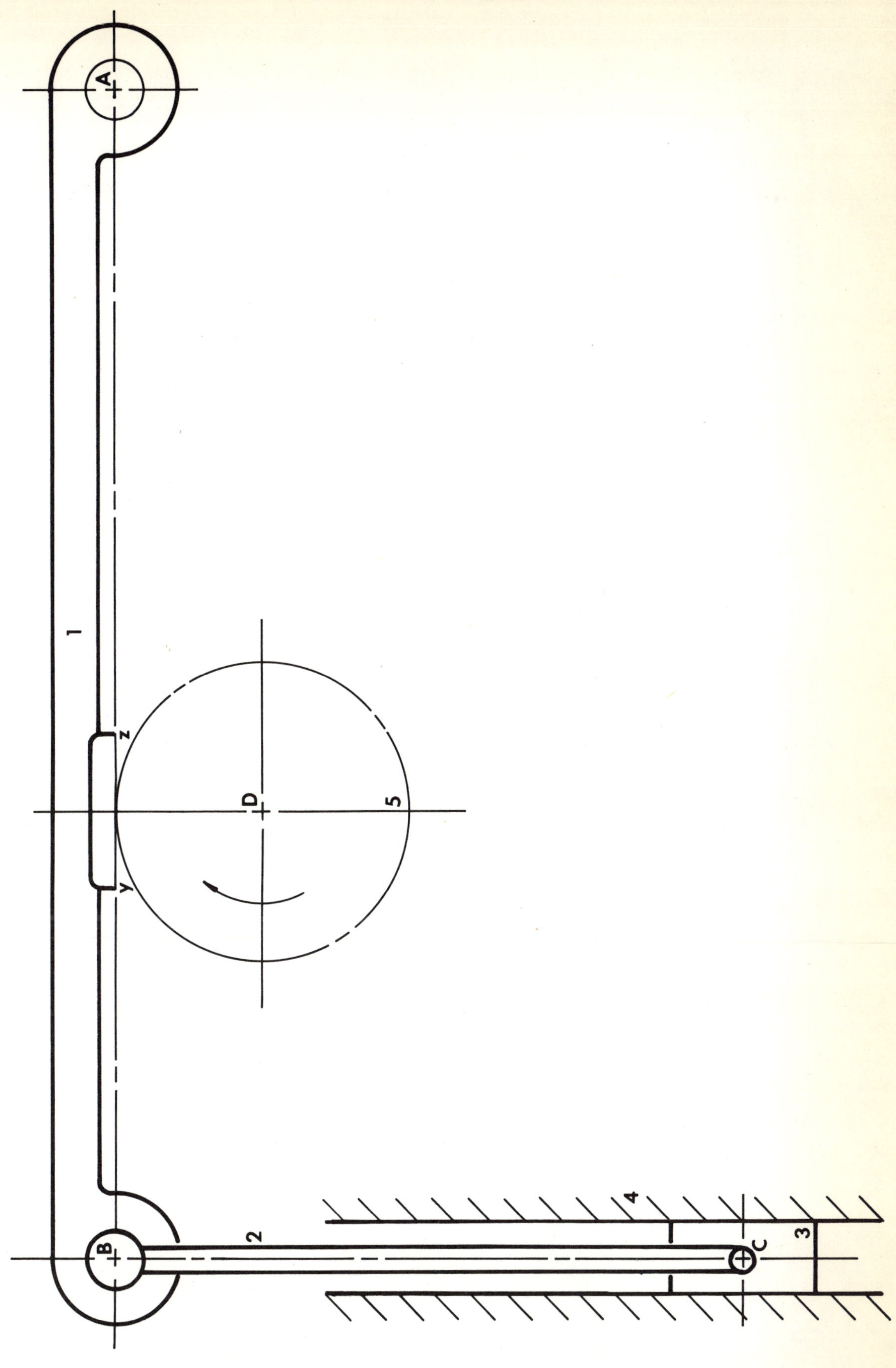

MECHANISM DRAFTING AND DESIGN	NAME: SECT. NO.: FILE NO.: DATE:	GRADE	PROBLEM 7J

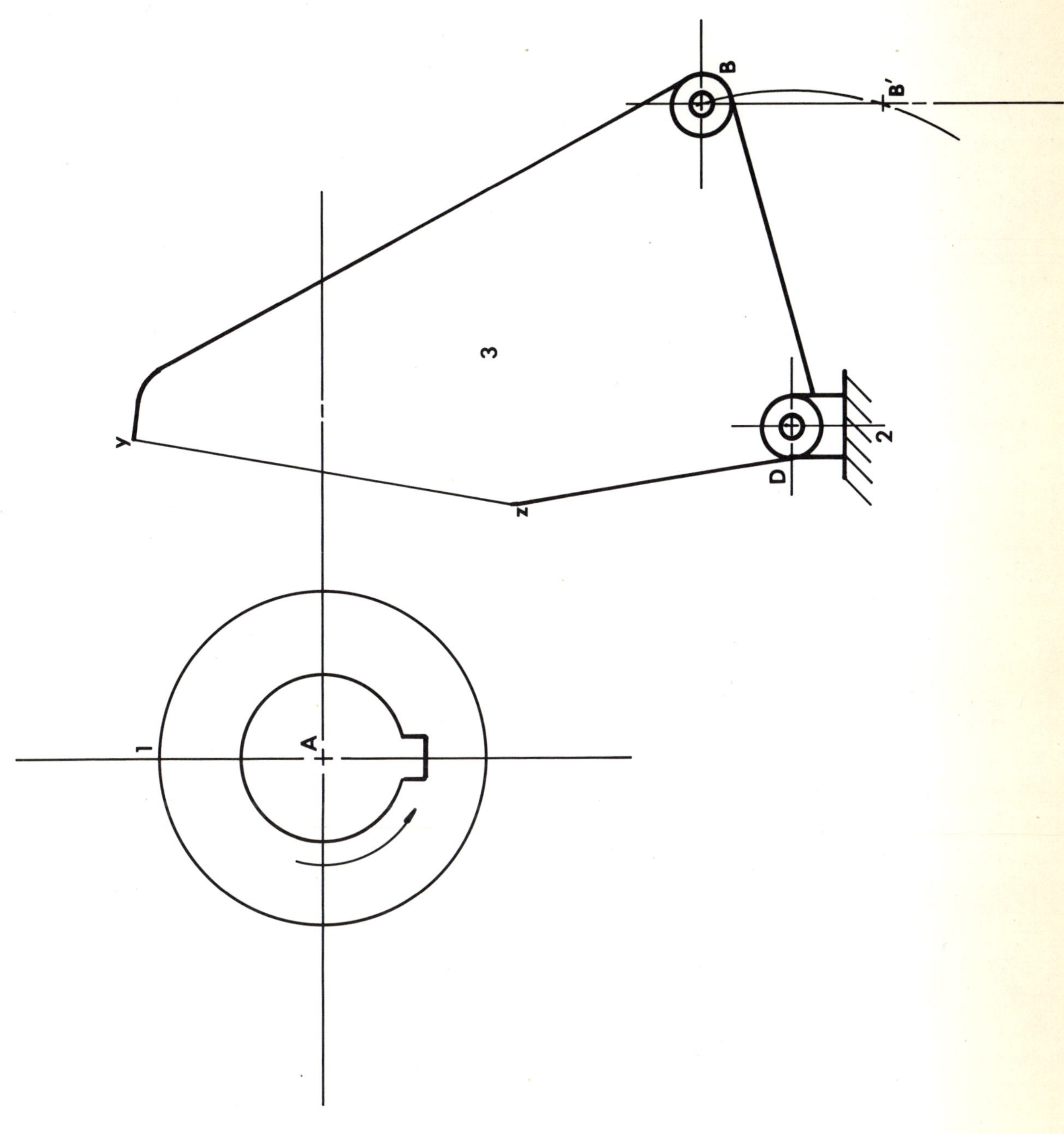

MECHANISM DRAFTING AND DESIGN	NAME: SECT. NO.: FILE NO.: DATE:	GRADE	PROBLEM 7K

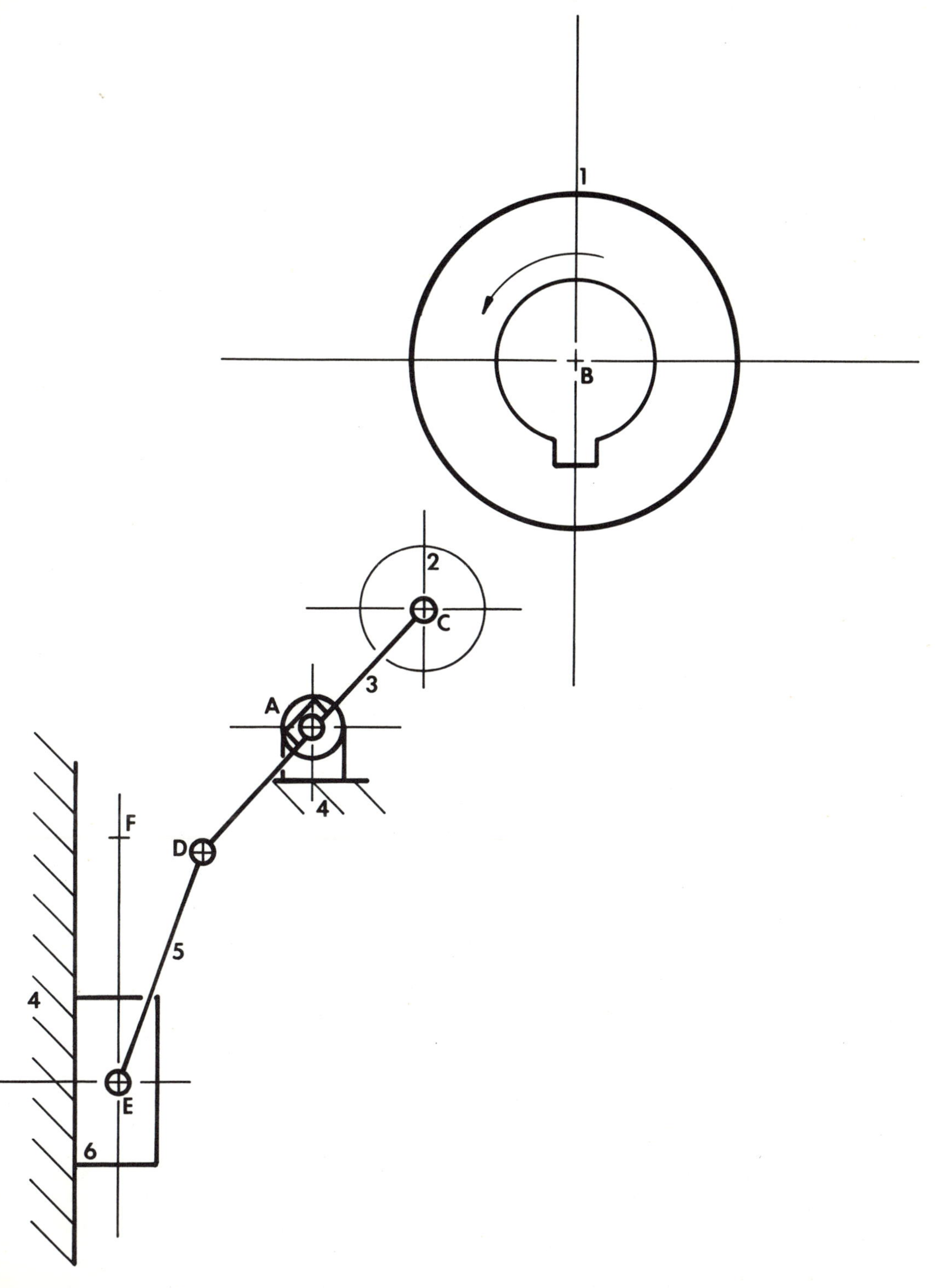

MECHANISM DRAFTING AND DESIGN	NAME: SECT. NO.: FILE NO.: DATE:	GRADE	PROBLEM 7L

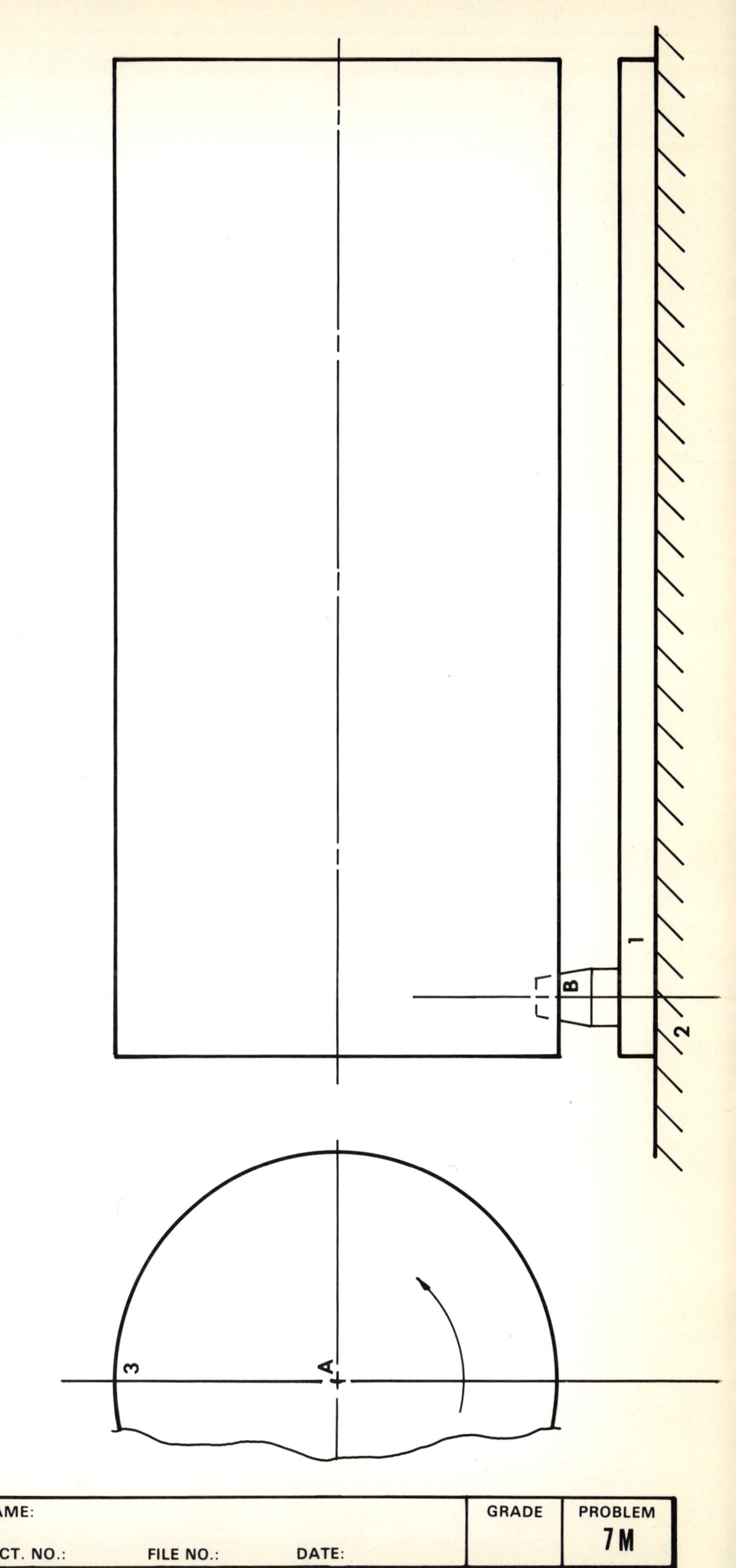

MECHANISM DRAFTING AND DESIGN	NAME: SECT. NO.: FILE NO.: DATE:	GRADE	PROBLEM 7M

1

2

3

4

5

MECHANISM DRAFTING AND DESIGN	NAME: SECT. NO.: FILE NO.: DATE:	GRADE	PROBLEM 8A

(1)

(2)

(3)

(4)

(5)

MECHANISM DRAFTING AND DESIGN	NAME: SECT. NO.: FILE NO.: DATE:	GRADE	PROBLEM 8B

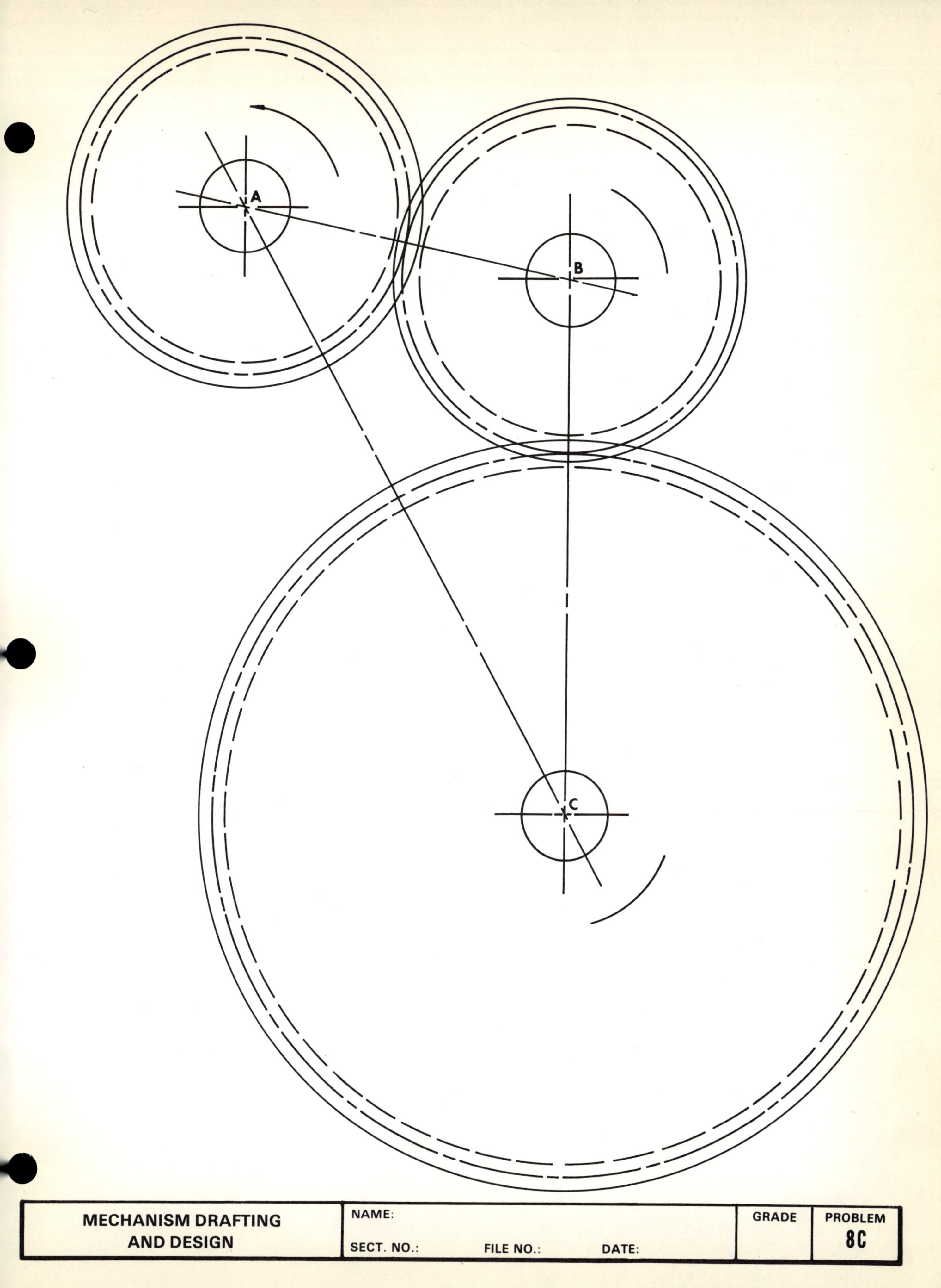

MECHANISM DRAFTING AND DESIGN	NAME: SECT. NO.: FILE NO.: DATE:	GRADE	PROBLEM 8C

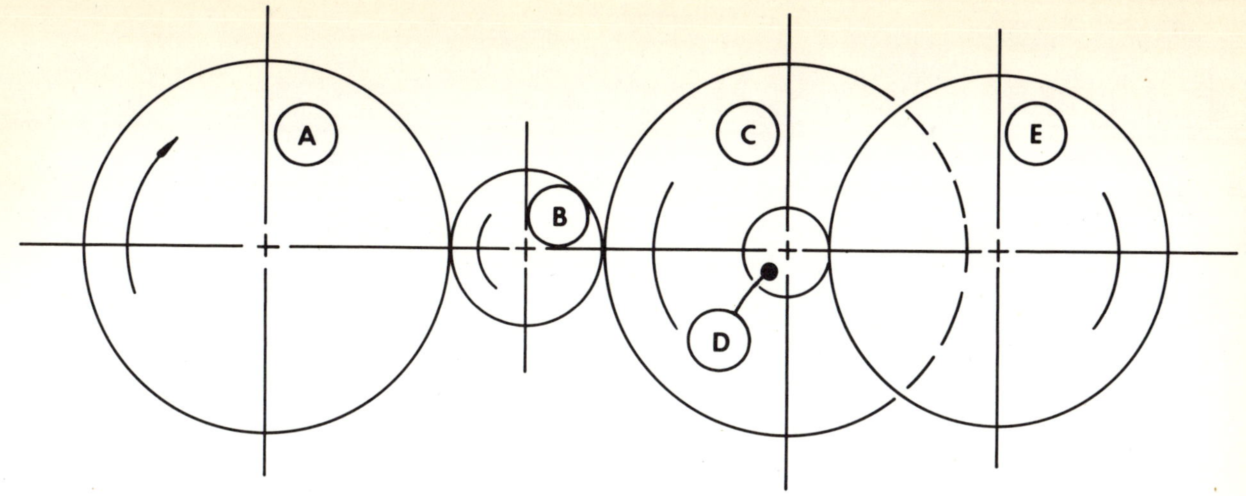

GEAR	RPM	DP	PD	N
A	100	2		50
B				20
C			25	50
D		3	5	15
E				60

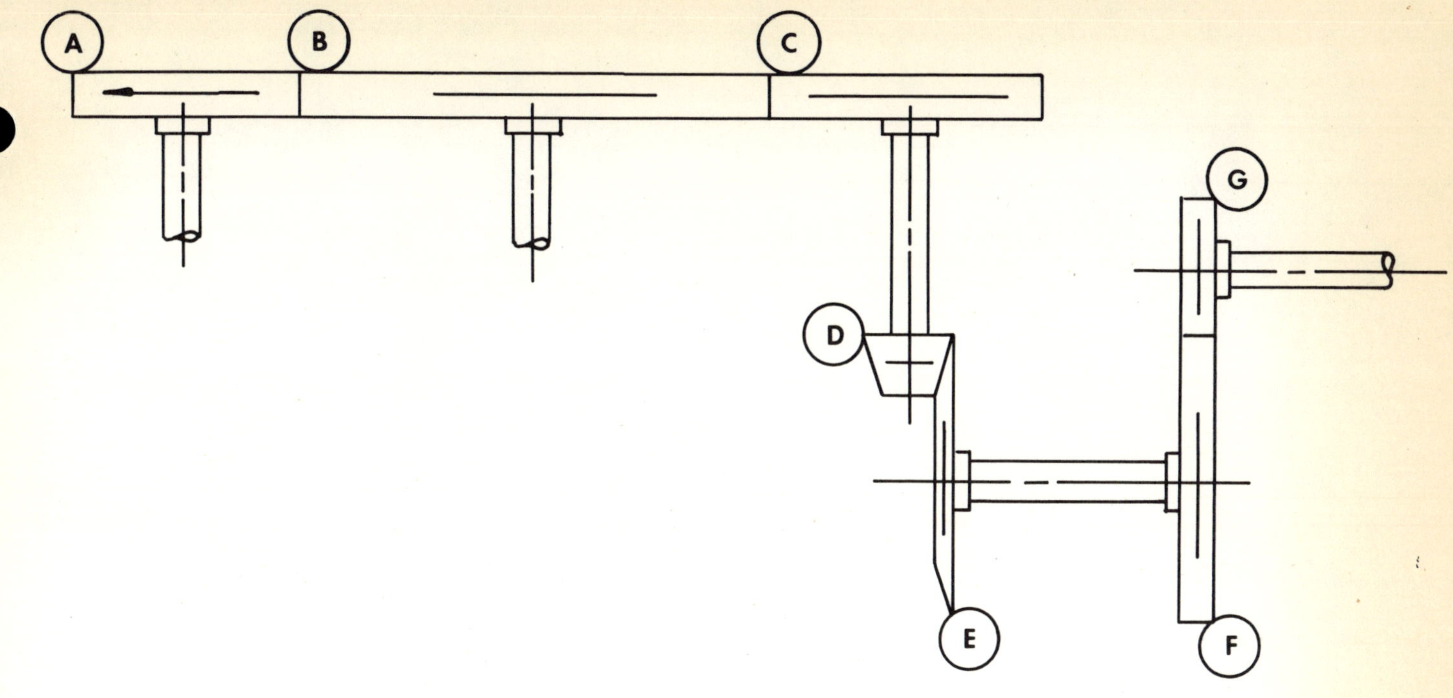

GEAR	RPM	DP	PD	N
A	500		5	20
B			10	40
C				24
D				16
E				48
F				80
G		8	4	32

MECHANISM DRAFTING AND DESIGN	NAME: SECT. NO.: FILE NO.: DATE:	GRADE	PROBLEM 8E

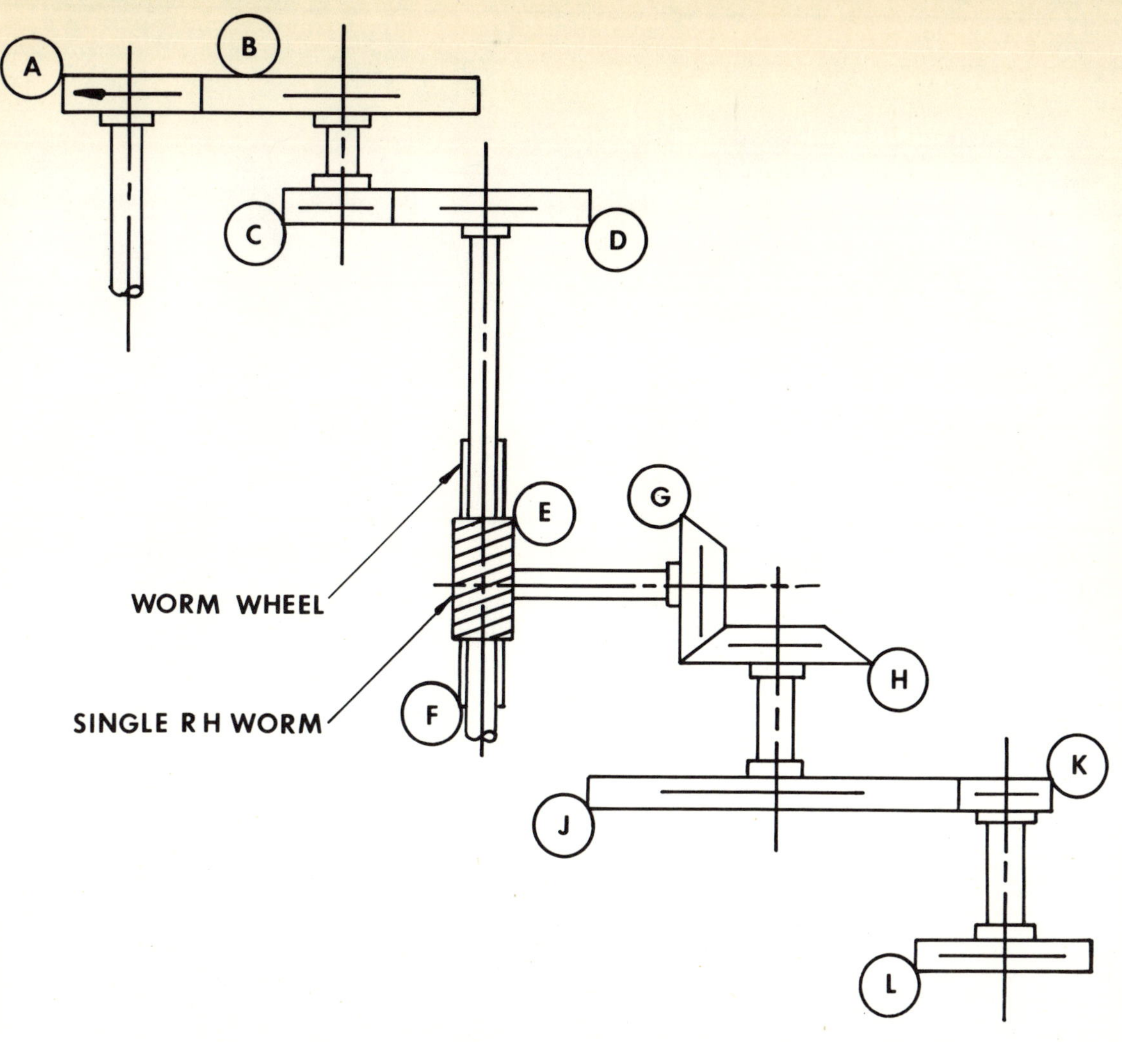

GEAR	RPM	PD	N
A	1750		8
B			16
C		10	
D		18	
WORM E			1
WHEEL F			72
G			16
H			22
J		20	
K		5	
L			20

MECHANISM DRAFTING AND DESIGN	NAME: SECT. NO.: FILE NO.: DATE:	GRADE	PROBLEM 8F

P
MECHANISM DRAFTING AND DESIGN
NAME:
SECT. NO.:
FILE NO.:
DATE:
GRADE
PROBLEM
8G

P

MECHANISM DRAFTING AND DESIGN	NAME: SECT. NO.: FILE NO.: DATE:	GRADE	PROBLEM 8H

P

MECHANISM DRAFTING AND DESIGN	NAME: SECT. NO.: FILE NO.: DATE:	GRADE	PROBLEM 81

MECHANISM DRAFTING AND DESIGN	NAME: SECT. NO.: FILE NO.: DATE:	GRADE	PROBLEM 8J

℄ PINION

GEAR ℄

MECHANISM DRAFTING AND DESIGN	NAME: SECT. NO.: FILE NO.: DATE:	GRADE	PROBLEM 8K

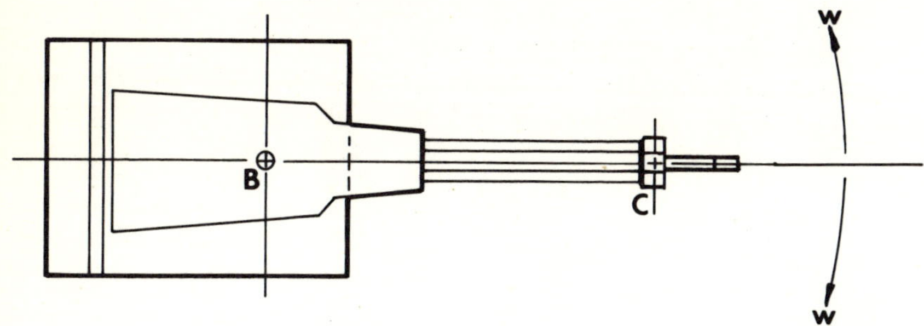

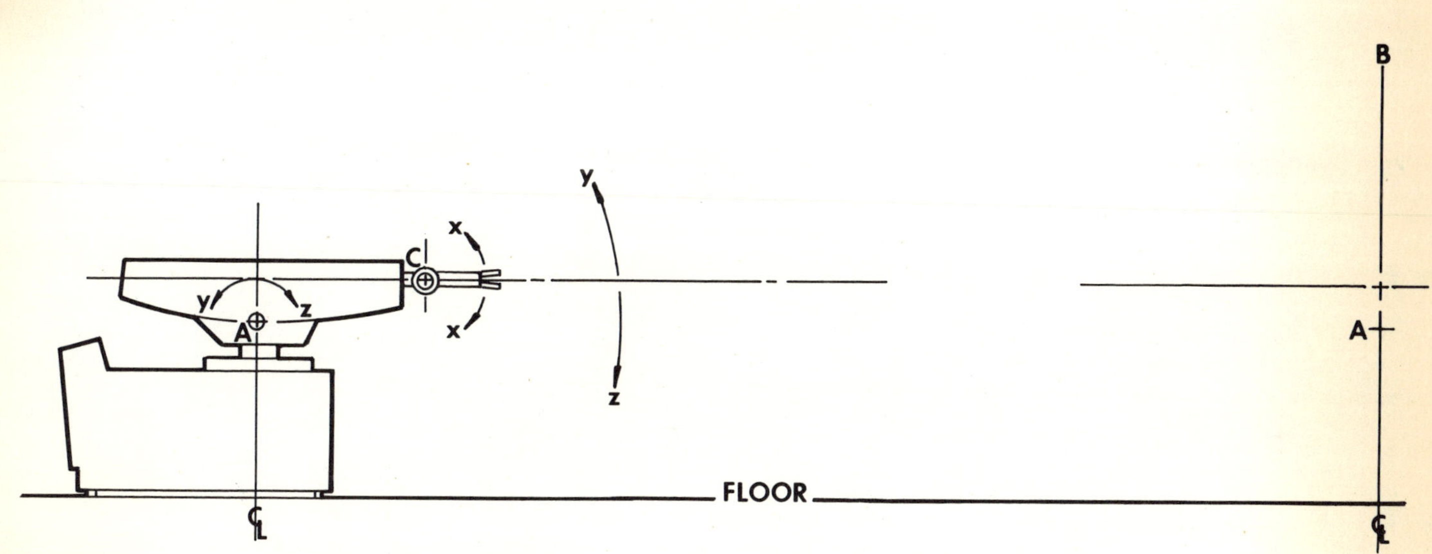

MECHANISM DRAFTING AND DESIGN	NAME: SECT. NO.: FILE NO.: DATE:	GRADE	PROBLEM 9A

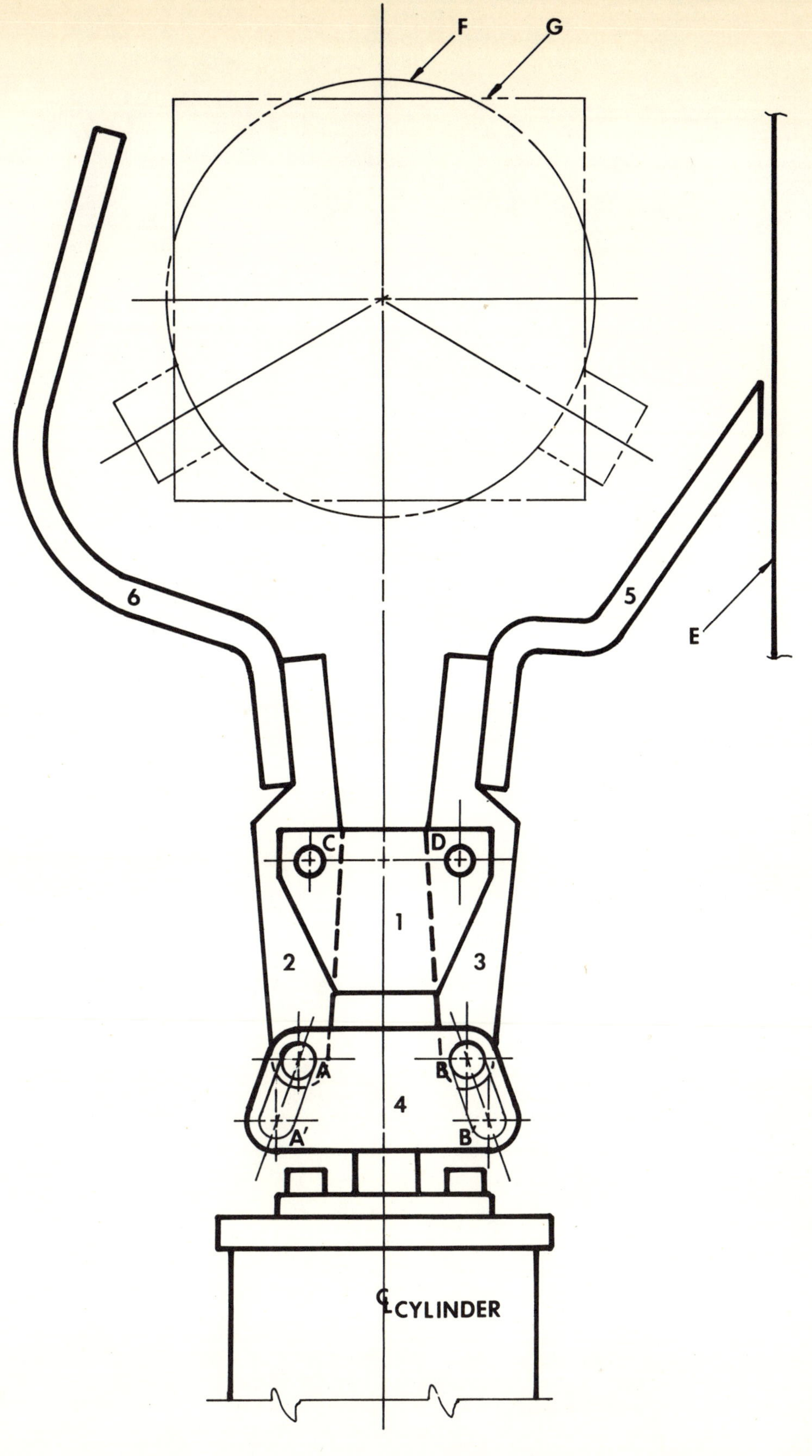

MECHANISM DRAFTING AND DESIGN	NAME: SECT. NO.: FILE NO.: DATE:	GRADE	PROBLEM 9B

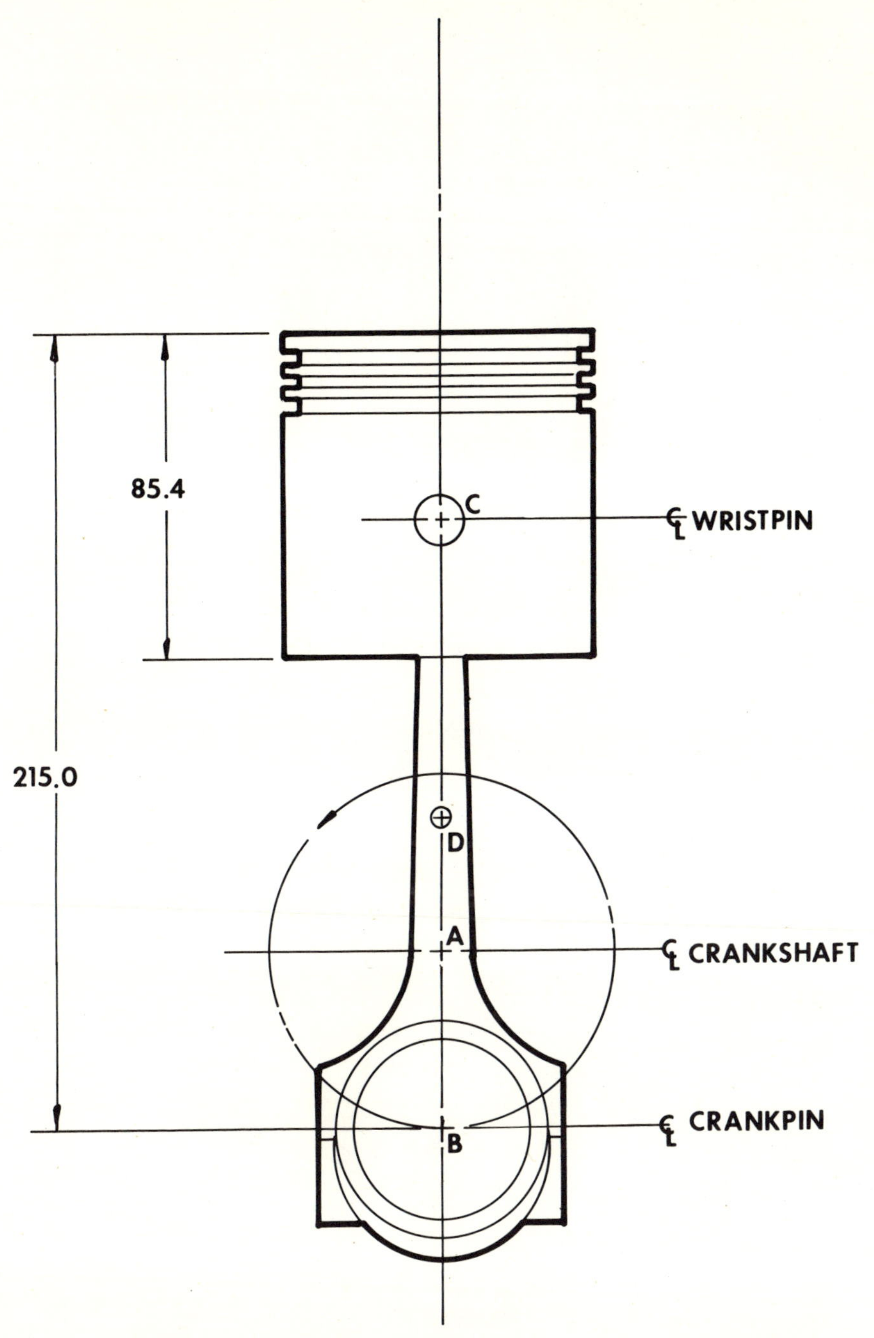

MECHANISM DRAFTING AND DESIGN	NAME: SECT. NO.: FILE NO.: DATE:	GRADE	PROBLEM 10 A

MECHANISM DRAFTING AND DESIGN	NAME: SECT. NO.: FILE NO.: DATE:	GRADE	PROBLEM 10 B

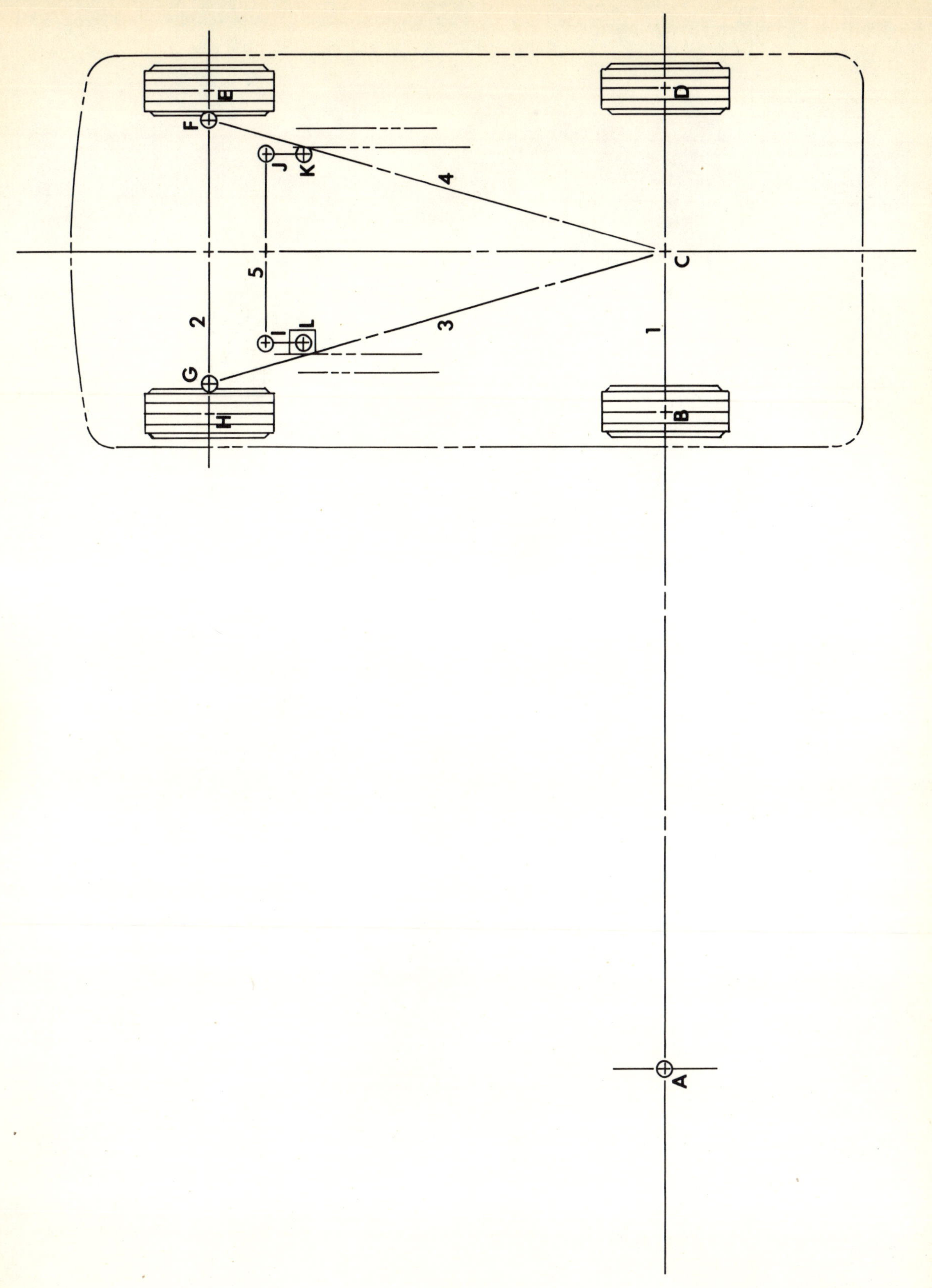

MECHANISM DRAFTING AND DESIGN	NAME: SECT. NO.: FILE NO.: DATE:	GRADE	PROBLEM 11A

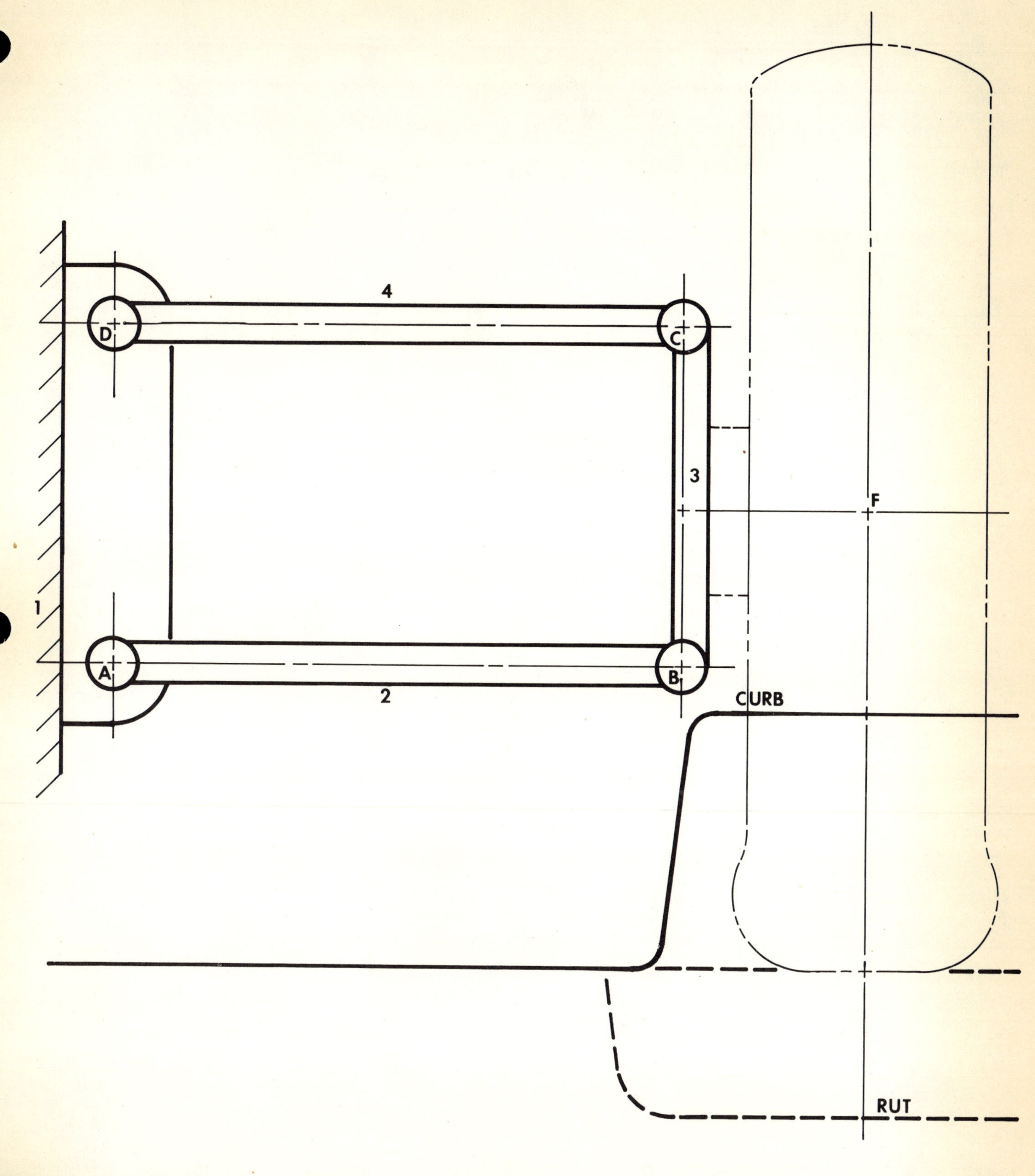

MECHANISM DRAFTING AND DESIGN	NAME: SECT. NO.: FILE NO.: DATE:	GRADE	PROBLEM 11B

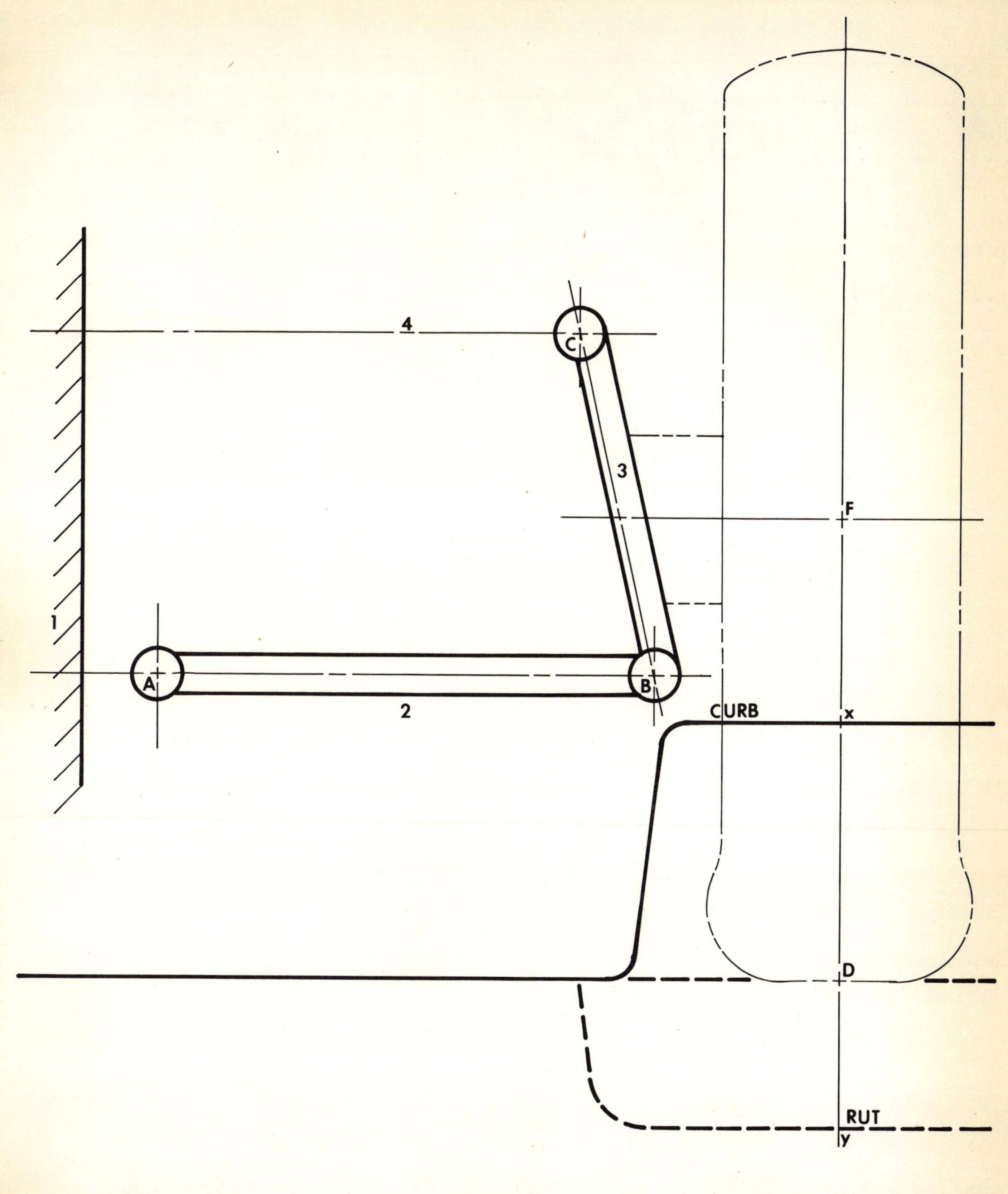

MECHANISM DRAFTING AND DESIGN	NAME: SECT. NO.: FILE NO.: DATE:	GRADE	PROBLEM 11C

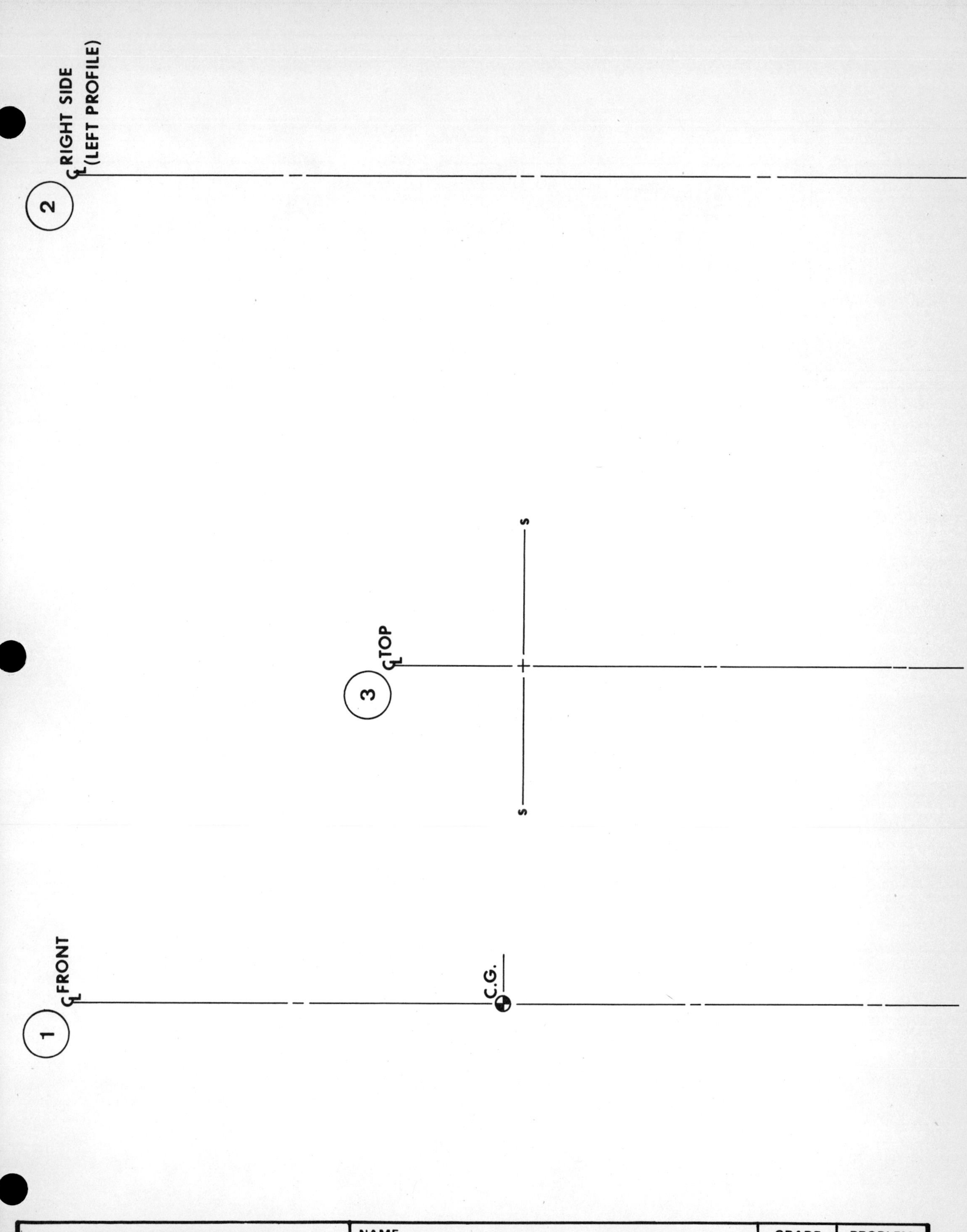

MECHANISM DRAFTING AND DESIGN	NAME: SECT. NO.: FILE NO.: DATE:	GRADE	PROBLEM 12A

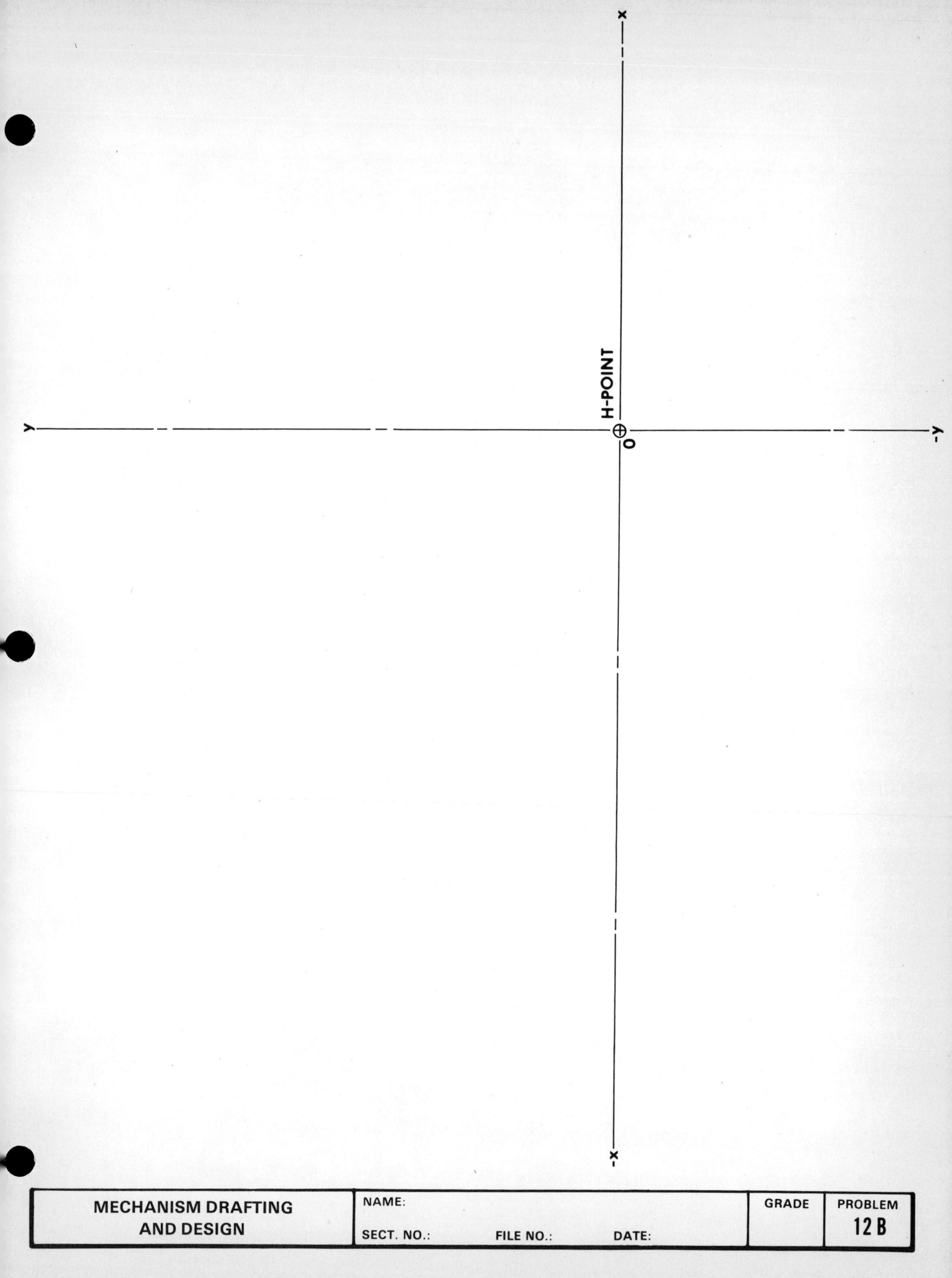

MECHANISM DRAFTING AND DESIGN	NAME: SECT. NO.: FILE NO.: DATE:	GRADE	PROBLEM 12 B

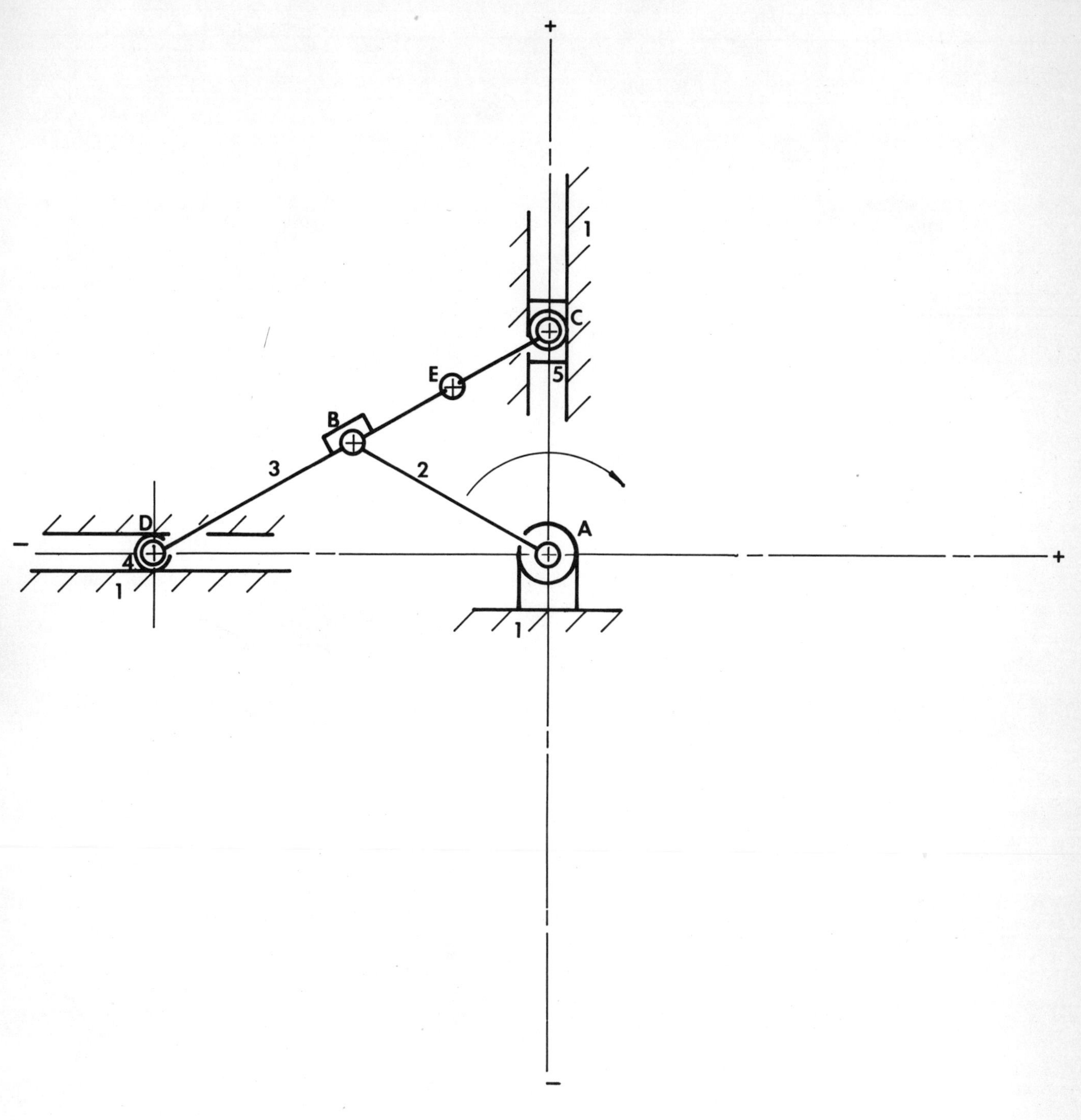

MECHANISM DRAFTING AND DESIGN	NAME: SECT. NO.: FILE NO.: DATE:	GRADE	PROBLEM 13A

360°

0°

MECHANISM DRAFTING AND DESIGN	NAME: SECT. NO.: FILE NO.: DATE:	GRADE	PROBLEM 13B

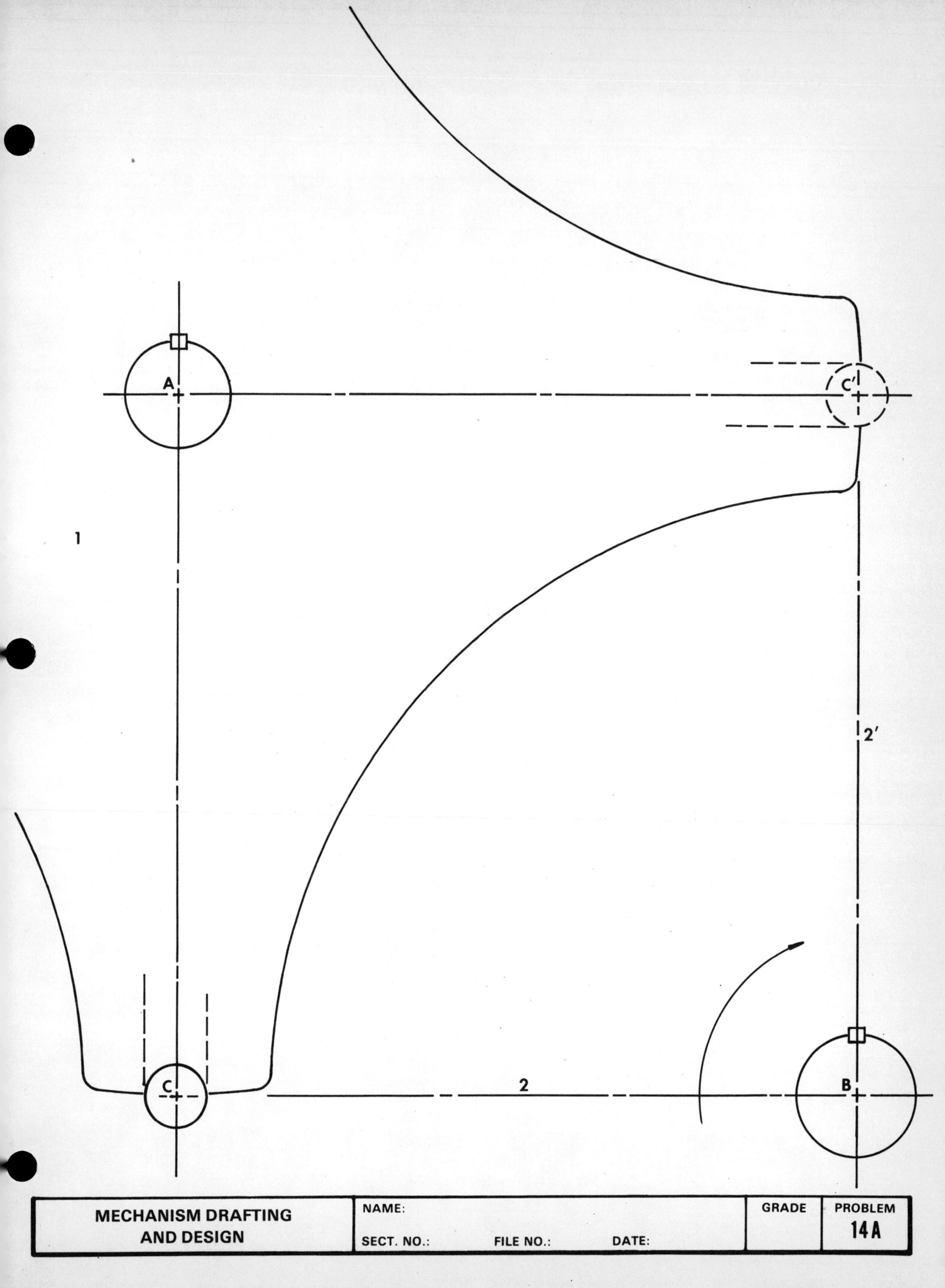

MECHANISM DRAFTING AND DESIGN	NAME:	GRADE	PROBLEM
	SECT. NO.: FILE NO.: DATE:		14 A

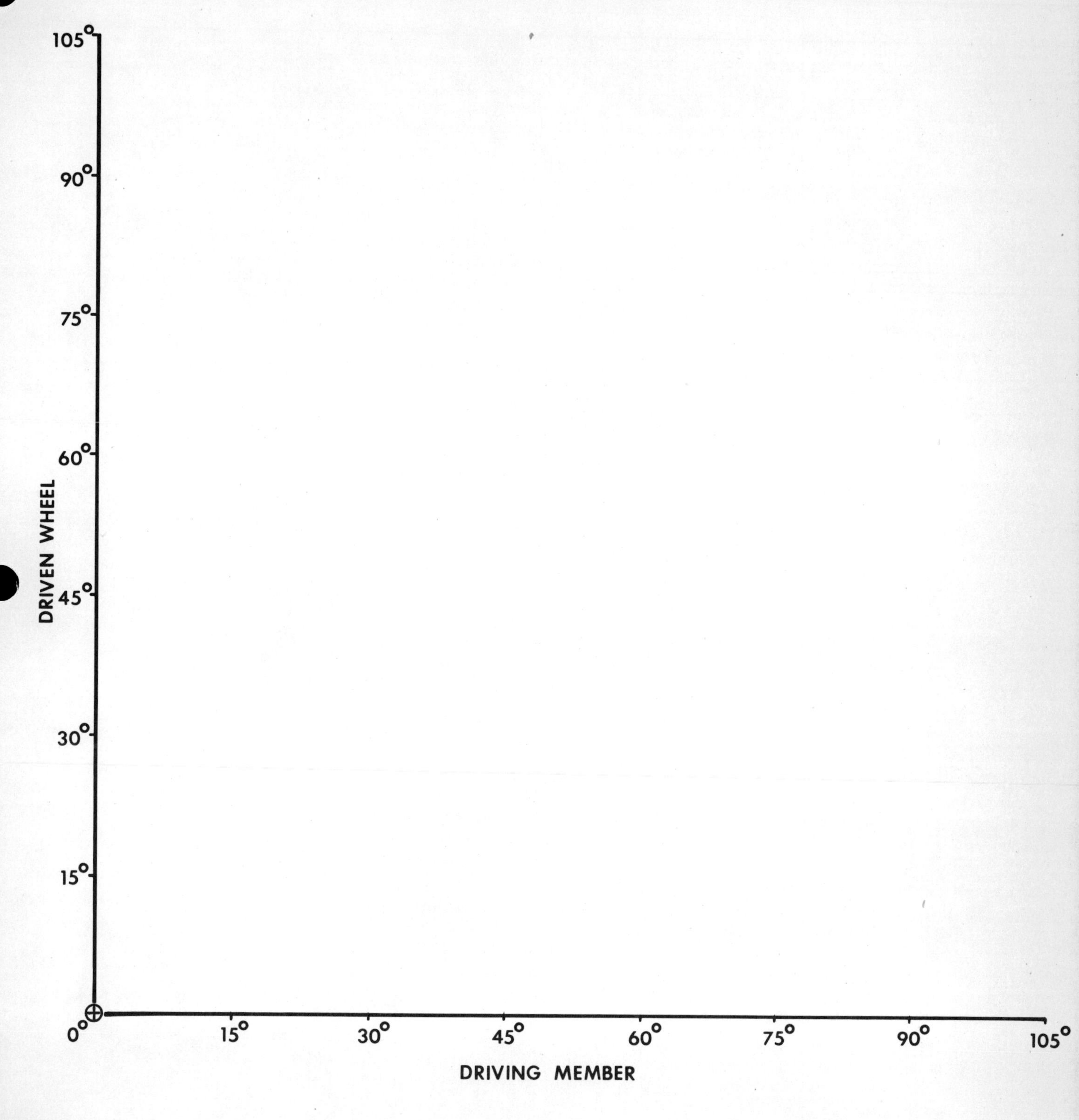

MECHANISM DRAFTING AND DESIGN	NAME: SECT. NO.: FILE NO.: DATE:	GRADE	PROBLEM 14 B

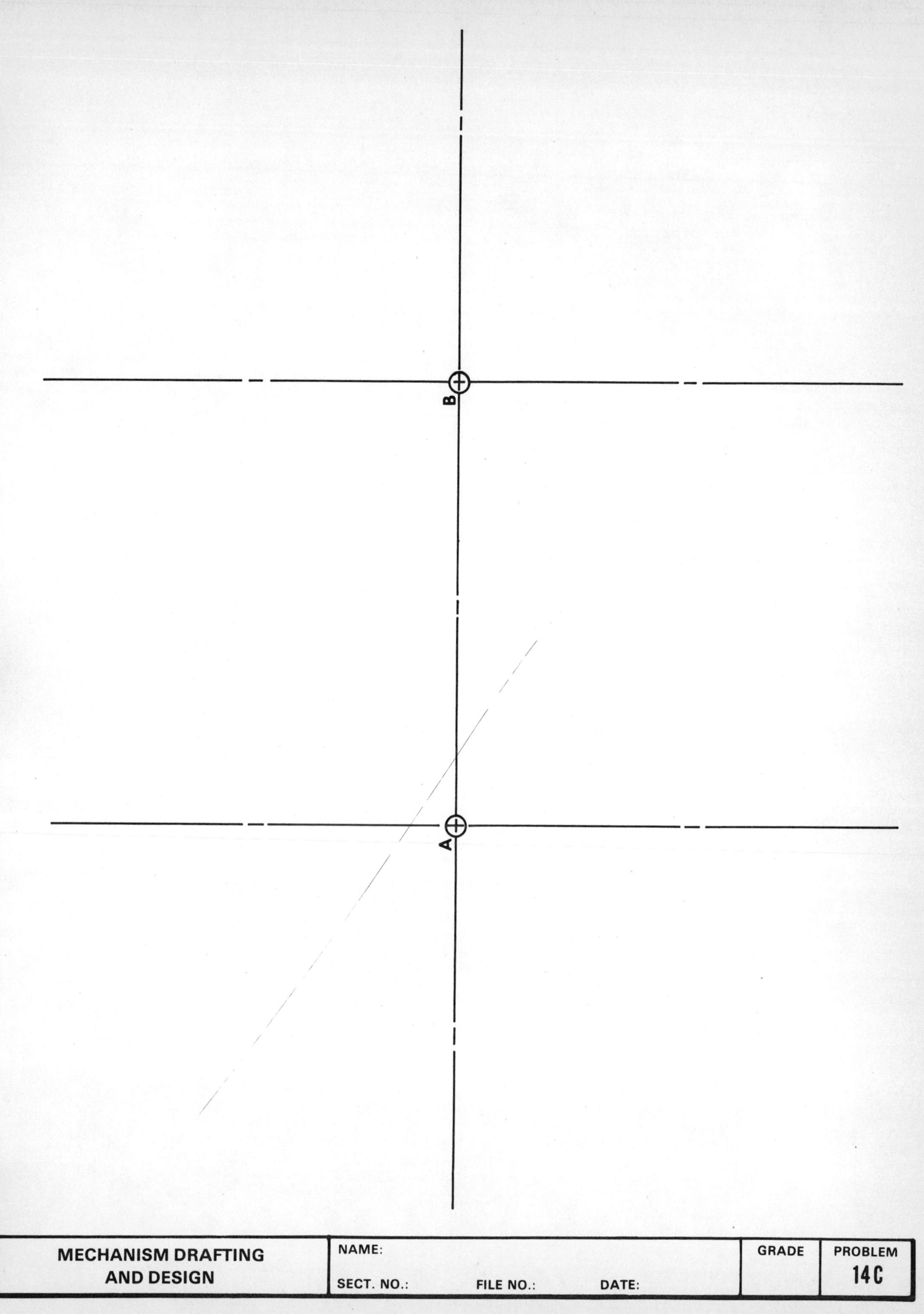

MECHANISM DRAFTING AND DESIGN	NAME: SECT. NO.: FILE NO.: DATE:	GRADE	PROBLEM 14C

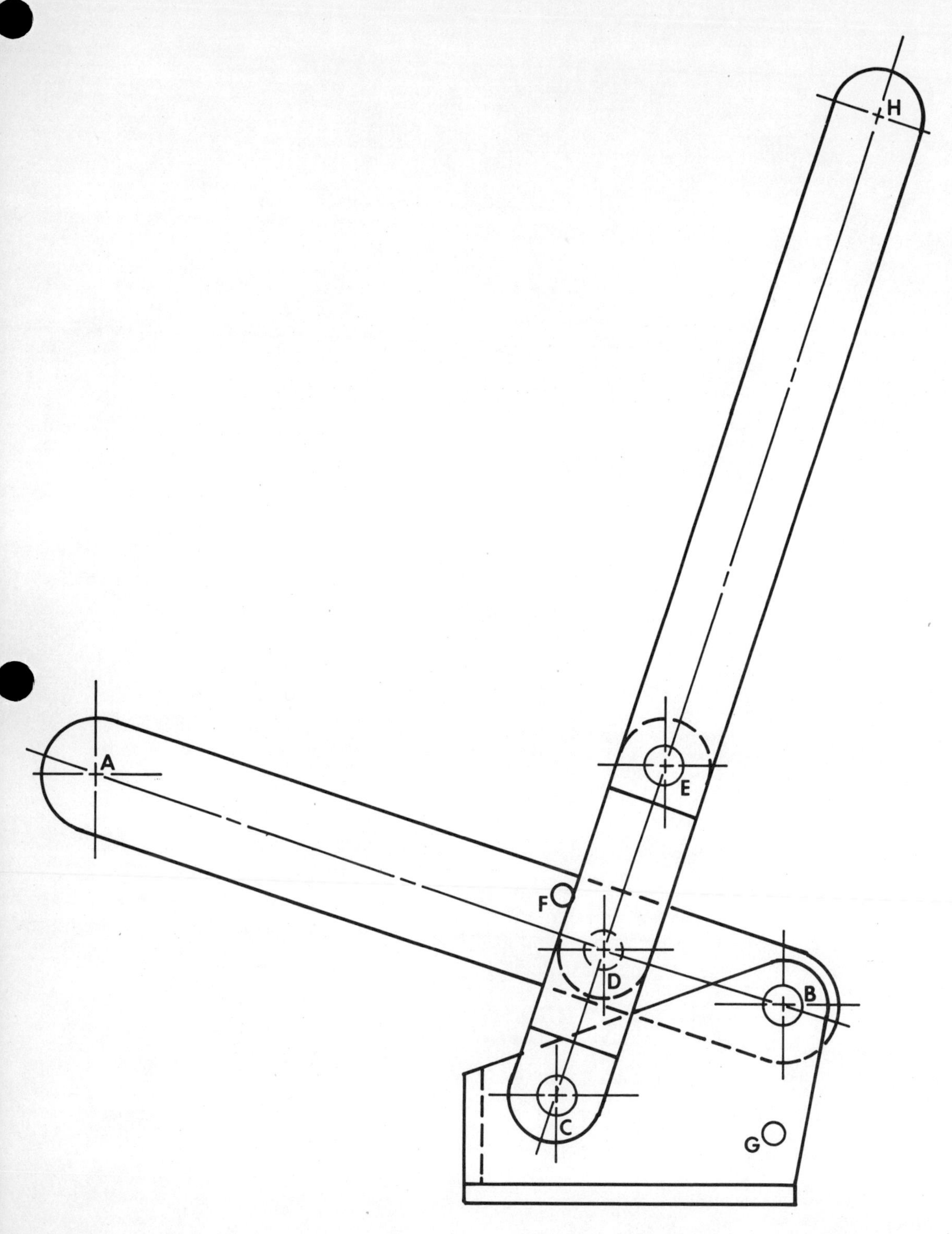

MECHANISM DRAFTING AND DESIGN	NAME:	GRADE	PROBLEM 15A
	SECT. NO.: FILE NO.: DATE:		

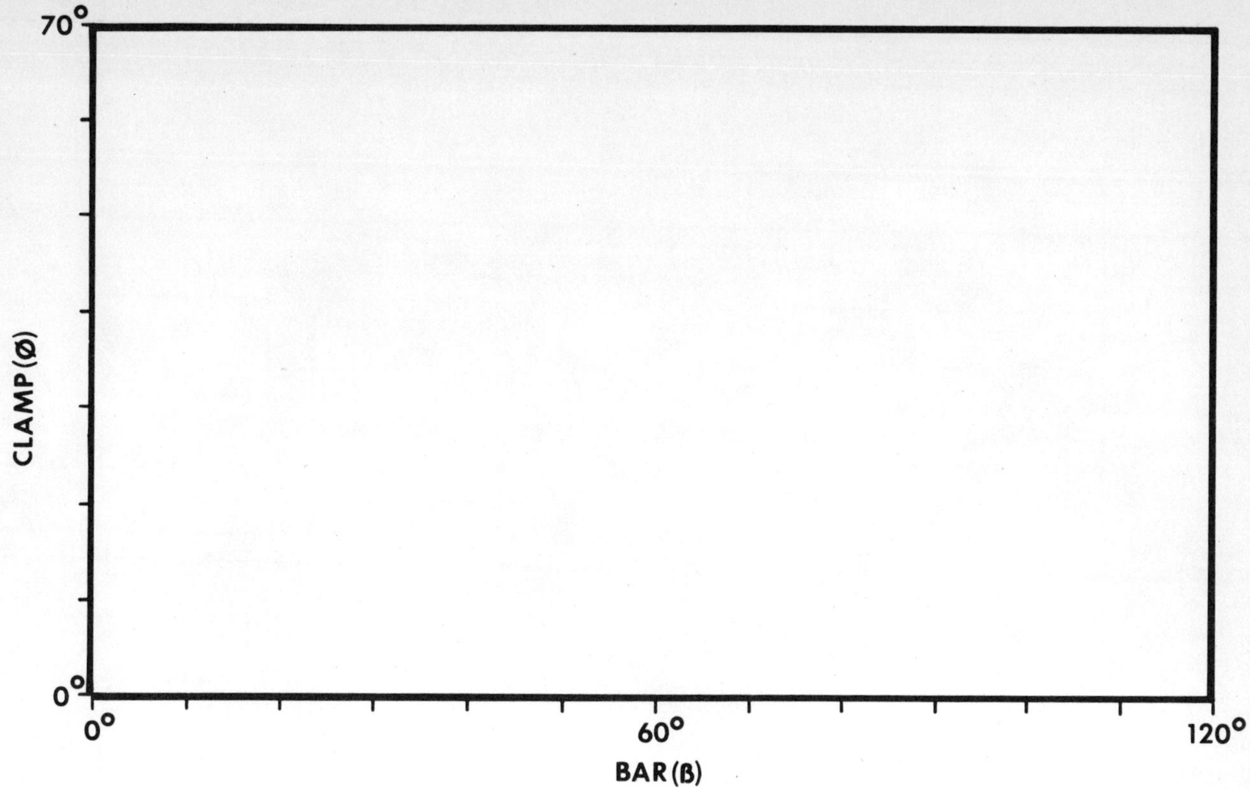

MECHANISM DRAFTING AND DESIGN	NAME: SECT. NO.: FILE NO.: DATE:	GRADE	PROBLEM 15B

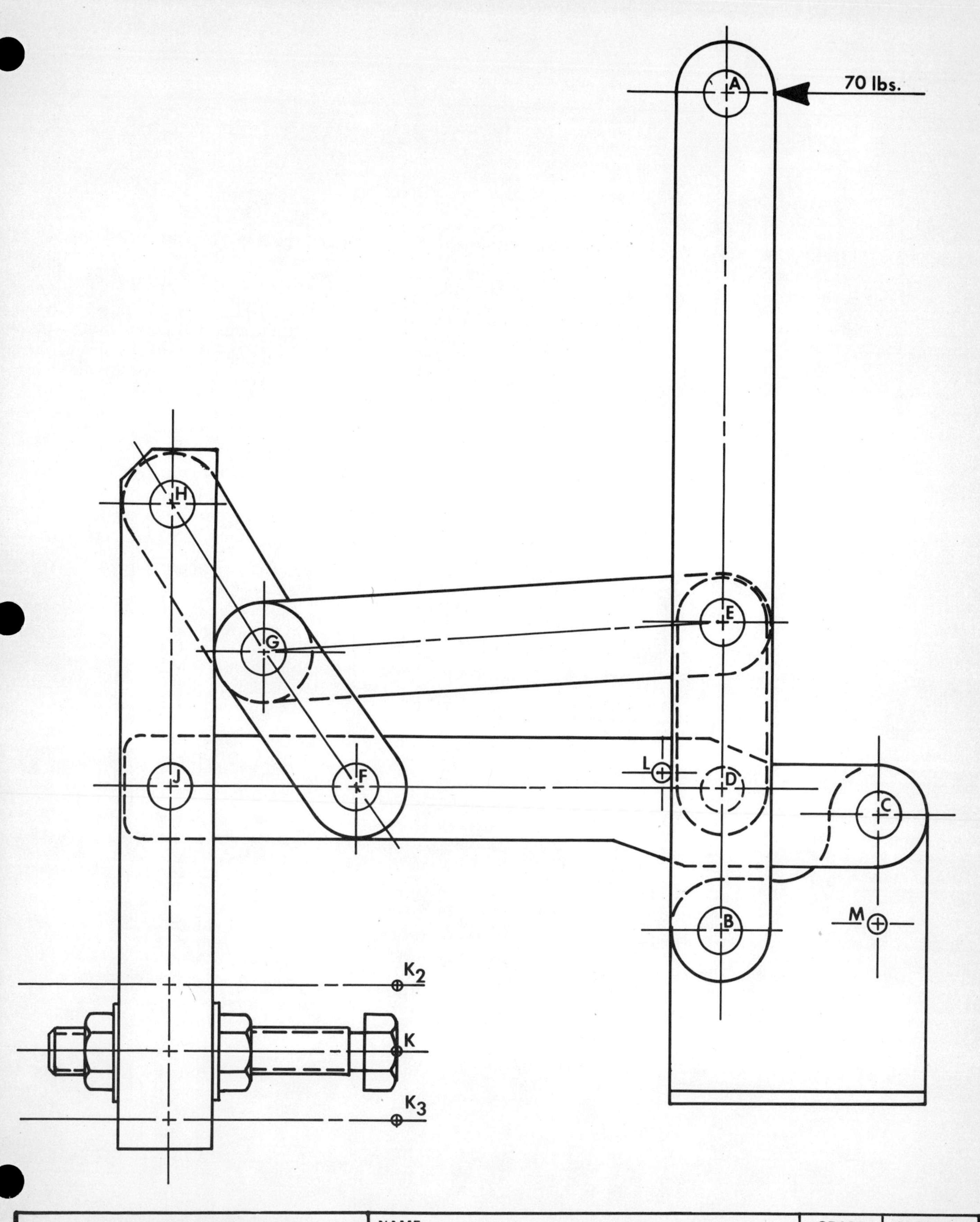

MECHANISM DRAFTING AND DESIGN	NAME: SECT. NO.: FILE NO.: DATE:	GRADE	PROBLEM 15C

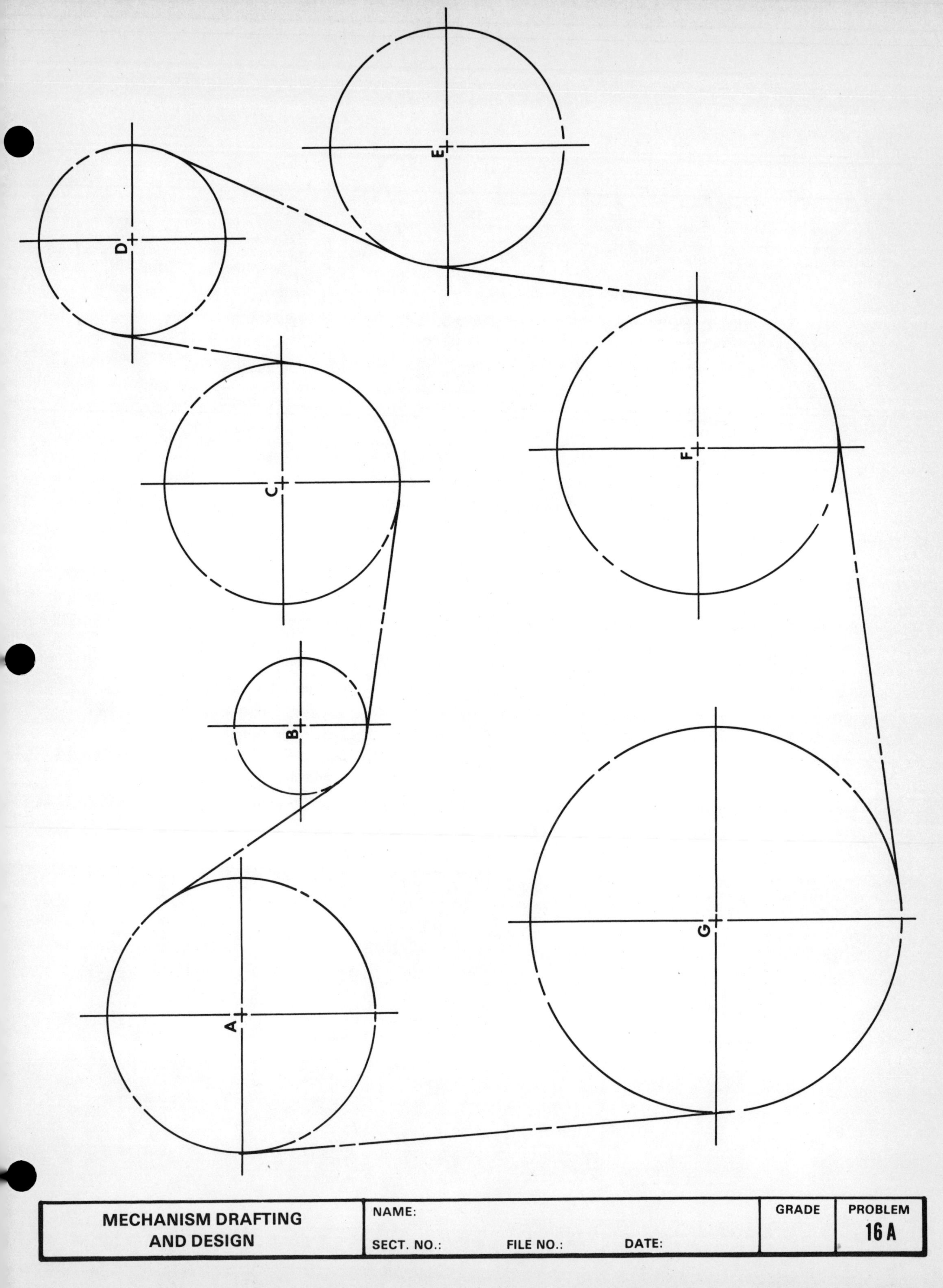

D
E
C
F
B
G
A
MECHANISM DRAFTING AND DESIGN
NAME:
SECT. NO.:
FILE NO.:
DATE:
GRADE
PROBLEM
16 A

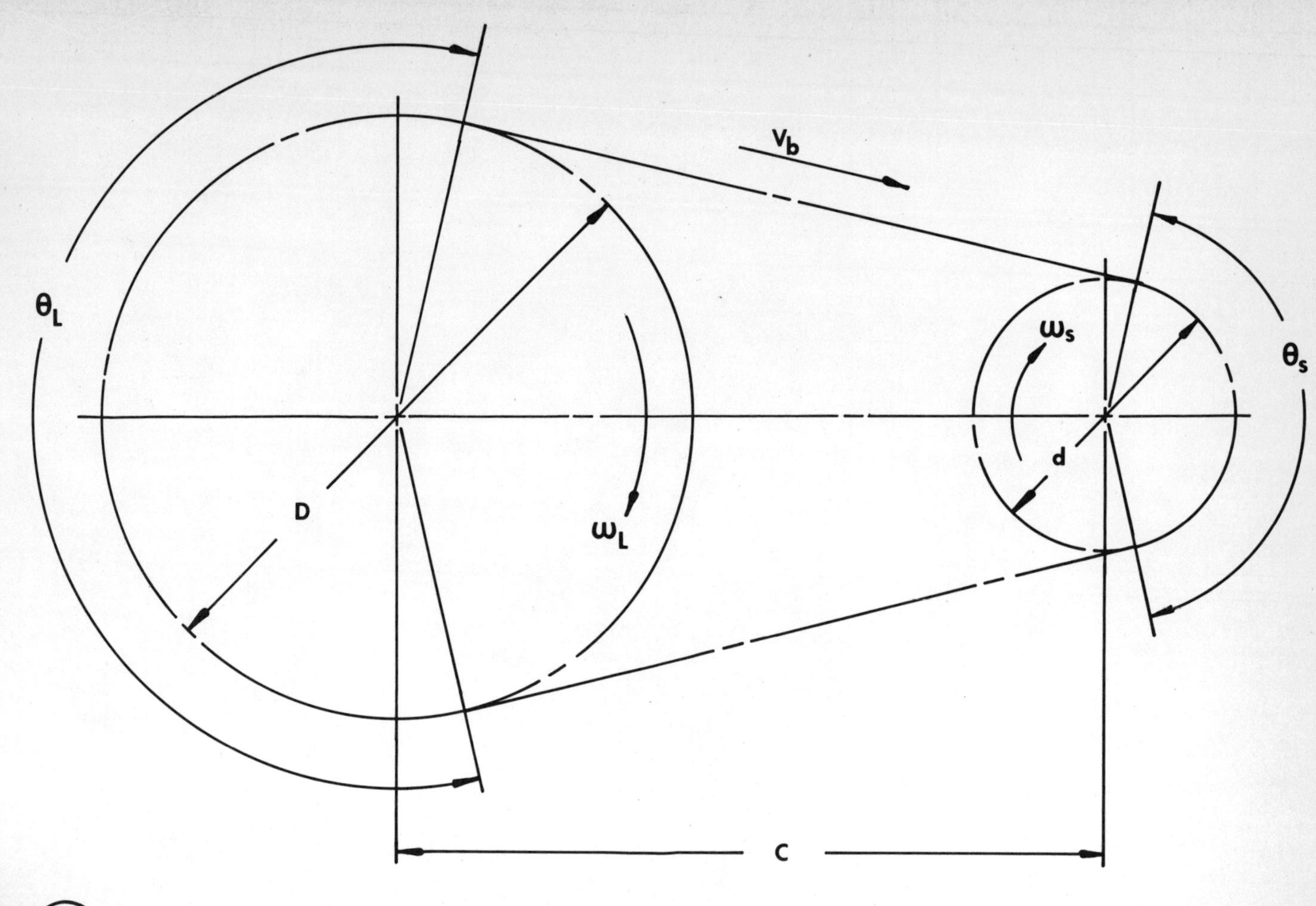

(1)

(2)

MECHANISM DRAFTING AND DESIGN	NAME: SECT. NO.: FILE NO.: DATE:	GRADE	PROBLEM 16 B

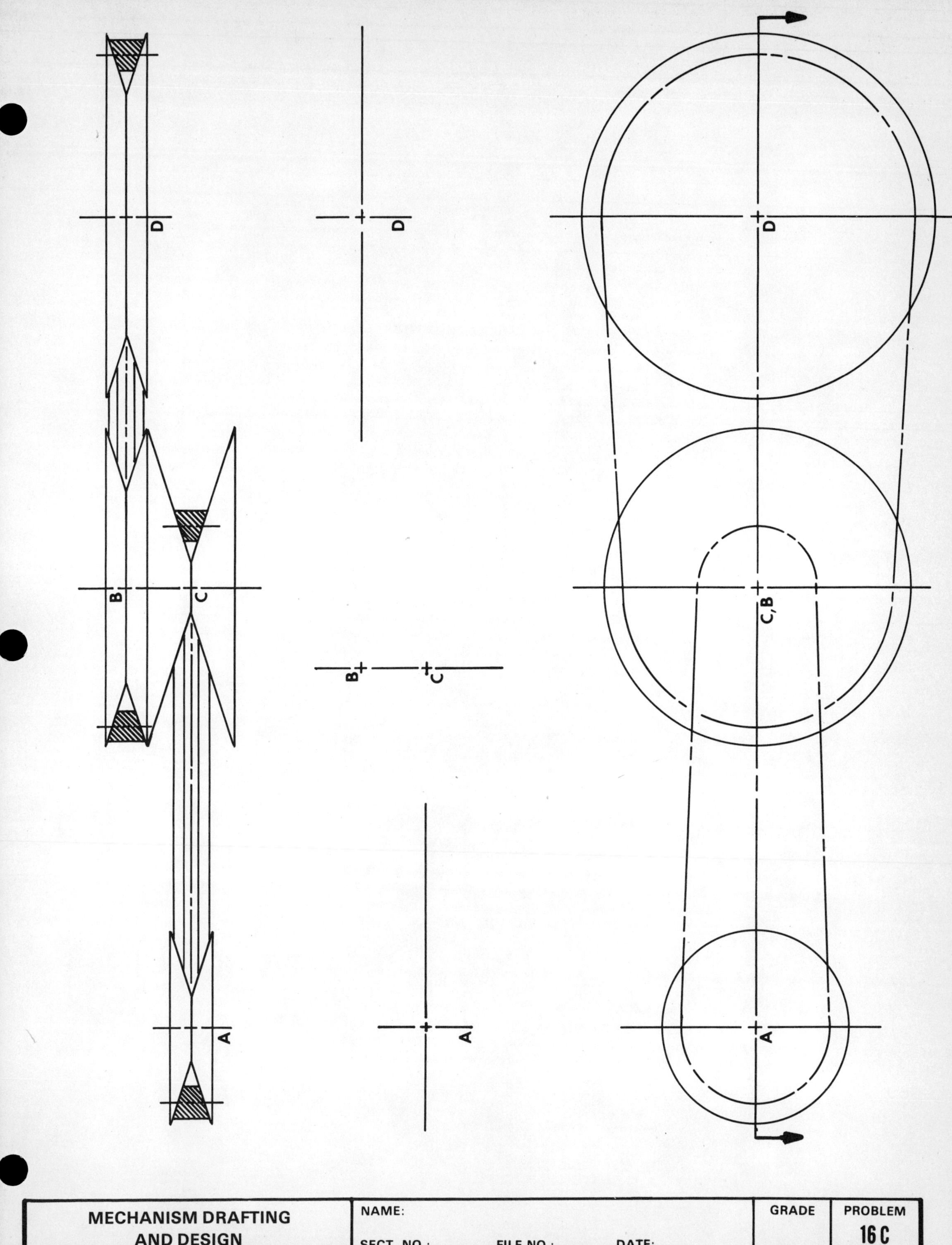

MECHANISM DRAFTING AND DESIGN	NAME: SECT. NO.: FILE NO.: DATE:	GRADE	PROBLEM 16 C

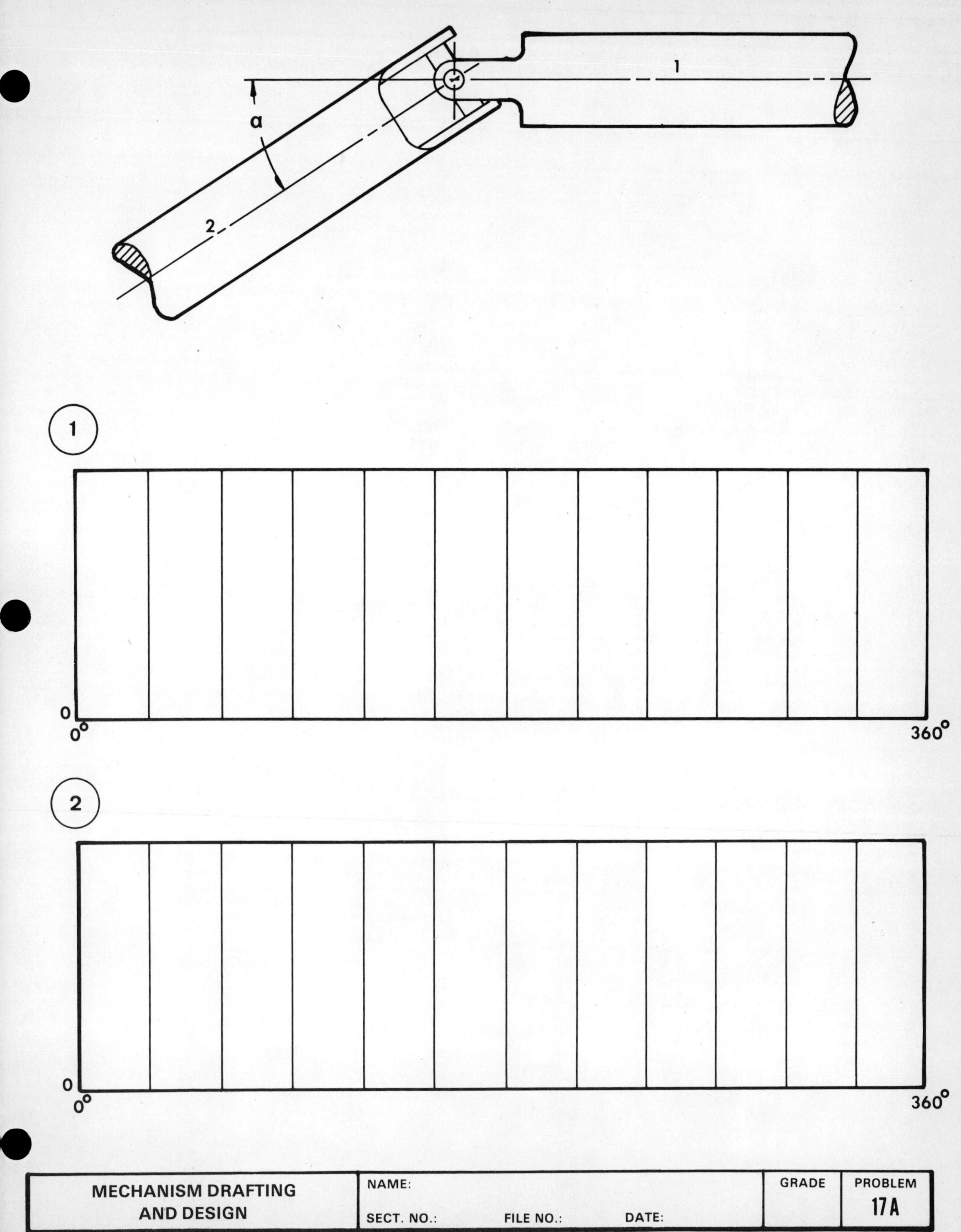
1
α
2
1
0
0°
360°
2
0
0°
360°
MECHANISM DRAFTING AND DESIGN
NAME:
SECT. NO.:
FILE NO.:
DATE:
GRADE
PROBLEM
17A

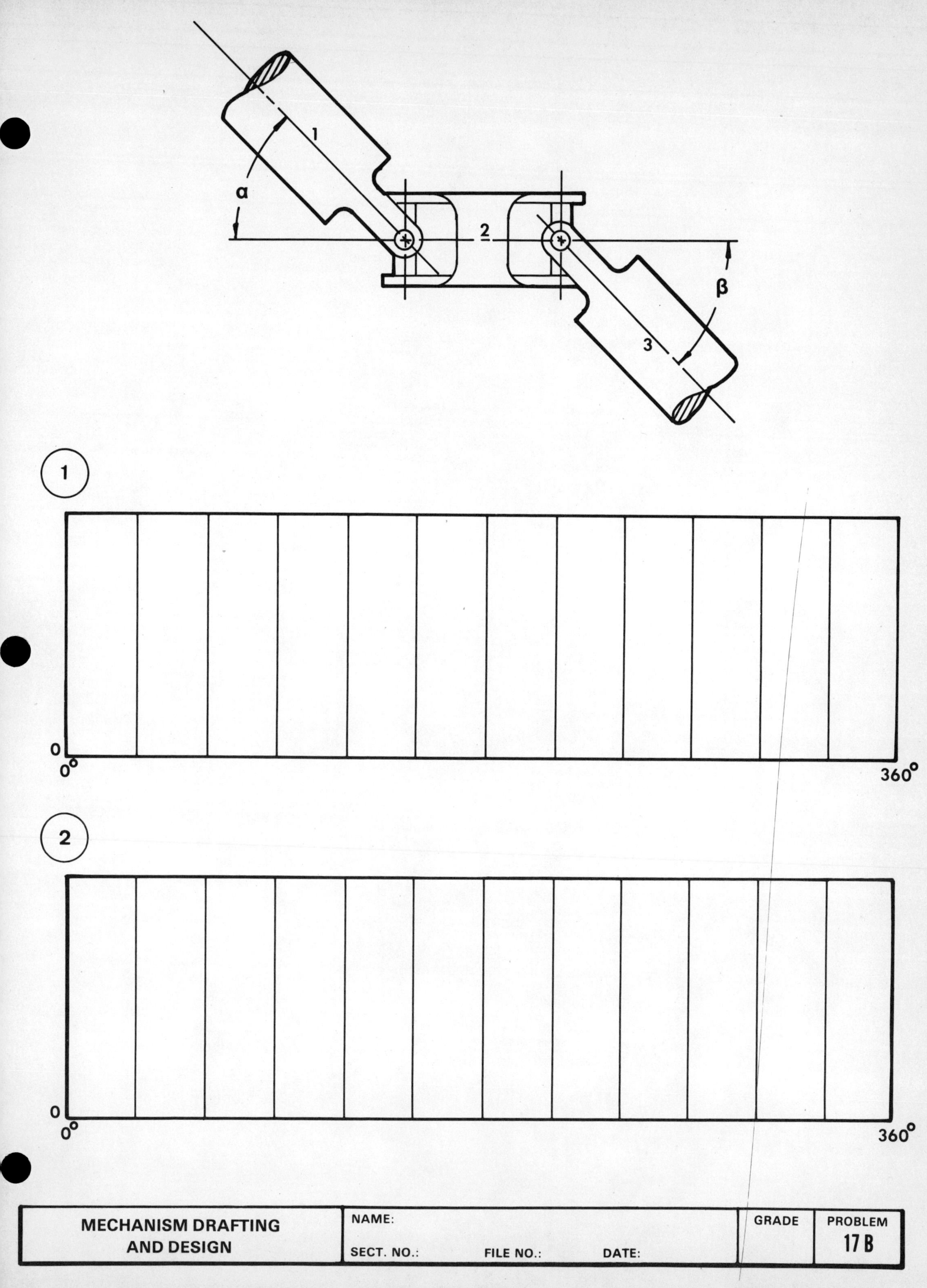

1
α
2
β
3
1
0
0°
360°
2
0
0°
360°
MECHANISM DRAFTING AND DESIGN
NAME:
SECT. NO.:
FILE NO.:
DATE:
GRADE
PROBLEM
17 B

1

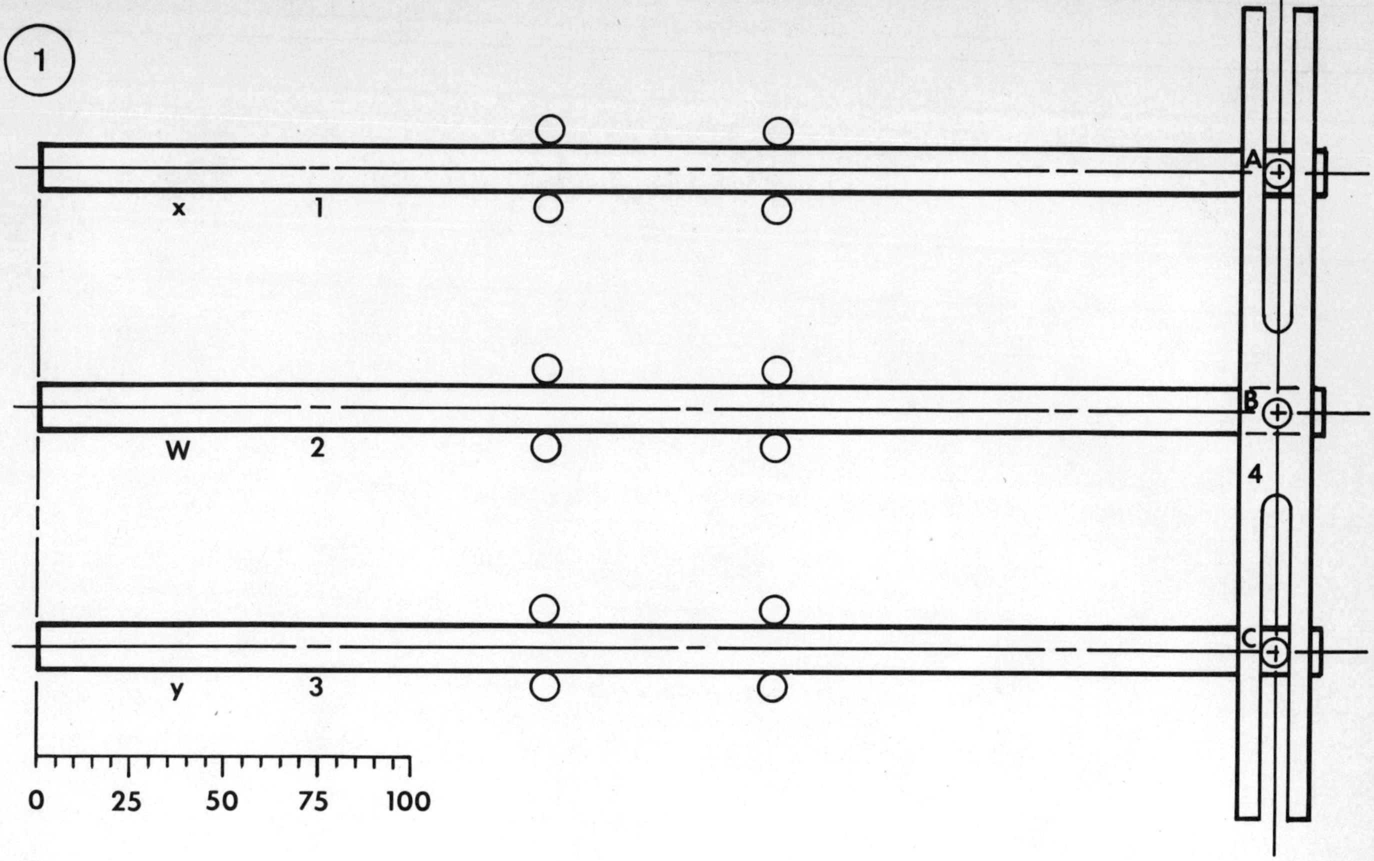

A
x 1
B
W 2
4
C
y 3

0 25 50 75 100

2

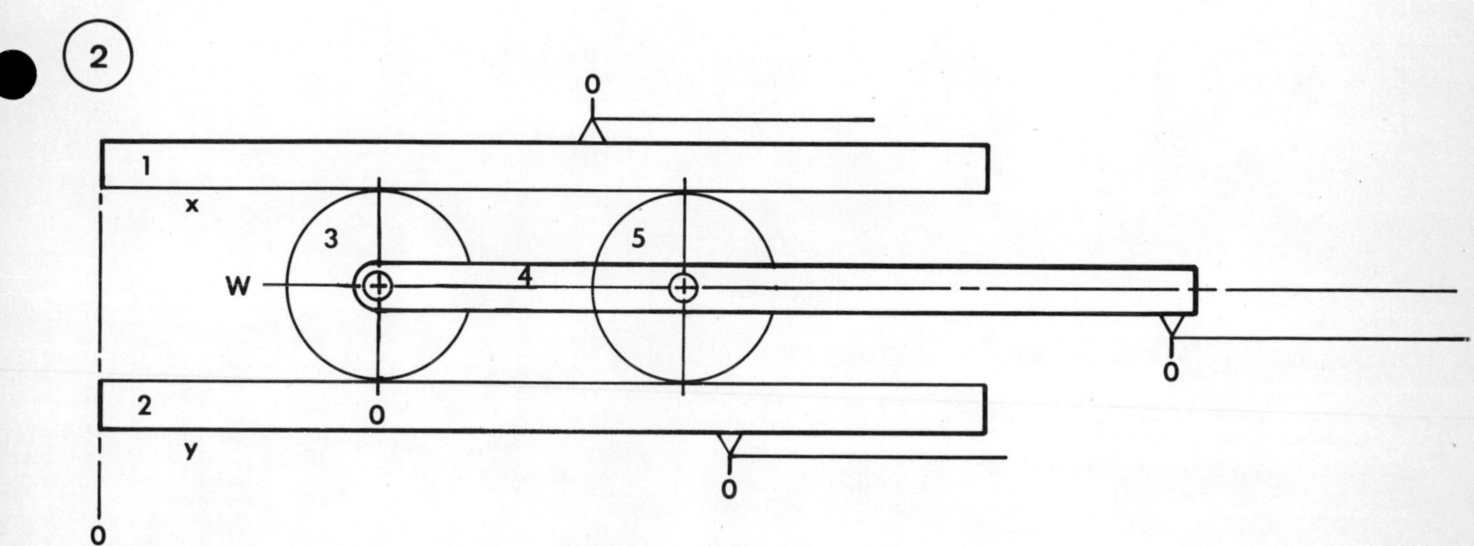

MECHANISM DRAFTING AND DESIGN	NAME: SECT. NO.: FILE NO.: DATE:	GRADE	PROBLEM 18 A

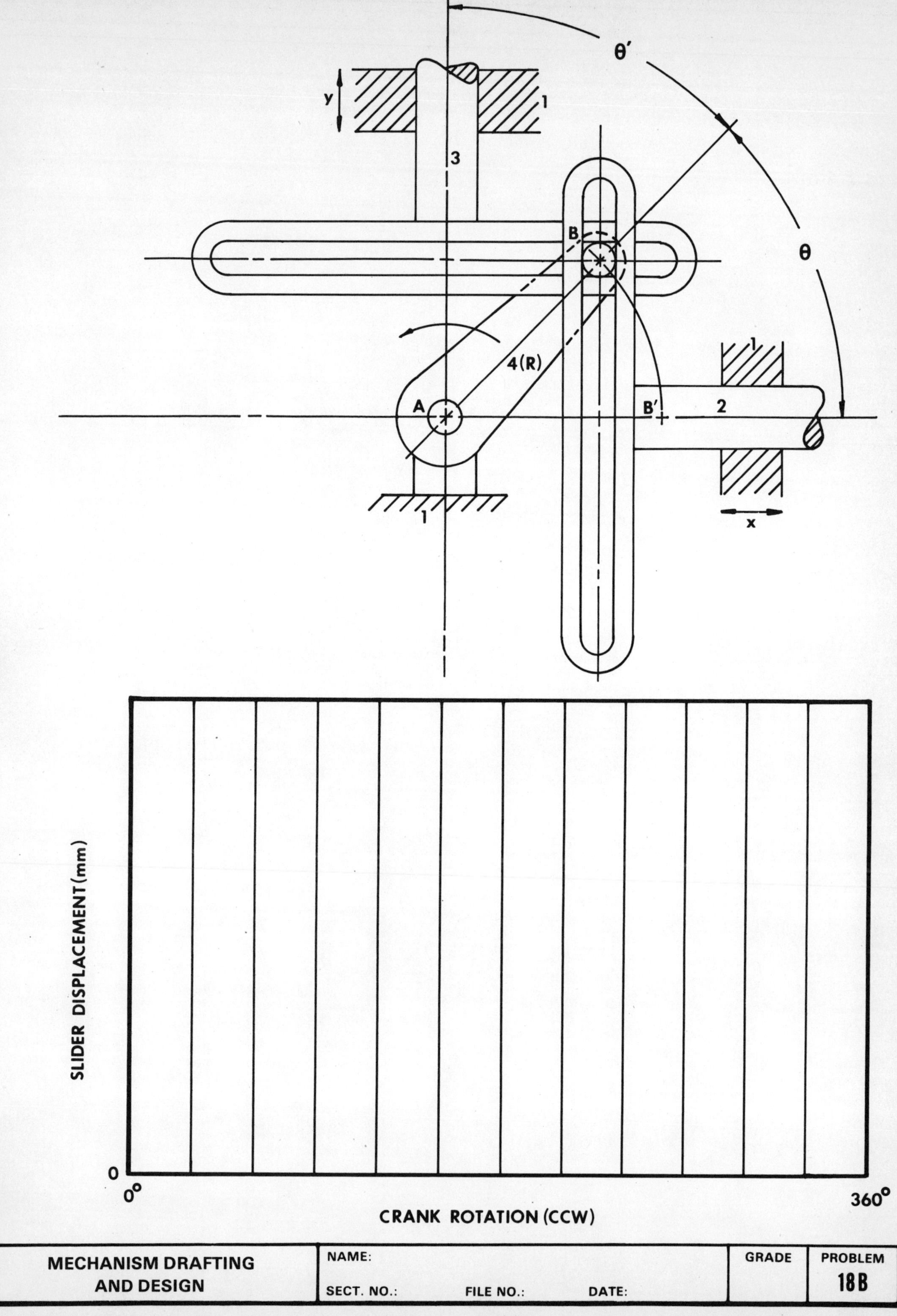
θ′
θ
y
1
3
B
4(R)
A
B′
2
1
1
x
SLIDER DISPLACEMENT (mm)
0
0°
360°
CRANK ROTATION (CCW)
MECHANISM DRAFTING AND DESIGN
NAME:
SECT. NO.:
FILE NO.:
DATE:
GRADE
PROBLEM
18B

(This form may be duplicated for use as an extra worksheet.)

MECHANISM DRAFTING AND DESIGN	NAME:	GRADE	PROBLEM
	SECT. NO.: FILE NO.: DATE: